Microbial Biocatalysis

Microbial Biocatalysis

Editors

Zhilong Wang
Tao Pan

MDPI • Basel • Beijing • Wuhan • Barcelona • Belgrade • Manchester • Tokyo • Cluj • Tianjin

Editors

Zhilong Wang
State Key Laboratory of Microbial Metabolism
Shanghai Jiao Tong University
Shanghai
China

Tao Pan
Jiangxi Province Key Laboratory of Mining and Metallurgy Environmental Pollution Control
Jiangxi University of Science and Technology
Ganzhou
China

Editorial Office
MDPI
St. Alban-Anlage 66
4052 Basel, Switzerland

This is a reprint of articles from the Special Issue published online in the open access journal *Catalysts* (ISSN 2073-4344) (available at: www.mdpi.com/journal/catalysts/special_issues/microbes_biocatal).

For citation purposes, cite each article independently as indicated on the article page online and as indicated below:

LastName, A.A.; LastName, B.B.; LastName, C.C. Article Title. *Journal Name* **Year**, *Volume Number*, Page Range.

ISBN 978-3-0365-7191-1 (Hbk)
ISBN 978-3-0365-7190-4 (PDF)

Contents

About the Editors

Zhilong Wang

Zhilong Wang received his bachelor degree of environmental engineering from Xiangtan University in 1992, master degree of chemical engineering from Tianjin University in 1998, and doctorate degree of biomedical engineering from Shanghai Jiao Tong University in 2004. Now, he is a professor of School of Pharmacy, and a core member of State Key Laboratory of Microbial Metabolism, Shanghai Jiao Tong University (SJTU).

The water-insoluble substrate, product inhibition and further degradation often occur in chemical reactions catalyzed by living microorganisms as biocatalysts (biotransformation, biodegradation and fermentation). His interest is using chemical technology (cloud point system, inverse phase transfer catalysis, and Pickering emulsion, as well as designing of biocompatible chemical reaction) to alleviate the above problems. Focusing on those topics, nearly 70 reviewed international articles and one monograph of *Extractive Microbial Transformation* were published.

Tao Pan

Tao Pan received his PhD in Microbiology from the South China University of Technology and joined Jiangxi University of Science and Technology in 2013 as deputy director of the Department of Environmental and Bioengineering there. In 2018, he was promoted to associate professor. Over the years, he has been committed to research on enhanced biodegradation technologies for hydrophobic organic pollutants and has received two grants from the National Natural Science Foundation of China. He is currently conducting research on the interfacial biodegradation of crude oil with the aim of addressing the risk of dispersant application.

catalysts

Editorial

Microbial Biocatalysis

Tao Pan [1,*] and Zhilong Wang [2,*]

[1] Jiangxi Province Key Laboratory of Mining and Metallurgy Environmental Pollution Control, School of Resource and Environmental Engineering, Jiangxi University of Science and Technology, Ganzhou 341000, China
[2] State Key Laboratory of Microbial Metabolism, School of Pharmacy, Shanghai Jiao Tong University, Shanghai 200240, China
* Correspondence: t.pan@jxust.edu.cn (T.P.); zlwang@sjtu.edu.cn (Z.W.)

Citation: Pan, T.; Wang, Z. Microbial Biocatalysis. *Catalysts* **2023**, *13*, 629. https://doi.org/10.3390/catal13030629

Received: 16 March 2023
Accepted: 20 March 2023
Published: 21 March 2023

Biocatalysis, which can be performed by whole cells and isolated enzymes, has become a topic of public interest for its potential use in the chemical industry in manufacturing, monitoring, and waste management. Enzymes are proteins that organisms produce to catalyze the biochemical reactions needed for life. However, the isolation and purification of enzymes may be costly and time-consuming, and cofactors may need to be added or recovered. An alternative approach is to use whole cells as "Microbial Biocatalysts" to perform multiple enzyme reactions in a single strain and regenerate cofactors internally. Whole-cell biocatalysts can be used for different types of processes, such as biotransformation and fermentation. They involve one or more steps of biocatalysis to produce valuable chemicals through biosynthesis/biotransformation or degrade organic pollutants completely.

Fermentation is a whole-cell biosynthesis process that involves multiple enzymes and native pathways. α-Ketoglutaric acid (KGA) is a valuable compound that can be produced by *Yarrowia lipolytica* CBS146773 using a mixed carbon source of glycerol and rapeseed oil [1]. This strain requires thiamine for growth and overexpresses genes encoding glycerol kinase, citrate synthase, and the mitochondrial acid transporter. By optimizing the fermentation conditions, the KGA yield reached 82.4 g/L. In contrast to the pure compound KGA, extracellular polymers (EPS) are complex secondary metabolites produced by microorganisms. They consist of proteins, polysaccharides, humic acids, and nucleic acids. A Cordyceps strain C058 and its bioaugmented biofloc, named mycelium biofloc (MBF), have shown high water purification efficiency due to their high EPS production [2]. MBF was constructed by both fungi and bacteria, with C058 being the main contributor to EPS synthesis. Multifunctional enzymes (MFEs), including various synthetases and post-modification enzymes, are involved in the biosynthesis of such secondary metabolites [3]. The review summarized the research advances of MFEs such as polyketides, non-ribosomal peptides, terpenoids, and a wide range of cytochrome P450s that participate in secondary metabolite synthesis [3].

Whole-cell biocatalysis relies on finding suitable biocatalysts for specific conversion reactions. A team from Shenyang Pharmaceutical University reported three examples of whole-cell biotransformation reactions in three articles: the 7α- and 7β-hydroxylation of Dehydroepiandrosterone by *Gibberella* sp. CICC 2498 and *Absidia coerulea* CICC 41050 [4]; the O-demethylation of rabeprazole by *Cunninghamella blakesleeana* 3.970 [5]; and the non-oxidative deamination of L-phenylalanine and L-tyrosine by recombinant *Escherichia coli* BL21 [6]. These reactions are useful for the synthesis of important pharmaceutical intermediates. α-Amylase is an important industrial enzyme that has not been extensively studied in marine sources. The α-amylase from the symbiotic bacteria *Bacillus* sp. HR13 and HR16 isolated from the intestines of *Sillago sihamas* and *Rastrelliger Kanagurta* exhibited high thermostability at 60 °C [7]. This suggests that they have potential applications in food and detergent industries. Phenol-degrading bacteria have been widely reported, but little is known about phenol degradation by cold-tolerant strains in extreme environments such as Antarctica. Two cold-adapted Arthrobacter strains that only possessed catechol 1,2-dioxygenase were able to degrade phenol via an ortho-cleavage pathway at temperatures between 10 and 15 °C [8].

In order to improve the efficiency of whole-cell catalysis, the catalyst can be modified by mutation, directed evolution, and immobilization. Random mutagenesis was the first technology that enabled the efficient generation of large diversity and is still used in many laboratories. For example, UV mutagenesis of the nystatin-producing strain Streptomyces noursei D-3-14 resulted in the mutant strain 72-22-1, with a high yield of polyfungin B and high fungicidal activities [9]. The mutant strain showed a 1.58 and 1.91-fold increase in chemical and biological potency, respectively, and had stable genetic characteristics. However, random mutagenesis has some limitations, such as the incompleteness of the diversity introduced. This can be overcome by combining it with other technologies to achieve the directed evolution of catalysts. A case in point is the use of error-prone PCR for random mutagenesis and a suitable and efficient high-throughput screening method to enhance the acetophenone tolerance of short-chain dehydrogenase/reductase (SDR) from *Empedobacter brevis* ZJUY-1401 [10]. The mutant M190V exhibited a 74.8% activity improvement compared with the wild-type when using 200 mM acetophenone as the substrate. Another way to improve whole-cell catalysis efficiency is to immobilize cells to keep them alive, stabilize their catalytic efficiency, and enable their reuse. This also simplifies cell recycling and downstream processing. For instance, immobilized whole *E. coli* pRSF-AfNit2 cells were able to effectively catalyze the hydrolysis of 3-cyanopyridine to nicotinic acid in a semi-continuous packed-bed bioreactor (sPBR) using recombinant nitrilase [11]. The conversion rate remained at 100% after repeating the operation for 41 batches of sPBR. A review of diclofenac biodegradation by microorganisms and immobilized systems was presented [12]. It showed that immobilized fungal and bacterial systems can achieve complete degradation of diclofenac by a metabolic relay that avoids the accumulation of toxic intermediates.

Process intensification can be an effective way to improve the efficiency of microbial biocatalysis. For example, in the water-organic solvent two-phase system, hydrophobic *Mycolicibacterium* can act as emulsifiers to stabilize the Pickering emulsion. This can enhance the interfacial biotransformation of phytosterols by increasing the substrate concentration to cause phase inversion and overcome substrate inhibition [13]. However, this strategy may not work for all biocatalytic reactions. In the same water-organic solvent two-phase system, the biodegradation of atrazine was inhibited by the organic solvent [14]. This could be because the atrazine-degrading bacterium is hydrophilic and the organic solvent limits substrate transfer between oil and water. The biodegradation process of pollutants in natural environments is also influenced by various environmental factors. Bioplastics are a potential alternative to petroleum-based polymers that can be degraded in compost, soil, and aquatic environments [15]. The review focused on the intensification of bioplastic biodegradation through composting technology.

To conclude, this Special Issue on "Microbial Biocatalysis" provides a comprehensive overview of the recent developments in catalyst discovery, its modification, and process intensification for whole-cell catalysis in the fermentation, biotransformation, or biodegradation processes. We hope that this collection of key studies will inspire further research in this field.

We would like to express our sincere gratitude to all authors for their valuable contributions, as well as to the editorial team of *Catalysts*, especially Miss Cathy Yang, for their kind support.

Author Contributions: The contribution of T.P. and Z.W. is equal. All authors have read and agreed to the published version of the manuscript.

Conflicts of Interest: The authors declare no conflict of interest.

References

1. Tomaszewska-Hetman, L.; Rywińska, A.; Lazar, Z.; Rymowicz, W. Enhancement of α-Ketoglutaric Acid Production by *Yarrowia Lipolytica* Grown on Mixed Renewable Carbon Sources through Adjustment of Culture Conditions. *Catalysts* **2023**, *13*, 14. [CrossRef]
2. Li, Y.; Luo, W.; Liu, W.; Yang, Y.; Lei, Z.; Tao, X.; Wang, B. C058 and Other Functional Microorganisms Promote the Synthesis of Extracellular Polymer Substances in Mycelium Biofloc. *Catalysts* **2022**, *12*, 693. [CrossRef]
3. Wang, J.-T.; Shi, T.-T.; Ding, L.; Xie, J.; Zhao, P.-J. Multifunctional Enzymes in Microbial Secondary Metabolic Processes. *Catalysts* **2023**, *13*, 581. [CrossRef]
4. Song, M.; Fu, R.; Cai, S.; Jiang, X.; Wang, F.; Xu, W.; Xu, W. 7α and 7β Hydroxylation of Dehydroepiandrosterone by *Gibberella* sp. and *Absidia Coerulea* Biotransformation. *Catalysts* **2023**, *13*, 272. [CrossRef]
5. Song, M.; Zhu, H.; Wang, J.; Xu, W.; Xu, W. An O-Demethylation Metabolite of Rabeprazole Sulfide by *Cunninghamella Blakesleeana* 3.970 Biotransformation. *Catalysts* **2023**, *13*, 15. [CrossRef]
6. Liu, Y.; Xu, W.; Xu, W. Production of *Trans*-Cinnamic and *p*-Coumaric Acids in Engineered *E. coli*. *Catalysts* **2022**, *12*, 1144. [CrossRef]
7. Erfanimoghadam, M.R.; Homaei, A. Identification of New Amylolytic Enzymes from Marine Symbiotic Bacteria of *Bacillus* Species. *Catalysts* **2023**, *13*, 183. [CrossRef]
8. Lee, G.L.Y.; Zakaria, N.N.; Futamata, H.; Suzuki, K.; Zulkharnain, A.; Shaharuddin, N.A.; Convey, P.; Zahri, K.N.M.; Ahmad, S.A. Metabolic Pathway of Phenol Degradation of a Cold-Adapted Antarctic Bacteria, *Arthrobacter* sp. *Catalysts* **2022**, *12*, 1422. [CrossRef]
9. Song, M.; He, W.; Cai, S.; Wang, F.; Xu, W.; Xu, W. Nysfungin Production Improvement by UV Mutagenesis in *Streptomyces Noursei* D-3-14. *Catalysts* **2023**, *13*, 247. [CrossRef]
10. Zhang, H.; Wang, B.; Yang, S.; Yu, H.; Ye, L. Enhancing Acetophenone Tolerance of Anti-Prelog Short-Chain Dehydrogenase/Reductase EbSDR8 Using a Whole-Cell Catalyst by Directed Evolution. *Catalysts* **2022**, *12*, 1071. [CrossRef]
11. Liu, X.-J.; Ma, B.-D.; Wu, X.-M.; Xu, Y. Highly Efficient Biosynthesis of Nicotinic Acid by Immobilized Whole Cells of *E. Coli* Expressing Nitrilase in Semi-Continuous Packed-Bed Bioreactor. *Catalysts* **2023**, *13*, 371. [CrossRef]
12. Wojcieszyńska, D.; Łagoda, K.; Guzik, U. Diclofenac Biodegradation by Microorganisms and with Immobilised Systems—A Review. *Catalysts* **2023**, *13*, 412. [CrossRef]
13. Zhao, W.; Xie, H.; Zhang, X.; Wang, Z. Correlation Relationship between Phase Inversion of Pickering Emulsions and Biocatalytic Activity of Microbial Transformation of Phytosterols. *Catalysts* **2023**, *13*, 72. [CrossRef]
14. Zhang, J.; Yu, Z.; Gao, Y.; Wang, M.; Wang, K.; Pan, T. Biodegradation of Crystalline and Nonaqueous Phase Liquid-Dissolved ATRAZINE by *Arthrobacter* sp. ST11 with Cd2+ Resistance. *Catalysts* **2022**, *12*, 1653. [CrossRef]
15. Ahsan, W.A.; Hussain, A.; Lin, C.; Nguyen, M.K. Biodegradation of Different Types of Bioplastics through Composting—A Recent Trend in Green Recycling. *Catalysts* **2023**, *13*, 294. [CrossRef]

catalysts MDPI

Review

Multifunctional Enzymes in Microbial Secondary Metabolic Processes

Jun-Tao Wang, Ting-Ting Shi, Lin Ding, Juan Xie and Pei-Ji Zhao *

State Key Laboratory for Conservation and Utilization of Bio-Resources in Yunnan, School of Life Sciences, Yunnan University, Kunming 650091, China
* Correspondence: pjzhao@ynu.edu.cn

Abstract: Microorganisms possess a strong capacity for secondary metabolite synthesis, which is represented by tightly controlled networks. The absence of any enzymes leads to a change in the original metabolic pathway, with a decrease in or even elimination of a synthetic product, which is not permissible under conditions of normal life activities of microorganisms. In order to improve the efficiency of secondary metabolism, organisms have evolved multifunctional enzymes (MFEs) that can catalyze two or more kinds of reactions via multiple active sites. However, instead of interfering, the multifunctional catalytic properties of MFEs facilitate the biosynthetic process. Among the numerous MFEs considered of vital importance in the life activities of living organisms are the synthases involved in assembling the backbone of compounds using different substrates and modifying enzymes that confer the final activity of compounds. In this paper, we review MFEs in terms of both synthetic and post-modifying enzymes involved in secondary metabolic biosynthesis, focusing on polyketides, non-ribosomal peptides, terpenoids, and a wide range of cytochrome P450s(CYP450s), and provide an overview and describe the recent progress in the research on MFEs.

Keywords: multifunctional enzymes; secondary metabolic; synthases; post-modifying enzymes

Citation: Wang, J.-T.; Shi, T.-T.; Ding, L.; Xie, J.; Zhao, P.-J. Multifunctional Enzymes in Microbial Secondary Metabolic Processes. *Catalysts* **2023**, *13*, 581. https://doi.org/10.3390/catal13030581

Academic Editors: Zhilong Wang and Tao Pan

Received: 31 December 2022
Revised: 8 March 2023
Accepted: 9 March 2023
Published: 14 March 2023

1. Introduction

The life activities of organisms involve a large number of functionally rich enzymes, among which multifunctional enzymes (MFEs) are a class of enzymes with two or more catalytic functions. The polypeptide chains of MFEs are linked together by covalent linkages, which help maintain the homeostasis of MFEs. In addition, the function of polypeptide chains in MFEs is not affected by the separation of polypeptide chains, which is an important way to distinguish them from multi-enzyme complexes [1]. There are also a large number of MFEs in the biosynthesis pathway of microbial secondary metabolites, especially synthetases and modifying enzymes. To assemble the basic skeleton of the products, synthases use specific substrates, which are the basis upon which the products are distinguished [2,3]. The most representative synthetic enzymes are polyketide synthases (PKSs), non-ribosomal peptide synthases (NRPSs), and terpene synthases (TPSs). The modification enzymes, which are the main contributors to the abundance of secondary metabolites and the key to their ultimate activity, modify the product skeleton, resulting in products with different functions. As an important modifying enzyme, cytochrome P450s (CYP450s) participate in the post-modification process, and the abundant number and function of CYP450s indicates that they are multifunctional [4]. Hence, it is reasonable to speculate that MFEs can make full use of energy and substances in vivo and in vitro, enabling a faster and more accurate response when regulation occurs and facilitating the vital activities of the organism. Therefore, exploring MFEs in microbial secondary metabolic processes is a meaningful task to more clearly demonstrate the process of certain important biosynthetic products (Figure 1) and to provide precise data for application.

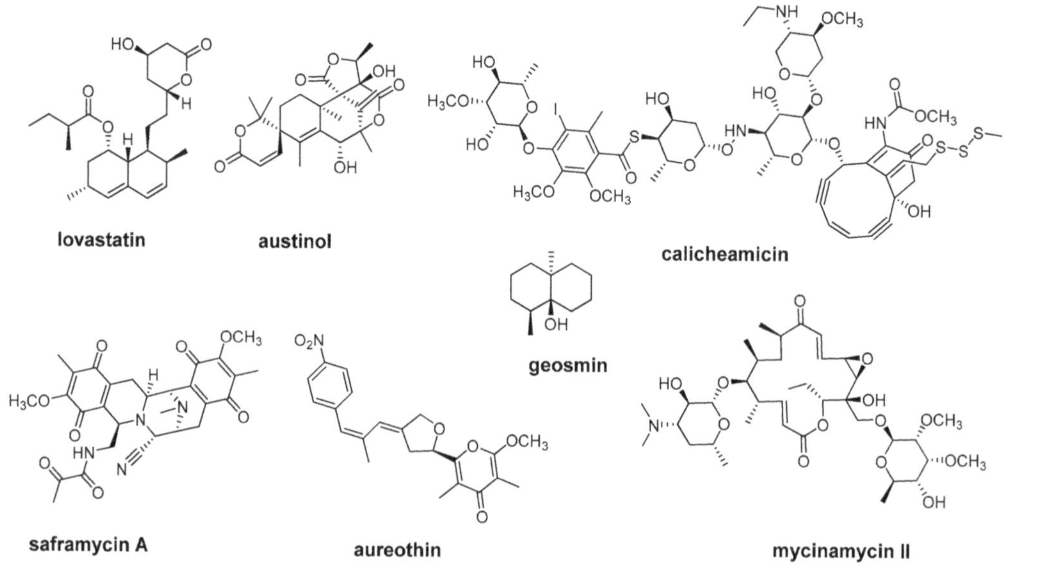

Figure 1. The main structures mentioned in this paper; their partial biosynthetic processes are catalyzed by MFEs.

2. Multifunctional Synthases in the Synthesis of Secondary Metabolic Processes

Polyketides, non-ribosomal peptides (NRP), and terpenoids are secondary metabolites widely found in microorganisms [5,6] and have been widely applied in antibiotic, anti-infective, antineoplastic, and immunization applications due to their excellent activity [3,7,8]. In this section, we mainly introduce the MFEs that are typically representative secondary metabolites (Figure 1). Interestingly, PKSs and NRPSs are analogous in many aspects concerning the function of domains and the biosynthesis process. On the one hand, the basic extension domains of PKSs are the acyltransferase (AT), acyl carrier protein (ACP), and condensation (KS) domains, which play the role of substrate loading, substrate transport, and condensation, respectively [9–12], while in NRPSs, these roles are performed by the adenylation domain A (A), peptidyl carrier protein (PCP) domain, and condensation domain C (C), respectively [13–17]. In the context of the biosynthesis process, the ACP (PCP) of PKSs (NRPSs) transports the substrate that is loaded and activated by AT (A), and finally KS (C) catalyzes the attachment of the substrate to the intermediate until the completion of all catalytic processes, after which the compound skeleton is formed. The backbone is extended via catalysis by the structural domain, which is eventually released by the terminating domain (TE) and modified by a post-modifying enzyme, resulting in the final active compound [13,14,16–31]. MFEs are ubiquitous, especially according to the available classification, and type I PKSs and type I NRPS are all MFEs. They are further divided into two categories, modular and iterative, where we focus more on the elaboration of iterative PKSs (IPKSs) and iterative NRPSs (INRPSs). Compared with modular enzymes, iterative enzymes contain structural domains that are reused in multiple rounds to complete product extension, although they have only one functional module. Interestingly, despite the smaller size and fewer structural domains of iterative enzymes, their synthesis capacity is not weaker than that of modular enzymes and, moreover, has the potential to be developed. Regarding multifunctional TPSs, their multifunctionality is more often seen in bifunctional enzymes with both isoprenoid transferase (PT) and TPS activities—for example, bifunctional terpene synthases (BTSs), which are bifunctional enzymes that continuously catalyze the proceeding key steps in terpene biosynthesis and have excellent terpenoid synthesis capabilities. In addition, there exist some TPSs with modifying roles that are directly involved in the post-modification process in addition to cyclization, all of which will be elaborated in this section.

2.1. Multifunctional Polyketide Synthases

6-Methylsalicylic acid synthase (6-MSAS), first isolated from *Penicillium patulum*, is a representative of IPKS [32]. 6-MSAS was found to be involved in the biosynthesis of 6-methylsalicylic acid (6-MSA), which is one of the simplest polypeptides produced [33]. The single protein of 6-MSAS contains all the catalytically active domains required for the synthesis of 6-MSA [34]. 6-MSAS catalyzes the biosynthesis of 6-MSA from one acetyl coenzyme A and three malonyl coenzyme A units by successive decarboxylation condensation [35–37]. 6-MSAS initiates the reaction by loading an acetyl starting unit and a malonyl extension unit, and the growing polyketide midbody is bound to the ACP until the end of synthesis. Usually, PKSs releases products through the termination domain (TE), which is absent in 6-MSAS. Thus, the question of what 6-MSAS releases its products through has persisted for some time. There are specialized domains for product release in other IPKSs, such as the Claisen cyclase domain, as well as the R domain [38,39]. The results of Tomomi et al. suggest that the product release of 6-MSAS is inextricably linked to its dehydratase domain (DH), called the thioester hydrolase domain (TH), with which 6-MSAS hydrolyzes the thioester bond of the tetraketone intermediate linked to the ACP for product release (Figure 2) [40].

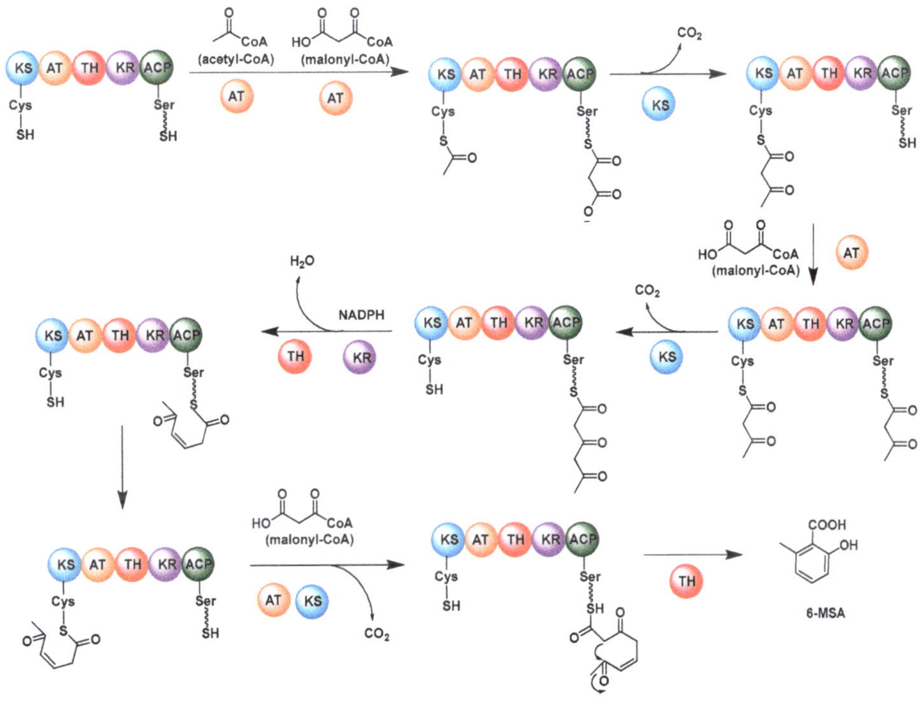

Figure 2. The representative IPKS 6-MSAS is involved in the biosynthesis of 6-MSA. 6-MSAS repeatedly uses a module to complete the biosynthesis, and each domain is reused in this process. This is a typical way for IPKSs to synthesize products.

In addition, IPKSs responsible for the biosynthesis of aromatic polyketides also occur in bacteria, where they may even be more prevalent. As the first bacterial 6-MSAS identified for 6-MSA biosynthesis, ChlB1, first isolated from *Streptomyces antibioticus* Tü99, is involved in the biosynthesis of the antibiotic chlorothricin (CHL) [41]. ChlB1 is responsible for the synthesis 6-MSA, which is methylated and chlorinated to give the 2-methoxy-5-chloro-6-methylsalicylic acid fraction of CHL [42,43]. In the biosynthesis of maduropeptin (MDP), an enediene from actinomyces, MdpB, similar to ChlB1, is involved in the biosynthesis of part of 6-MSA, namely the part that is modified and used as a component of MDP [44]. PtmQ from *Streptomyces pactum* ATCC 27456, which is highly similar to ChlB1, is thought to be involved in the biosynthesis of 6-MSA in this organism [45]. As the first discovered bacterial 6-MSAS, ChlB1 is homologous to the fungal 6-MSAS in organization and sequence, and phylogenetic analysis shows that ChlB1 is genetically closer to AviM, CalO5, and NNS, which are bacterial IPKSs [43].

The representative IPKS LovB is responsible for the biosynthesis of lovastatin, a cholesterol-lowering drug formed from dihydromonacolin L (DML) [46] found in *Aspergillus oryzae*. LovB contains an enoyl reductase (ER)-like structural domain that lacks the corresponding activity and therefore requires the assistance of a separate enoyl reductase, LovC, for the biosynthesis of lovastatin [47,48]. The entire process of lovastatin biosynthesis involves eight cycles of synthesis, and throughout the biosynthesis of lovastatin, the individual structural domains of LovB are used in selective combinations: the MT–KR–DH–ER domains combination are used in the third cycle; the KR–DH–ER domains combination are used in the fourth and sixth cycles; the KR–DH domains combination are used in the first, second, and fifth cycles; and only the KR domain is used in the seventh and eighth cycles (Figure 3). The study by Suzanne et al. demonstrated that the C-terminus of LovB contains neither the TE nor other common unloading domains. When using LovB and LovC to produce polyketides in vitro, the polyketides remain attached to LovB and, therefore, the other structural domains release DML [49]. A study by Xu et al. identified that the gene *lovG*, located between lovB and lovC in the biosynthetic gene cluster of lovastatin, belongs to the esterase–lipase family of serine hydrolases and is involved in the release of LovB products, which is in agreement with the results that the presence of a LovB-independent structure is involved in DML release [50].

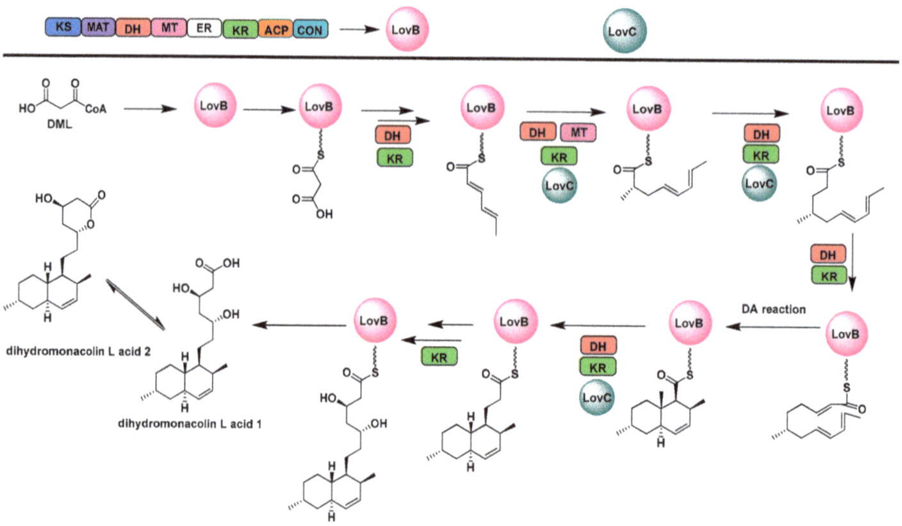

Figure 3. LovB, as an IPKS, is involved in the biosynthesis of lovastatin. LovB reuses its domain combinations to participate in different stages of lovastatin synthesis, and different domain combinations give lovastatin a complex structure and excellent activity. The methyltransferase domain (MT) is active only during the fourth cycle, producing the intermediate α-methyl 3-ACP, a reaction that follows chain elongation catalyzed by ketone synthase (KS) and precedes the function of the keto reductase (KR), dehydratase (DH), and alkenyl reductase (ER) domains.

IPKSs play an important role in the formation of polyketone structures during the biosynthesis of a noteworthy class of antibiotics, the enediynes. Calicheamicin, a non-pigmented enediyne from *Micromonospora* that can be used as an antitumor agent, possesses two specific polyketide structural elements, PKS E and PKS O, of which PKS E is known as the calicheamicin warhead and is an important functional structure of calicheamicin. Two genes in the biosynthetic gene cluster of calicheamicin, *calE8* and *calO5*, encode two independent IPKSs, CalE8 and CalO5, respectively. CalO5 shares a high degree of similarity to the IPKS-AviM responsible for avilamycin synthesis. Joachim found that the CalE8 of a *Micromonospora* disruption mutant produced neither calicheamicin nor enediene activity nor compounds that would be produced by disruption of the PKS O synthesis machinery, consistent with disruption of PKS E, suggesting a role for CalE8 on PKS E [51]. Another study showed that an IPKS-SgcE from *Streptomyces globisporus* is involved in the biosynthesis of the slug portion of the enediene antibiotic C-1027, which, unlike calicheamicin, is a class of pigment protein enediene [52]. By comparing the slug structures of C-1027, calicheamicin, and the fellow nonchromoprotein dynemicin, it was hypothesized that a highly conserved PKS is responsible for the biosynthesis of the slug structures of both the pigmented and nonchromoprotein enediynes [51]. In addition, it was shown that the naphthalene fraction of enediyne, another enediyne antitumor antibiotic, is synthesized via condensation of six intact acetate units by naphthalene synthase, an IPKS that, similarly to CalO5 above, shares a high degree of sequence homology with AviM [53].

2.2. Multifunctional Non-Ribosomal Peptide Synthase

INRPSs are often found to be involved in the biosynthesis of NRPs with repetitive sequence composition, and these NRPSs often reuse a domain, especially the TE. Many microbial siderophores are synthesized by NRPSs [54], such as coelichelin, fuscachelin C, etc. Hai et al. proposed the INRPS involved in fuscachelin C biosynthesis and demonstrated a mechanism for fungal NRPS in assembling iron carriers [55]. Enterobactin NRPS, a typical representative INRPS, is a siderophore found in Gram-positive bacteria that promote cellular iron transport. Enterobactin is a circular trimer consisting of a dihydroxybenzoyl group and serine. Prior to biosynthesis, conversion of dihydroxybenzoyl from an aromatic amino acid precursor branching acid to an isobranching acid is first required, then to 2,3-dihydro-2,3-dihydroxybenzoate, and finally to 2,3-dihydroxybenzoic acid (DHB) [56]. Amide linkage of DHB and L-serine is catalyzed by EntD, EntE, EntF, and the C-terminal aryl carrier of EntB [57], in which EntF consists of four domains, C, A, PCP, and TE. L-Serine is activated by domain A and subsequently bound to PCP in the form of an acyl-*S*-pantothiamine intermediate awaiting binding to DHB [58], which is activated by EntE and binds to EntB [59]. The formation of a peptide bond is catalyzed by domain C of EntF, which binds DHB and L-serine, forming a dihydroxybenzoyl–serine unit. This is followed by the most interesting process in the biosynthesis of enterobactin, wherein the dihydroxybenzoyl–serine unit is transferred to the TE of EntF, leaving the site open for assembly of the next unit (Figure 4). This process is repeated three times and, finally, the cyclization on TE forms a cyclic trimer of dihydroxybenzoyl–serine units, which are finally released by hydrolysis [60]. Even more interesting is that there are countless examples of such iterative use of the TE in such modules as 2,3-dihydroxybenzoate (DHB)-glycine, an iron carrier produced by *Bacillus subtilis* during iron deficiency that is very similar to enterobactin and is deduced to be a DHB–Gly–Thr trimer with the same DHB portion as enterobactin. Similarly, 2,3-dihydroxybenzoate (DHB)–glycine uses the TE at the end of the C-terminus of INRPS for iterative repeated triple condensation to form the final product. The proteins DhbE, DhbB, and DhbF are involved in the whole biosynthesis process, among which DhbE and DhbB function similarly to EntE and EntB in enterobactin for loading and activating DHB; DhbF is different from EntF, being divided into DhbF–Gly and DhbF–Thr, which are correspondingly responsible for loading Gly and Thr to form the DHB–Gly–Thr monomer, after which TE plays an iterative role to complete biosynthesis of the product [58]. The biosynthetic processes of enedimycin, valinomycin, echinocandin, and gramicin, compounds of bacterial origin, are similar in that they are all formed from repetitive condensation of TE monomers [61–65], which seems to foreshadow the similarity in the biosynthesis of this class of compounds in bacteria.

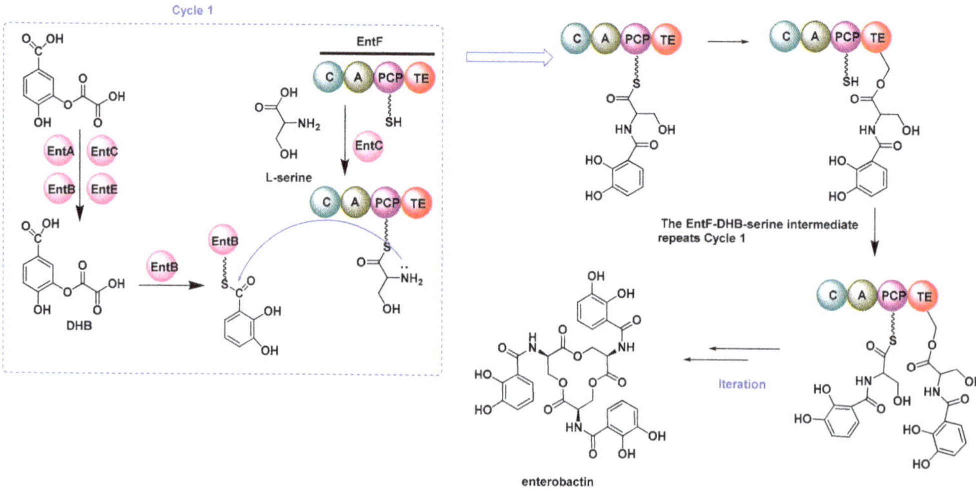

Figure 4. DHB is synthesized under the action of EntC, EntB, EntA, and EntE, in which EntE (a bifunctional enzyme) repeatedly plays a role. EntF is an INRPS that binds L-serine and catalyzes its binding to DHB. This process occurs three times, during which the intermediate product binds to the TE domain, and the final product, enterobactin, is released under the catalysis of TE.

In 2012, Oikawa et al. performed in vitro expression and biochemical testing of the NRPS for saframycin A biosynthesis and proposed that NRPS SfmC plays an iterative role in completing the final assembly of saframycin A, which is a potent antitumor antibiotic. SfmC plays an iterative role by add two tyrosine derivative units to its skeleton. Unlike the biosynthesis process of NRP, such as for enterobactin described above, the C-terminal structural domain R of SfmC, according to the in vitro reaction results, catalyzes multiple steps of other reduction reactions during this biosynthesis in addition to playing a role in catalyzing the reductive cleavage and release of the mature polypeptide chain from the last PCP structural domain adjacent to the R structural domain [66].

2.3. Multifunctional Terpene Synthases

Two key enzymes play a role in terpenoid biosynthesis: isoprenoid transferases (PTs), which are responsible for the production of C5 precursors, and terpene synthases (TPSs), which catalyze the formation of the basic terpene backbone by cyclizing these chain precursors synthesized by PTs. Generally speaking, PTs and TPSs are two separate enzymes; however, PTs and TPSs have increasingly been found to coexist in the same protein as separate active sites involved in catalytic sequential reactions, and these bifunctional enzymes are called bifunctional terpene synthases (BTSs) [67,68]. Type I BTSs contain $\alpha\alpha$ structures, such as the fusicoccadiene synthase PaFS from *Phomopsis amygdali*, while type II BTSs have $\alpha\beta\gamma$ structures, such as the copalyl diphosphate synthase (CDS) from *Penicillium verruculosum* [69–71]. In recent years, both PaFS and CDS have been researched relatively in-depth by David's group [72]. PaFS is the first BTS identified to catalyze the first two steps of diterpene glycoside fusicoccadiene biosynthesis [68,73]. Its C-terminal PT structural domain generates geranylgeranyl diphosphate (GGPP), which is used by the N-terminal class I cyclase structural domain to generate the tricyclic hydrocarbon backbone of fusicoccadiene. CDS was the first BTS found to possess both PT and class II TPS activity [74]. The α structural domain of PT catalyzes the condensation of C5 dimethylallyl diphosphate with three successive additions of C-5 isopentenyl diphosphate (IPP) to form C-20 GGPP, which is then cyclized in the N-terminal class II cyclase structural domain to form copalyl diphosphate [75]. Based on the attraction of the efficient biological activities of terpenoids, various techniques are being actively attempted in the expectation of improving product yields or obtaining structurally novel structures. The yields achieved using ophiobolin synthase, a BTS that is highly homologous to PaFS, have been increased more than 100-fold by constructing a high-yield chassis, a method that can efficiently increase synthetic yields [76]. It was demonstrated that the depth of the PT active site determines the final length of the product [77], and fixed-point mutagenesis experiments showed that the length of the product chain can be changed by altering the depth of the active site pocket [78]. Thus, David et al. demonstrated that CDS can be modified to produce sesquiterpenes and that PT partially produces farnesyl diphosphate (FPP) instead of GGPP [75]. Extending the biosynthetic potential of BTS through combinatorial biology is a good option, but this aspect remains to be explored in relation to PKSs and NRPSs. Some studies have made some progress in demonstrating the applicability of combinatorial biology approaches to the development of TPS [79].

In another study, Tiangang Liu selected two TPSs with a substrate admixture by constructing a phylogenetic tree, and their broad substrate specificity was verified by modifying the mevalonate pathway such that it could provide sufficient amounts of C5 precursors. Six BTSs were constructed by combining the screened TPSs with PT for efficient synthesis of FPPS, GGPPS, and GFPPS, resulting in seven new terpenoids with three new backbones [80]. This modular-like combination seems to be a feasible way to exploit the potential of TPSs and rapidly expand the number of terpenoids. However, because of limited substrate specificity, there is still a need to discover new TPSs. New compounds can be obtained by constructing multifunctional TPS using combinatorial biology methods [81].

Genome-based mining for BTS is a fast and efficient route, and new enzymes are constantly being discovered [80,82–86]. Two BTSs, PbSS and PvPS, were presumably obtained by analyzing the genomes of *Penicillium brasilianum* NBRC 6234 and *Penicillium verruculosum* TPU1311, and their cyclization mechanisms were investigated using isotope labeling experiments [87]. Genome mining of *Aspergillus ustus* 094102 led to identification of a BTS, AuAS, whose heterologous expression resulted in five new diester terpenoids, while coexpression of AuAS and the upstream CYP450 (AuAP450) yielded four new diester terpene alcohols [88]. Subsequently, AcAS from *Aspergillus calidoustus*, a homologue of AuAS, was heterologously expressed and found to be involved in the biosynthesis of six ester terpenoids, one of which is the new skeleton ester terpenoid calidoustene, and the other five are the five diester terpenoids found after the aforementioned heterologous expression of AuAS (Figure 5) [89]. Liu et al. made remarkable progress by systematically analyzing the genomes of 477 fungal species and found a total of 227 homologous BTS genes, which includes 20 PTTS genes whose functions are known. In order to obtain more information on the catalytic functions of these BTSs, 74 were selected based on the results of functional identification. Yeast engineered strains of 74 BTSs were constructed using a high dipterpene producing yeast chassis, of which 34 new functional BTS were successfully identified. This study presents some new evidence on the origin of BTSs and the consistency of their functions, as well as the convergent evolution of gene functions in fungi, plants, and bacteria [90].

In addition to the BTSs mentioned above, there are also some multifunctional TPSs that tend to have some additional modifying effects, and these additional modification activities are often conducive to the formation of the final products. The volatile sesquiterpene derivative geosmin from *Streptomyces coelicolor* is synthesized by geosmin synthase (GS), the first bifunctional TPS found in bacteria, and is present in various Gram-positive soil bacteria. GS is a bifunctional TPS with two α structural domains: the N-terminal α structural domain of type I TPS catalyzes the FPP reaction to form cyclized sesquiterpenes, which are precursors of geosmin, and the C-terminal α structural domain catalyzes the cyclization break reaction to produce hyoscyamine and acetone [91–93]. A bifunctional synthase encoded by the gene *crtYB* was found to synthesize octahydro lycopene and cyclize lycopene to β-carotene. First, the octahydro lycopene-β-carotene synthase active domain condenses two GGPP molecules to produce octahydro lycopene, and four desaturation reactions subsequently take place to produce lycopene. Finally, lycopene cyclase activity results in cyclization of both ends of lycopene to produce β-carotene [94].

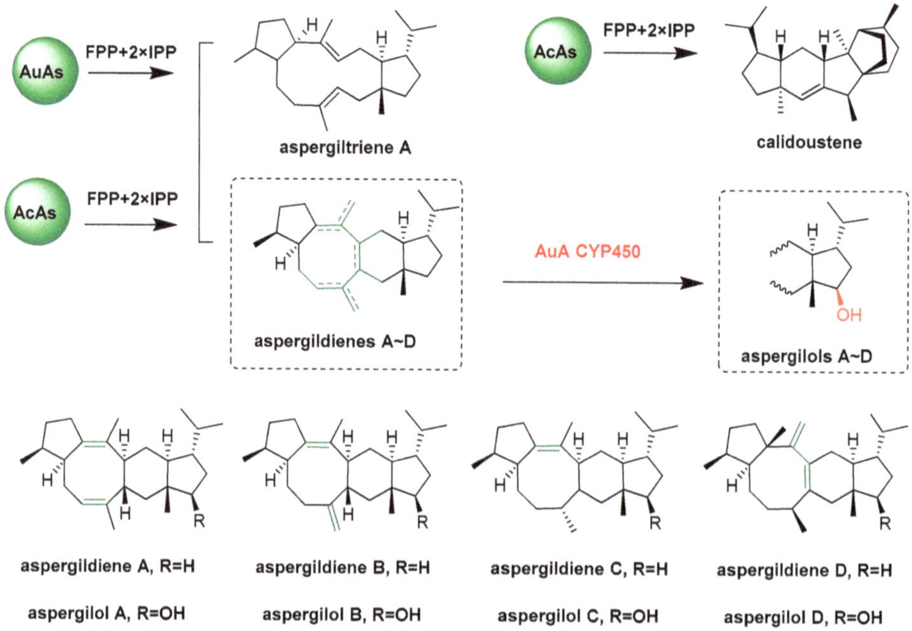

Figure 5. AuAs and AcAs are homologous BTSs containing PT and TPS active domains, and they can use the same substrates to synthesize five different terpenoids: aspergitriene A and aspergidiene A, B, C, and D. AuAs and CYP450 can use aspergidiene A, B, C, and D to produce the corresponding oxidation products aspergilol A, B, C, and D. In addition, AcAs were found to synthesize calidoustens, another compound with a new skeleton.

3. Multifunctional Post-Modifying Enzymes in the Synthesis of Secondary Metabolic Processes

After the synthases complete the assembly of the compound backbone, modifying enzymes, such as oxidases, methyltransferases, glycosyltransferases, and halogenases, catalyze highly diverse and complex structural modifications, a critical step in conferring biological activity to the compounds [2,3]. Among them, oxidases are extensively involved in structural modifications to obtain complex and highly oxidized structures, which are essential for the products to acquire biological activity. The main oxidases commonly found in microbial secondary metabolism are CYP450s, flavin-dependent monooxygenases (FMO), and dioxygenases represented by non-heme α-ketoglutarate-dependent dioxygenases, which present multifunctional activity.

3.1. *Multifunctional Cytochrome P450*

As an essential modifier enzyme in many secondary metabolic processes, the CYP450 gene is typically integrated in biosynthetic clusters [95] and functions to hydroxylate hydrocarbon bonds through insertion of oxygen atoms to produce hydroxylated metabolites and H_2O [96], which can catalyze region-specific and stereospecific oxidation of precursors, allowing for structural diversity and enhanced biological activity. More and more

MFEs are being discovered, and many CYP450s from fungi and bacteria are found to be multifunctional.

The same CYP450 enzymes can catalyze different biosynthetic processes or different stages of the same biosynthetic process. Most of the multifunctional CYP450s function in the same biosynthetic process. In contrast to pure hydroxylation and epoxidation reactions, these multifunctional CYP450s are usually responsible for sequential reactions, catalyzing sequential hydroxylation or epoxidation combinatorial reactions, directly using the products of the previous reaction as substrates [97,98]. The most typical example of this is the synthesis of gibberellins (GAs) in fungi, for example, a gibberellic acid called GA3, a tetracyclic dihydroxy gamma-lactone acid containing two ethylene bonds and a free carboxylic acid group. In fungi, such as *Fusarium fujikuroi*, the main final product of the gibberellin synthesis pathway is gibberellic acid, which is synthesized by seven genes located in a gene cluster. During subsequent modification, four CYP450 genes encode four multifunctional P450s, namely P450-1, P450-2, P450-3, and P450-4 [99]. Five oxidases participate in the GA biosynthetic pathway, including P450-1, P450-2, P450-3, P450-4, and a 2-oxoglutarate-dependent dioxygenase DES. The role played by oxidases in the biosynthesis of GA3 is well elucidated: P450-4 is responsible for the oxidation of ent-kaurene to ent-kaurenoic acid. P450-1 oxidizes ent-kaurenoic acid to GA14 and catalyzes four sequential steps in the GA biosynthetic pathway: 7β-hydroxylation, oxidation of C-6 to give a contractile ring B, 3β-hydroxylation, and C-7 oxidation [100]; P450-2 functions as a GA20-oxidase, oxidizing GA14 to GA4, and is able to oxidize GA12 to GA9 [101]; P450-3 encodes a 13-hydroxylase responsible for the conversion of GA7 to GA3 and also catalyzes the 13-hydroxylation of GA4 to GA1 [102]; and 2-oxoglutarate-dependent dioxygenase DES is responsible for desaturation and conversion of GA4 to GA7 (Figure 6) [102,103].

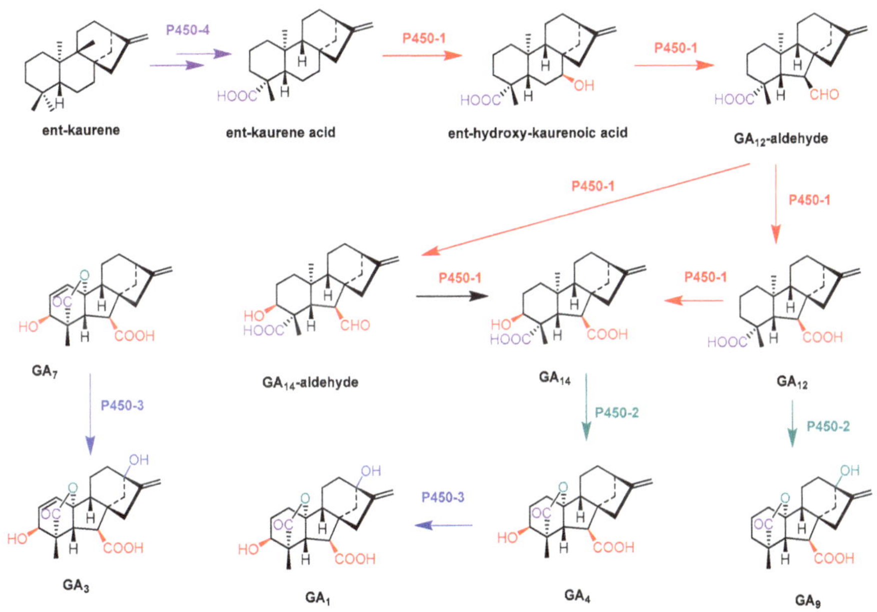

Figure 6. Four CYP450s, P450-1, P450-2, P450-3, and P450-4, are involved in post-modification in the process of GA biosynthesis. All are MFEs and play a variety of oxidative functions, resulting in a variety of products. Among them, P450-1 is the most typical, which can catalyze four oxidation reactions: 7β-hydroxylation, oxidation of C-6 to obtain a contractile ring B, 3β-hydroxylation, and C-7 oxidation.

A multifunctional CYP450 is involved in the biosynthesis of a well-known compound, lovastatin, and its backbone synthesis is described in the IPKS section above. However, there is no clear explanation of how the cyclic ninhydrin DML is converted to monacolin J and of how the double bond and hydroxylation are formed. Both in vivo and in vitro characterizations by Jorge Barriuso and Don T. Nguyen showed that LovA can catalyze the regio- and stereospecific hydroxylation of monacolin L-acid to produce monacolin J acid. MycG, a multifunctional CYP450 found in *Micromonospora griseorubida* A11725, which is involved in the biosynthesis of mycinamycin II, a 16-membered macrolide antibiotic. Mycinamycin II contains two types of sugars, deglycosamine and mycinamycin, at C5 and C21, respectively. Genetic complementation analysis and in vitro characterization of MycG confirmed that it is a multifunctional CYP450 that catalyzes sequential hydroxylation and epoxidation of substrates [104]. During the biosynthesis of mycinamycin II, MycG can continuously catalyze the hydroxylation of C14 and the epoxidation of C12 and C13 of macrolides on the substrate mycinamicin IV to obtain mycinamycin II [105,106]. AurH, a CYP450 from *Streptomyces thioluteus*, catalyzes two sequential oxidation reactions in the biosynthesis of the polyketide antibiotic aureothin, in which AurH first catalyzes hydroxylation of the allyl groups and then oxidation of the methyl groups to generate tetrahydrofuran rings, which are important active structures of aureothin [107,108]. The researchers who discovered this multifunctional CYP450 subsequently revealed the structural and biochemical basis of AurH-catalyzed tetrahydrofuran ring formation [109]. In addition, the multifunctional CYP450 Fma-P450 is involved in three successive oxidation steps of the antiparasitic and antiangiogenic drug, namely the terpenoid fumagillin from *Aspergillus fumigatus* [110]. CYP450 RosC is responsible for the formation of hydroxymethyl, formyl, and carboxyl groups during the biosynthesis of rosamicin (16-membered macrolide) produced by *Micromonospora rosaria* IFO 13697 [111–113]. Similar examples have been described elsewhere [114–117]. Aflatoxin is a class of toxic and carcinogenic microbial secondary metabolites, among which the most famous and toxic is aflatoxin B1. To obtain the final chemical structure in the biosynthesis of aflatoxin B1, two successive oxidation reactions are required. These two successive oxidation reactions have been shown to be catalyzed by a CYP450 OrdA encoded by a single gene. The synthesis intermediate o-methyl-sterimatocystin is first catalyzed by OrdA, which adds a hydroxyl group to the A-ring of *O*-methyl-sterimatocystin and then continues to produce an unstable seven-membered lactone ring under OrdA catalysis. After deacidification and rearrangement, it is converted to aflatoxin B1. *ordA*-encoded proteins have previously been found to be involved in the biosynthesis of aflatoxins B1, G1, B2, and G2 by functioning as a CYP450s, and they have shown the ability to convert OMST to aflatoxin B1 independently (Figure 7) [118,119].

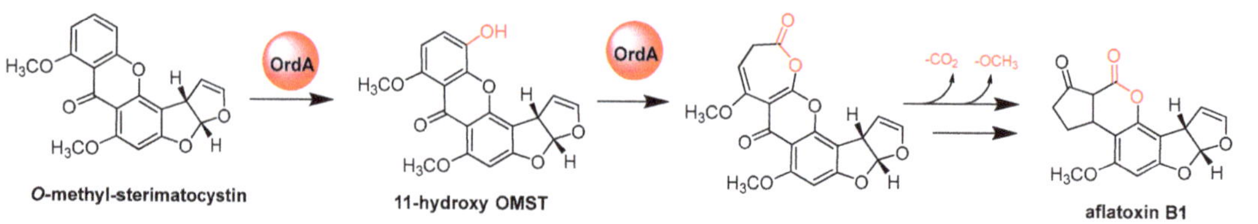

Figure 7. OrdA functions as a multifunctional CYP450 in the modification of aflatoxin B1. OrdA catalyzes a two-step oxidation reaction in the post-modification of aflatoxin B1. Firstly, a hydroxyl group is added to the A-ring of *O*-methyl-sterimatocystin, and the unstable seven-membered lactone is then generated under OrdA catalysis. Subsequent deacidification and rearrangement are also thought to involve CYP450, a putative formation step shown in the figure.

A fraction of multifunctional CYP450 can function in different synthetic processes due to their wide range of substrate properties. For example, the first reported CYP450 hydroxylase, PikC, can accept macrolide substrates of different ring sizes, catalyzing the addition of hydroxyl groups at two different positions on the macrolide. In *Streptomyces venezuelae*, it is the enzyme responsible for hydroxylation during the biosynthesis of the 12-membered ring compounds methymycin and neomethymycin and the 14-membered ring ketolide pikromycin [120]. CYP170A is a bifunctional enzyme that catalyzes the last two consecutive allylic oxidations in the biosynthesis of albaflavenone [121]. CYP170A1 first catalyzes the hydroxylation reaction to generate a heteromeric mixture of albaflavenols and then converts these heteromeric albaflavenols to albaflavenone [122]. Crystal structure and mutagenesis analysis revealed that CYP170A1 also has a separate non-CYP450 active site, a novel TPS active site that converts FPP into a mixture of farnesene isomers [123].

3.2. Other Oxygenases

FMO is a common enzyme for biochemical Baeyer–Villiger oxidation and also accomplishes a variety of oxidation reactions, including epoxidation, hydroxylation, oxidative decarboxylation, halogenation, and sulfonation oxidation, and MFEs exist for this class of oxidases [124–126]. XanO4 is an MFE responsible for catalyzing the oxidative substitution of carbonyl and methoxy in anthraquinone compounds during the biosynthesis of yellow phospholipids, leading to the formation of xanthenone rings and the replacement of methoxy with hydroxyl groups. Through isotopic labeling experiments, it has been shown that the reaction involves the sequential insertion of two oxygen atoms and is accompanied by demethoxylation [127]. Cytochalasin, an angiogenesis inhibitor produced by *Aspergillus oryzae*, contains a vinyl carbonate fraction, and CcsB is an FMO responsible for the introduction of the carbonate fraction via two oxidations of the ketomacrocyclic precursor [128]. Bradley and colleagues have shown that the FMO EncM catalyzes two sequential oxidations of the polyketide midbody, followed by an unusual Favorskii-type rearrangement to produce enteromycin [129,130].

Non-heme α-ketoglutarate-dependent dioxygenases are representatives of dioxygenases that are widely distributed in nature and usually use α-ketoglutarate as a co-substrate and ferrous iron as a cofactor to oxidize the substrate through decarboxylation of α-ketoglutarate into succinic acid with subsequent oxidative coupling of the substrate. Hydroxylation is probably the most common reaction performed by these enzymes, but they can also catalyze other oxidation reactions [131–133]. Austinol is a fungal diterpene derived from 3,5-dimethylmustardic acid with a unique chemical structure and a remarkable and unusual spirolactone ring system, whose formation requires the α-ketoglutarate-dependent dioxygenase AusE, along with two FMOs for catalysis. AusE is responsible for iterative oxidation steps, including oxidation of the spirocyclic ring formation reactions, to produce the austinol scaffold [134]. This is the first example of an α-ketoglutarate-dependent dioxygenase that catalyzes the formation of spirolactone rings. PrhA from *Penicillium brasilianum* has high homology with AusE, and the two enzymes can catalyze the same substrate via a similar catalysis process, except that AusE first desaturates at C1-C2 to form preaustinoid A2 and then rearranges, leading to the formation of spirolactone in preaustinoid A3, whereas PrhA first desaturates at C5-C6 to form berkeleyone B and then rearranges the A/B ring to form the cycloheptadiene portion of berkeleydione [135,136]. Interchanging the non-conserved residues on AusE and PrhA by targeted mutagenesis was used to demonstrate the possibility of functional conversion. After interchanging residues Val150 and Ala232 in PrhA and the corresponding residues Leu150 and Ser232 in AusE, both mutants lost their original function. The results demonstrate that the catalytic function of non-heme iron α-ketoglutarate-dependent oxygenases can be altered by a certain degree of substitution with non-conserved residues, and that subtle differences in the active site structure can lead to dramatic changes in the reaction results, serving as an example for the modification of such enzymes [137].

Other multifunctional non-heme iron α-ketoglutarate-dependent oxygenases, such as CAS, catalyze three oxidation reactions in the production of clavulanic acid, an important β-lactamase inhibitor [138,139]. In cephalosporin C biosynthesis, DAOC, a bifunctional enzyme present in eukaryotes, such as *Cephalosporium acremonium*, is responsible for the sequential oxidative ring expansion of penicillin N, to produce deacetoxycephalosporin C, and its hydroxylation to form deacetylcephalosporin C [140,141]. A bifunctional α-KG oxygenase, carbapenem synthase (CarC), catalyzes differential isomerization and desaturation reactions in the formation of carbapenem-3-carboxylate, a carbapenem antibiotic [142].

3.3. Post-Modifying Enzymes in the Synthesis of Lanthipeptides

Peptide secondary metabolites of microorganisms are a class of natural products with a wide range of biological activities, typically represented by NRP, and another class represented by ribosomally synthesized post-translationally modified peptides (RiPPs) [143,144], which result from the ribosomal synthesis of polypeptide chains followed by post-modification. The biosynthesis of RiPPs is usually accompanied by a critical hydrolysis step that separates the N-terminal lead peptide from the C-terminal core peptide [145], of which the core peptide is the biologically active RiPPs. Lanthipeptides comprise one of the largest subfamilies of RiPPs and are widely utilized as an antibiotic [145]. According to the biosynthesis pathway, lanthipeptides are classified into four categories [146], and with the exception of type I lanthipeptides, the post-modification enzymes of the other three types of lanthipeptides are MFEs. Although these enzymes are involved in post-modification, they are, in most cases, referred to as lanthipeptide synthases.

Type I lanthipeptides are post-modified by two enzymes: LanB, a dehydratase, and LanC, a cyclase. Type II lanthipeptides are synthesized by LanM, a bifunctional enzyme with a dehydratase domain at the N-terminal end and a cyclase domain at the C-terminal end. Type III and IV lanthipeptides are post-modified by the trifunctional enzymes LanKC and LanL, both of which contain an N-terminal cleavage domain and a central kinase domain with a LanC-like domain at the C-terminal end [147]. However, different structures exist at the end of their C-termini, with the C-terminal domain of LanL containing a conserved zinc-binding motif, and such motifs are not found in the C-terminal cyclase domain of LanKC [148,149]. Notably, new lanthipeptides are still being discovered, and the first V-type lanthipeptide cacaoidin, a glycosylated lanthipeptide, was discovered in 2020 [150,151].

All lanthipeptides contain either the cyclic amino acid lanoline (Lan) or the methyl lanoline (MeLan), the formation of which requires the dehydration of Ser and Thr residues into dehydroalanine (Dha) and dehydrobutyrine (Dhb), followed by a subsequent cyclization reaction in which the Cys residues undergo conjugate addition to Dha and Dhb, corresponding to the formation of thioether cross-linked lanoline and methyllanoline, respectively [152]. In addition, the dehydration mechanism for Ser and Thr residues is the main way to distinguish lanthipeptides [153]. Phylogenomic studies have shown that although these enzymes have very similar cyclase structural domains, the three classes of LanM, LanKC, and LanL have evolved independently. At present, there are a few studies on the structures of these post-modifying enzymes. Only the crystal structure of CylM, which is involved in type II lanthipeptides synthesis, has been revealed and shows the expected structure [154].

The discovery of type III lanthipeptides was relatively late. The first type III lanthipeptide synthase identified was RamC, which is involved in the biosynthesis of the lanthipeptide SapB. A sequence analysis showed that this enzyme contains a Ser/Thr protein kinase structural domain and a C-terminal structural domain [155]. Shortly afterward, a new lanthipeptide synthase with homology to the N-terminus of RamC was discovered, which was shown to contain three catalytic structural domains, including a kinase-like domain responsible for phosphorylation of Ser/Thr residues, an N-terminal lyase domain responsible for phosphate elimination, and a C-terminal cyclase domain. The truncated protein without the C-terminal cyclase domain can also independently catalyze the dehydration reaction. The C-terminal cyclase domain has the same cyclization strategy as LanM and LanC proteins, except that this enzyme contains a conserved zinc-binding motif, which shows that its C-terminal cyclase domain is zinc ion-dependent, whereas RamC-type enzymes do not have a zinc-binding motif [156]. Type III and IV lanthipeptides synthases differ in that the C-terminal cyclization sequence of type III lanthipeptides synthase has no zinc-binding motif and is known as LanKC, whereas the C-terminal cyclization sequence of type IV lanthipeptide synthase contains a zinc-binding motif and is known as LanL [157]. The mechanism of LanL cyclization is similar to that of cyclase LanC due to the similarity of the cyclization structural domain. The cyclization mechanism of LanKC is the most puzzling because their cyclization structural domains do not have a conserved zinc-binding motif. In addition, the presence of a lanthionin or labionin carbon ring in type III lanthipeptides indicates the presence of another type of cyclization, and there are no clear studies showing how this occurs.

3.4. Other Multifunctional Modifying Enzymes

The introduction of fluorine, chlorine, bromine, and iodine substituents into compounds by halogenases can lead to organohalides with high biological activity, due to the significant electronic and spatial properties conferred by the halogen part [158,159]. Of the nearly 5000 or so halogenated natural products identified, chlorinated and brominated metabolites predominate, with iodinated and fluorinated metabolites being fairly rare [160]. A new halogenase, AoiQ, has been identified in *Aspergillus oryzae* and is representative of multifunctional halogenases. It is involved in the synthesis of the halogenated polyketide compound dichlorodiaporthin and is responsible for catalyzing the highly regiospecific sequential aliphatic dichloride of independent polyketide substrates. Prior to this, only a few aliphatic halogenases have been biochemically characterized compared with the large number of well-studied aromatic halogenases, and all are Fe II ketoglutarate-dependent enzymes. This is the first flavin-dependent halogenase (FDH) for fungal aliphatic halogenation. Bioinformatic analysis and functional genetics indicate that AoiQ is a bifunctional enzyme. Successful recombination of AoiQ in vivo and in vitro demonstrates its ability to progressively add gem-dichloride inactive carbon atoms onto independent substrates. cDNA sequencing confirmed that AoiQ has 1014 amino acids and contains two distinct functional domains: one with FDH activity for catalytic halogenation and the other with a conserved S-adenosylmethionine (SAM) binding domain with methyltransferase activity, suggesting it is an unidentified halogenase–methyltransferase heterodimer. Furthermore, the halogenase site catalyzes an unactivated methyl group, whereas almost all other in vitro reconstituted FDHs are involved in aromatic substitution reactions using Cl^+ equivalents, and achieving regioselective functionalization of inactive carbon atoms, such as aliphatic halogenation, is a major synthetic challenge. The discovery of the first characterized fungal aliphatic halogenase, AoiQ, provided a solution for aliphatic halogenation [161,162]. XanH is a bifunctional protein with both reducing and chlorinating functions, but the reduced products cannot be fully and efficiently utilized. By constructing the fusion enzyme FDR-XanH, additional flavin reductase (FDR) can be added to facilitate the halogenation reaction, and its activity has been verified [163,164].

Carbamoylation is ubiquitous in secondary metabolism, causing oxygen and nitrogen atoms to undergo *O*-carbamoylation and *N*-carbamoylation, and this post-modification often imparts antibiotic activity and cytotoxicity to its products. The biosynthesis of ansamitocin, an antitumor agent similar to maytansinoid [165], involves the addition of seven polyketide units that undergo a series of post-modifications, including carbamoylation [166]. After some genes were shown to be involved in ansamycin modification [167–169], carbamoylated *asm21* was presumed to be involved, and the experimental results also indicate that *asm21* encodes a carbamoyltransferase. 4″-*O*-Aminoformylansamycin was isolated and identified as a novel ansamitocin with an aminoformylated polyketoskeleton and glycosylated moiety. As the only carbamoyltransferase gene present in the ansamiline biosynthetic gene cluster, Asm21 is thought to be involved in carbamylation of the glycosyl fraction, a conjecture that was confirmed. The results indicate that ASM21 is a MFE with dual carbamoylation activity on both the polyketide backbone and the glycosyl fraction (Figure 8). Avermectin, a macrolide natural product containing a 6,6 helix ketone group, is an important antiparasitic drug, and a bifunctional post-modifying enzyme was also found to be involved in its biosynthesis. A specific protein, AveC, has no sequence homology to any enzyme of known function and was long thought to only be involved in dehydration. This protein was found to catalyze stereospecific spironoketosis and dehydration of the dihydroxyketone intermediate during the biosynthesis of avermectin, and the dehydration precedes the formation of helical ketones [170]. To some extent, it facilitates the progress of biosynthesis of spironoketone-containing compounds.

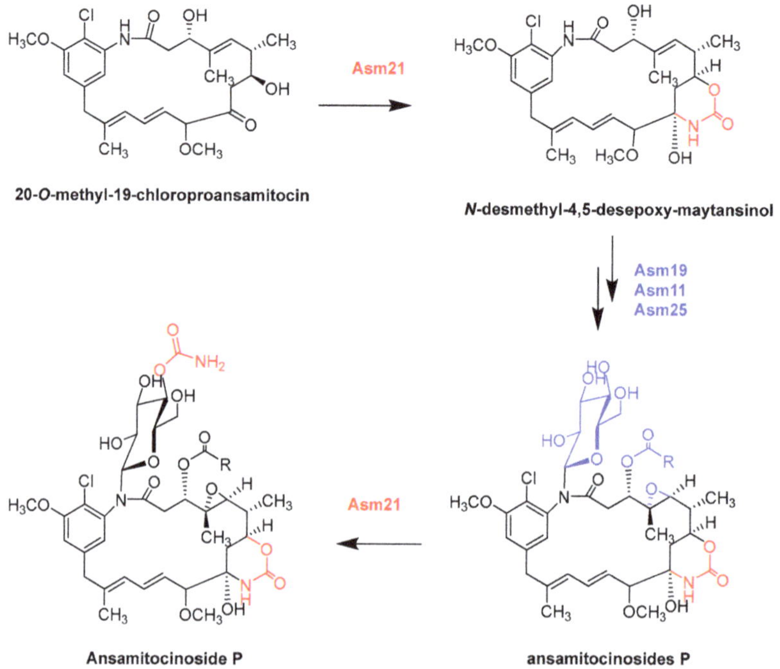

Figure 8. Asm21, as an MFE, participates in post-modification during ansamitocin biosynthesis. Asm21 catalyzes the carbamoylation of polyketoskeleton and the glycosylates part of a special ansamitocin (4″-*O*-aminoformylansamycin), indicating that it has double carbamoylated activity on the polyketoskeleton and glycosylated part, which sufficiently demonstrates that Asm21 is a multifunctional post-modifying enzyme.

4. Future Perspective

MFEs in secondary metabolic processes are highly efficient and versatile. It can be said that catalytic approaches based on MFEs can be used to increase the efficiency of synthesis and save space and substrate loss, as evidenced by their conservation in microorganisms. MFEs have been continuously and systematically studied, which allows us to keep abreast of their progress. In addition to the continuous search for microbial resources with production potential and fermentation isolation, various techniques have been actively applied to explore novel compounds, including genome mining, heterologous expression, and combinatorial biology, and these approaches are used in conjunction. The application of combinatorial biology in the modification of multifunctional synthetases can yield some unnatural products. It is of great value to design a perfect biosynthetic pathway by combining various methods and to exploit the functions of multifunctional enzymes using suitable biomaterials. In any case, the emergence of MFEs has improved the use of energy, matter, and space by organisms, and this advantage can be exploited by humans.

Author Contributions: J.-T.W., T.-T.S. and P.-J.Z. proposed the concept; J.-T.W. and P.-J.Z. drafted the manuscript, L.D. and J.X. participated in the data collection. All authors have read and agreed to the published version of the manuscript.

Funding: This research was supported by the National Natural Science Foundation of China (Grant No. 31970060 (P.-J.Z.) and 32270132 (P.-J.Z.)) and the Yunnan Science and Technology Special Project (202102AA100013 and 202001BB050061).

Data Availability Statement: Not applicable.

Conflicts of Interest: The authors declare no conflict of interest.

References

1. Stark, G.R. Multifunctional proteins: One gene—More than one enzyme. *Trends Biochem. Sci.* **1977**, *2*, 64–66. [CrossRef]
2. Hopwood, D.A. Genetic contributions to understanding polyketide synthases. *Chem. Rev.* **1997**, *97*, 2465–2498. [CrossRef]
3. Fischbach, M.A.; Walsh, C.T. Assembly-line enzymology for polyketide and nonribosomal Peptide antibiotics: Logic, machinery, and mechanisms. *Chem. Rev.* **2006**, *106*, 3468–3496. [CrossRef] [PubMed]
4. Guengerich, F.P.; Macdonald, T.L. Chemical mechanisms of catalysis by cytochromes-P-450-a unified view. *Acc. Chem. Res.* **1984**, *17*, 9–16. [CrossRef]
5. Grunewald, J.; Marahiel, M.A. Chemoenzymatic and template-directed synthesis of bioactive macrocyclic peptides. *Microbiol. Mol. Biol. Rev.* **2006**, *70*, 121–146. [CrossRef]
6. Velkov, T.; Lawen, A. Non-ribosomal peptide synthetases as technological platforms for the synthesis of highly modified peptide bioeffectors–Cyclosporin synthetase as a complex example. *Biotechnol. Annu Rev.* **2003**, *9*, 151–197. [CrossRef]
7. Koglin, A.; Walsh, C.T. Structural insights into nonribosomal peptide enzymatic assembly lines. *Nat. Prod. Rep.* **2009**, *26*, 987–1000. [CrossRef]
8. Christianson, D.W. Structural and chemical biology of terpenoid cyclases. *Chem. Rev.* **2017**, *117*, 11570–11648. [CrossRef]
9. Shen, B. Polyketide biosynthesis beyond the type I, II and III polyketide synthase paradigms. *Curr. Opin. Chem. Biol.* **2003**, *7*, 285–295. [CrossRef]
10. Sultana, A.; Kallio, P.; Jansson, A.; Wang, J.S.; Niemi, J.; Mantsala, P.; Schneider, G. Structure of the polyketide cyclase SnoaL reveals a novel mechanism for enzymatic aldol condensation. *EMBO J.* **2004**, *23*, 1911–1921. [CrossRef]
11. Chooi, Y.H.; Tang, Y. Navigating the fungal polyketide chemical space: From genes to molecules. *J. Org. Chem.* **2012**, *77*, 9933–9953. [CrossRef] [PubMed]
12. Piel, J. Biosynthesis of polyketides by trans-AT polyketide synthases. *Nat. Prod. Rep.* **2010**, *27*, 996–1047. [CrossRef] [PubMed]
13. Conti, E.; Stachelhaus, T.; Marahiel, M.A.; Brick, P. Structural basis for the activation of phenylalanine in the non-ribosomal biosynthesis of gramicidin S. *EMBO J.* **1997**, *16*, 4174–4183. [CrossRef] [PubMed]
14. May, J.J.; Kessler, N.; Marahiel, M.A.; Stubbs, M.T. Crystal structure of DhbE, an archetype for aryl acid activating domains of modular nonribosomal peptide synthetases. *Proc. Natl. Acad. Sci. USA* **2002**, *99*, 12120–12125. [CrossRef] [PubMed]
15. Stachelhaus, T.; Mootz, H.D.; Marahiel, M.A. The specificity-conferring code of adenylation domains in nonribosomal peptide synthetases. *CHEM Biol.* **1999**, *6*, 493–505. [CrossRef] [PubMed]
16. Stachelhaus, T.; Marahiel, M.A. Modular structure of peptide synthetases revealed by dissection of the multifunctional enzyme GrsA. *J. Biol. Chem.* **1995**, *270*, 6163–6169. [CrossRef]
17. Weissman, K.J. The structural biology of biosynthetic megaenzymes. *Nat. Chem. Biol.* **2015**, *11*, 660–670. [CrossRef]
18. Bevitt, D.J.; Cortes, J.; Haydock, S.F.; Leadlay, P.F. 6-deoxyerythronolide-b synthase-2 from *Saccharopolyspora-erythraea*-cloning of the structural gene, sequence-analysis and inferred domain-structure of the multifunctional enzyme. *Eur. J. Biochem.* **1992**, *204*, 39–49. [CrossRef]
19. Cortes, J.; Haydock, S.F.; Roberts, G.A.; Bevitt, D.J.; Leadlay, P.F. An unusually large multifunctional polypeptide in the erythromycin-producing polyketide synthase of *Saccharopolyspora-erythraea*. *Nature* **1990**, *348*, 176–178. [CrossRef]
20. Donadio, S.; Katz, L. Organization of the enzymatic domains in the multifunctional polyketide synthase involved in erythromycin formation in *Saccharopolyspora-erythraea*. *Gene* **1992**, *111*, 51–60. [CrossRef]
21. Donadio, S.; Staver, M.J.; McAlpine, J.B.; Swanson, S.J.; Katz, L. Modular organization of genes required for complex polyketide biosynthesis. *Science* **1991**, *252*, 675–679. [CrossRef] [PubMed]
22. Chen, A.Y.; Cane, D.E.; Khosla, C. Structure-based dissociation of a type I polyketide synthase module. *CHEM Biol.* **2007**, *14*, 784–792. [CrossRef] [PubMed]
23. Nivina, A.; Yuet, K.P.; Hsu, J.; Khosla, C. Evolution and diversity of assembly-line polyketide synthases. *Chem. Rev.* **2019**, *119*, 12524–12547. [CrossRef] [PubMed]
24. Stachelhaus, T.; Huser, A.; Marahiel, M.A. Biochemical characterization of peptides carrier protein (PCP), the thiolation domain of multifunctional peptide synthetases. *CHEM Biol.* **1996**, *3*, 913–921. [CrossRef] [PubMed]
25. vonDohren, H.; Keller, U.; Vater, J.; Zocher, R. Multifunctional peptide synthetases. *Chem. Rev.* **1997**, *97*, 2675–2705. [CrossRef]
26. Stachelhaus, T.; Mootz, H.D.; Bergendahl, V.; Marahiel, M.A. Peptide bond formation in nonribosomal peptide biosynthesis-Catalytic role of the condensation domain. *J. Biol. Chem.* **1998**, *273*, 22773–22781. [CrossRef]

27. Finking, R.; Marahiel, M.A. Biosynthesis of nonribosomal peptides. *Annu. Rev. Microbiol.* **2004**, *58*, 453–488. [CrossRef]
28. Takeda, K.; Kemmoku, K.; Satoh, Y.; Ogasawara, Y.; Shin-ya, K.; Dairi, T. N-phenylacetylation and nonribosomal peptide synthetases with substrate promiscuity for biosynthesis of heptapeptide variants, JBIR-78 and JBIR-95. *ACS Chem. Biol.* **2017**, *12*, 1813–1819. [CrossRef]
29. Miller, B.R.; Gulick, A.M. Structural biology of nonribosomal peptide synthetases. *Methods Mol. Biol.* **2016**, *1401*, 3–29. [CrossRef]
30. Walsh, C.T. Insights into the chemical logic and enzymatic machinery of NRPS assembly lines. *Nat. Prod. Rep.* **2016**, *33*, 127–135. [CrossRef]
31. Marahiel, M.A. A structural model for multimodular NRPS assembly lines. *Nat. Prod. Rep.* **2016**, *33*, 136–140. [CrossRef] [PubMed]
32. Spencer, J.B.; Jordan, P.M. Purification and properties of 6-methylsalicylic acid synthase from *Penicillium patulum*. *Biochem. J.* **1992**, *288*, 839–846. [CrossRef] [PubMed]
33. Beck, J.; Ripka, S.; Siegner, A.; Schiltz, E.; Schweizer, E. The multifunctional 6-methylsalicylic acid synthase gene of *Penicillium-patulum*-its gene structure relative to that of other polyketide synthases. *Eur. J. Biochem.* **1990**, *192*, 487–498. [CrossRef] [PubMed]
34. Dimroth, P.; Walter, H.; Lynen, F. Biosynthesis of 6-methylsalicylic acid. *Eur. J. Biochem.* **1970**, *13*, 98–110. [CrossRef]
35. Parascandolo, J.S.; Havemann, J.; Potter, H.K.; Huang, F.; Riva, E.; Connolly, J.; Wilkening, I.; Song, L.; Leadlay, P.F.; Tosin, M. Insights into 6-methylsalicylic acid bio-assembly by using chemical probes. *Angew. Chem.* **2016**, *55*, 3463–3467. [CrossRef]
36. Child, C.J.; Spencer, J.B.; Bhogal, P.; ShoolinginJordan, P.M. Structural similarities between 6-methylsalicylic acid synthase from *Penicillium patulum* and vertebrate type I fatty acid synthase: Evidence from thiol modification studies. *Biochemistry* **1996**, *35*, 12267–12274. [CrossRef]
37. Nicholson, T.P.; Rudd, B.A.M.; Dawson, M.; Lazarus, C.M.; Simpson, T.J.; Cox, R.J. Design and utility of oligonucleotide gene probes for fungal polyketide synthases. *CHEM Biol.* **2001**, *8*, 157–178. [CrossRef]
38. Fujii, I.; Watanabe, A.; Sankawa, U.; Ebizuka, Y. Identification of Claisen cyclase domain in fungal polyketide synthase WA, a naphthopyrone synthase of *Aspergillus nidulans*. *CHEM Biol.* **2001**, *8*, 189–197. [CrossRef]
39. Bailey, A.M.; Cox, R.J.; Harley, K.; Lazarus, C.M.; Simpson, T.J.; Skellam, E. Characterisation of 3-methylorcinaldehyde synthase (MOS) in *Acremonium strictum*: First observation of a reductive release mechanism during polyketide biosynthesis. *Chem. Commun.* **2007**, *39*, 4053–4055. [CrossRef]
40. Moriguchi, T.; Kezuka, Y.; Nonaka, T.; Ebizuka, Y.; Fujii, I. Hidden function of catalytic domain in 6-methylsalicylic acid synthase for product release. *J. Biol. Chem.* **2010**, *285*, 15637–15643. [CrossRef]
41. Gerhard, A.; Muntwyler, R.; Kellerschierlein, W. Metabolic products of microorganisms.147. unexpected transformation in chlorothricin series. *Helv. Chim. Acta.* **1975**, *58*, 1323–1338. [CrossRef] [PubMed]
42. Jia, X.Y.; Tian, Z.H.; Shao, L.; Qu, X.D.; Zhao, Q.F.; Tang, J.; Tang, G.L.; Liu, W. Genetic characterization of the chlorothricin gene cluster as a model for spirotetronate antibiotic biosynthesis. *Chem. Biol.* **2006**, *13*, 575–585. [CrossRef] [PubMed]
43. Shao, L.; Qu, X.D.; Jia, X.Y.; Zhao, Q.F.; Tian, Z.H.; Wang, M.; Tang, G.L.; Liu, W. Cloning and characterization of a bacterial iterative type I polyketide synthase gene encoding the 6-methylsalicyclic acid synthase. *Biochem. Biophys. Res. Commun.* **2006**, *345*, 133–139. [CrossRef] [PubMed]
44. Van Lanen, S.G.; Oh, T.J.; Liu, W.; Wendt-Pienkowski, E.; Shen, B. Characterization of the maduropeptin biosynthetic gene cluster from *Actinomadura* madurae ATCC 39144 supporting a unifying paradigm for enediyne biosynthesis. *J. Am. Chem. Soc.* **2007**, *129*, 13082–13094. [CrossRef]
45. Ito, T.; Roongsawang, N.; Shirasaka, N.; Lu, W.L.; Flatt, P.M.; Kasanah, N.; Miranda, C.; Mahmud, T. Deciphering pactamycin biosynthesis and engineered production of new pactamycin analogues. *ChemBioChem* **2009**, *10*, 2253–2265. [CrossRef]
46. Wang, J.L.; Liang, J.D.; Chen, L.; Zhang, W.; Kong, L.L.; Peng, C.; Su, C.; Tang, Y.; Deng, Z.X.; Wang, Z.J. Structural basis for the biosynthesis of lovastatin. *Nat. Commun.* **2021**, *12*, 867. [CrossRef] [PubMed]
47. Kennedy, J.; Auclair, K.; Kendrew, S.G.; Park, C.; Vederas, J.C.; Hutchinson, C.R. Modulation of polyketide synthase activity by accessory proteins during lovastatin biosynthesis. *Science* **1999**, *284*, 1368–1372. [CrossRef]
48. Ma, S.M.; Li, J.W.H.; Choi, J.W.; Zhou, H.; Lee, K.K.M.; Moorthie, V.A.; Xie, X.K.; Kealey, J.T.; Da Silva, N.A.; Vederas, J.C.; et al. Complete reconstitution of a highly reducing iterative polyketide synthase. *Science* **2009**, *326*, 589–592. [CrossRef]
49. Hendrickson, L.; Davis, C.R.; Roach, C.; Nguyen, D.K.; Aldrich, T.; McAda, P.C.; Reeves, C.D. Lavastatin biosynthesis in *Aspergillus terreus*: Characterization of blocked mutants, enzyme activities and a multifunctional polyketide synthase gene. *CHEM Biol.* **1999**, *6*, 429–439. [CrossRef]
50. Xu, W.; Chooi, Y.H.; Choi, J.W.; Li, S.; Vederas, J.C.; Da Silva, N.A.; Tang, Y. LovG: The thioesterase required for dihydromonacolinl release and lovastatin nonaketide synthase turnover in lovastatin biosynthesis. *Angew. Chem.* **2013**, *52*, 6472–6475. [CrossRef]
51. Ahlert, J.; Shepard, E.; Lomovskaya, N.; Zazopoulos, E.; Staffa, A.; Bachmann, B.O.; Huang, K.X.; Fonstein, L.; Czisny, A.; Whitwam, R.E.; et al. The calicheamicin gene cluster and its iterative type I enediyne PKS. *Science* **2002**, *297*, 1173–1176. [CrossRef] [PubMed]
52. Liu, W.; Christenson, S.D.; Standage, S.; Shen, B. Biosynthesis of the enediyne antitumor antibiotic C-1027. *Science* **2002**, *297*, 1170–1173. [CrossRef] [PubMed]
53. Sthapit, B.; Oh, T.J.; Lamichhane, R.; Liou, K.; Lee, H.C.; Kim, C.G.; Sohng, J.K. Neocarzinostatin naphthoate synthase: An unique iterative type I PKS from neocarzinostatin producer Streptomyces carzinostaticus. *FEBS Lett.* **2004**, *566*, 201–206. [CrossRef] [PubMed]
54. Barry, S.M.; Challis, G.L. Recent advances in siderophore biosynthesis. *Curr. Opin. Chem. Biol.* **2009**, *13*, 205–215. [CrossRef]

55. Hai, Y.; Jenner, M.; Tang, Y. Fungal siderophore biosynthesis catalysed by an iterative nonribosomal peptide synthetase. *Chem. Sci.* **2020**, *11*, 11525–11530. [CrossRef] [PubMed]
56. Raymond, K.N.; Dertz, E.A.; Kim, S.S. Enterobactin: An archetype for microbial iron transport. *Proc. Natl. Acad. Sci. USA* **2003**, *100*, 3584–3588. [CrossRef]
57. Gehring, A.M.; Bradley, K.A.; Walsh, C.T. Enterobactin biosynthesis in Escherichia coli: Isochorismate lyase (EntB) is a bifunctional enzyme that is phosphopantetheinylated by EntD and then acylated by EntE using ATP and 2,3-dihydroxybenzoate. *Biochemistry* **1997**, *36*, 8495–8503. [CrossRef]
58. Reichert, J.; Sakaitani, M.; Walsh, C.T. Characterization of EntF as a serine-activating enzyme. *Protein Sci.* **1992**, *1*, 549–556. [CrossRef]
59. Drake, E.J.; Nicolai, D.A.; Gulick, A.M. Structure of the EntB multidomain nonribosomal peptide synthetase and functional analysis of its interaction with the EntE adenylation domain. *CHEM Biol.* **2006**, *13*, 409–419. [CrossRef]
60. Shaw-Reid, C.A.; Kelleher, N.L.; Losey, H.C.; Gehring, A.M.; Berg, C.; Walsh, C.T. Assembly line enzymology by multimodular nonribosomal peptide synthetases: The thioesterase domain of E-coli EntF catalyzes both elongation and cyclolactonization. *CHEM Biol.* **1999**, *6*, 385–400. [CrossRef]
61. Haese, A.; Schubert, M.; Herrmann, M.; Zocher, R. Molecular characterization of the enniatin synthetase gene encoding a multifunctional enzyme catalyzing n-methyldepsipeptide formation in *Fusarium-scirpi*. *Mol. Microbiol.* **1993**, *7*, 905–914. [CrossRef] [PubMed]
62. Cheng, Y.Q. Deciphering the biosynthetic codes for the potent anti-SARS-CoV cyclodepsipeptide valinomycin in *Streptomyces tsusimaensis* ATCC 15141. *ChemBioChem* **2006**, *7*, 471–477. [CrossRef] [PubMed]
63. Watanabe, K.; Hotta, K.; Praseuth, A.P.; Koketsu, K.; Migita, A.; Boddy, C.N.; Wang, C.C.C.; Oguri, H.; Oikawa, H. Total biosynthesis of antitumor nonribosomal peptides in *Escherichia coli*. *Nat. Chem. Biol.* **2006**, *2*, 423–428. [CrossRef]
64. Kohli, R.M.; Trauger, J.W.; Schwarzer, D.; Marahiel, M.A.; Walsh, C.T. Generality of peptide cyclization catalyzed by isolated thioesterase domains of nonribosomal peptide synthetases. *Biochemistry* **2001**, *40*, 7099–7108. [CrossRef]
65. Hoyer, K.M.; Mahlert, C.; Marahiel, M.A. The iterative gramicidin S thioesterase catalyzes peptide ligation and cyclization. *CHEM Biol.* **2007**, *14*, 13–22. [CrossRef] [PubMed]
66. Koketsu, K.; Minami, A.; Watanabe, K.; Oguri, H.; Oikawa, H. Pictet-Spenglerase involved in tetrahydroisoquinoline antibiotic biosynthesis. *Curr. Opin. Chem. Biol.* **2012**, *16*, 142–149. [CrossRef]
67. Mitsuhashi, T.; Abe, I. Chimeric terpene synthases possessing both terpene cyclization and prenyltransfer activities. *ChemBioChem* **2018**, *19*, 1106–1114. [CrossRef]
68. Toyomasu, T.; Tsukahara, M.; Kaneko, A.; Niida, R.; Mitsuhashi, W.; Dairi, T.; Kato, N.; Sassa, T. Fusicoccins are biosynthesized by an unusual chimera diterpene synthase in fungi. *Proc. Natl. Acad. Sci. USA* **2007**, *104*, 3084–3088. [CrossRef]
69. Srere, P.A. Complexes of sequential metabolic enzymes. *Annu. Rev. Biochem.* **1987**, *56*, 89–124. [CrossRef]
70. Miles, E.W.; Rhee, S.; Davies, D.R. The molecular basis of substrate channeling. *J. Biol. Chem.* **1999**, *274*, 12193–12196. [CrossRef]
71. Huang, X.Y.; Holden, H.M.; Raushel, F.M. Channeling of substrates and intermediates in enzyme-catalyzed reactions. *Annu. Rev. Biochem.* **2001**, *70*, 149–180. [CrossRef] [PubMed]
72. Faylo, J.L.; Ronnebaum, T.A.; Christianson, D.W. Assembly-line catalysis in bifunctional terpene synthases. *Acc. Chem. Res.* **2021**, *54*, 3780–3791. [CrossRef] [PubMed]
73. Chen, M.B.; Chou, W.K.W.; Toyomasu, T.; Cane, D.E.; Christianson, D.W. Structure and function of fusicoccadiene synthase, a hexameric bifunctional diterpene synthase. *ACS Chem. Biol.* **2016**, *11*, 889–899. [CrossRef] [PubMed]
74. Mitsuhashi, T.; Okada, M.; Abe, I. Identification of chimeric alpha beta gamma diterpene synthases possessing both Type II terpene cyclase and prenyltransferase activities. *ChemBioChem* **2017**, *18*, 2104–2109. [CrossRef] [PubMed]
75. Ronnebaum, T.A.; Eaton, S.A.; Brackhahn, E.A.E.; Christianson, D.W. Engineering the prenyltransferase domain of a bifunctional assembly-line terpene synthase. *Biochemistry* **2021**, *60*, 3162–3172. [CrossRef]
76. Yuan, W.; Lv, S.; Chen, L.Y.; Zhao, Y.; Deng, Z.X.; Hong, K. Production of sesterterpene ophiobolin by a bifunctional terpene synthase in *Escherichia coli*. *Appl. Microbiol. Biotechnol.* **2019**, *103*, 8785–8797. [CrossRef]
77. Ronnebaum, T.A.; Gupta, K.; Christianson, D.W. Higher-order oligomerization of a chimeric alpha beta gamma bifunctional diterpene synthase with prenyltransferase and class II cyclase activities is concentration-dependent. *J. Struct. Biol.* **2020**, *210*, 107463. [CrossRef]
78. Chang, T.H.; Guo, R.T.; Ko, T.P.; Wang, A.H.J.; Liang, P.H. Crystal structure of type-III geranylgeranyl pyrophosphate synthase from *Saccharomyces cerevisiae* and the mechanism of product chain length determination. *J. Biol. Chem.* **2006**, *281*, 14991–15000. [CrossRef]
79. Andersen-Ranberg, J.; Kongstad, K.T.; Nielsen, M.T.; Jensen, N.B.; Pateraki, I.; Bach, S.S.; Hamberger, B.; Zerbe, P.; Staerk, D.; Bohlmann, J.; et al. Expanding the landscape of diterpene structural diversity through stereochemically controlled combinatorial biosynthesis. *Angew. Chem.* **2016**, *55*, 2142–2146. [CrossRef]
80. Bian, G.K.; Han, Y.C.; Hou, A.W.; Yuan, Y.J.; Liu, X.H.; Deng, Z.X.; Liu, T.G. Releasing the potential power of terpene synthases by a robust precursor supply platform. *Metab. Eng.* **2017**, *42*, 1–8. [CrossRef]
81. Yuan, Y.J.; Litzenburger, M.; Cheng, S.; Bian, G.K.; Hu, B.; Yan, P.; Cai, Y.S.; Deng, Z.X.; Bernhardt, R.; Liu, T.G. Sesquiterpenoids produced by combining two sesquiterpene cyclases with promiscuous myxobacterial CYP260B1. *ChemBioChem* **2019**, *20*, 677–682. [CrossRef] [PubMed]

82. Ye, Y.; Minami, A.; Mandi, A.; Liu, C.W.; Taniguchi, T.; Kuzuyama, T.; Monde, K.; Gomi, K.; Oikawa, H. Genome mining for sesterterpenes using bifunctional terpene synthases reveals a unified intermediate of di/sesterterpenes. *J. Am. Chem. Soc.* **2015**, *137*, 11846–11853. [CrossRef] [PubMed]
83. Narita, K.; Sato, H.; Minami, A.; Kudo, K.; Gao, L.; Liu, C.W.; Ozaki, T.; Kodama, M.; Lei, X.G.; Taniguchi, T.; et al. Focused genome mining of structurally related sesterterpenes: Enzymatic formation of enantiomeric and diastereomeric products. *Org. Lett.* **2017**, *19*, 6696–6699. [CrossRef] [PubMed]
84. Chiba, R.; Minami, A.; Gomi, K.; Oikawa, H. Identification of ophiobolin F synthase by a genome mining approach: A sesterterpene synthase from *Aspergillus clavatus*. *Org. Lett.* **2013**, *15*, 594–597. [CrossRef]
85. Matsuda, Y.; Mitsuhashi, T.; Quan, Z.Y.; Abe, I. Molecular basis for stellatic acid biosynthesis: A genome mining approach for discovery of sesterterpene synthases. *Org. Lett.* **2015**, *17*, 4644–4647. [CrossRef]
86. Jiang, L.; Zhang, X.; Sato, Y.; Zhu, G.L.; Minami, A.; Zhang, W.Y.; Ozaki, T.; Zhu, B.; Wang, Z.X.; Wang, X.Y.; et al. Genome-based discovery of enantiomeric pentacyclic sesterterpenes catalyzed by fungal bifunctional terpene synthases. *Org. Lett.* **2021**, *23*, 4645–4650. [CrossRef]
87. Mitsuhashi, T.; Rinkel, J.; Okada, M.; Abe, I.; Dickschat, J.S. Mechanistic characterization of two chimeric sesterterpene synthases from *Penicillium*. *Chemistry* **2017**, *23*, 10053–10057. [CrossRef] [PubMed]
88. Guo, J.J.; Cai, Y.S.; Cheng, F.C.; Yang, C.J.; Zhang, W.Q.; Yu, W.L.; Yan, J.J.; Deng, Z.X.; Hong, K. Genome mining reveals a multiproduct sesterterpenoid biosynthetic gene cluster in *Aspergillus ustus*. *Org. Lett.* **2021**, *23*, 1525–1529. [CrossRef]
89. Quan, Z.Y.; Hou, A.W.; Goldfuss, B.; Dickschat, J.S. Mechanism of the bifunctional multiple product sesterterpene synthase AcAS from *Aspergillus calidoustus*. *Angew. Chem. Int. Ed.* **2022**, *61*, e202117273. [CrossRef]
90. Chen, R.; Jia, Q.D.; Mu, X.; Hu, B.; Sun, X.; Deng, Z.X.; Chen, F.; Bian, G.K.; Liu, T.G. Systematic mining of fungal chimeric terpene synthases using an efficient precursor-providing yeast chassis. *Proc. Natl. Acad. Sci. USA* **2021**, *118*, e2023247118. [CrossRef]
91. Cane, D.E.; Watt, R.M. Expression and mechanistic analysis of a germacradienol synthase from *Streptomyces coelicolor* implicated in geosmin biosynthesis. *Proc. Natl. Acad. Sci. USA* **2003**, *100*, 1547–1551. [CrossRef] [PubMed]
92. Harris, G.G.; Lombardi, P.M.; Pemberton, T.A.; Matsui, T.; Weiss, T.M.; Cole, K.E.; Koksal, M.; Murphy, F.V.; Vedula, L.S.; Chou, W.K.W.; et al. Structural studies of geosmin synthase, a bifunctional sesquiterpene synthase with alpha alpha domain architecture that catalyzes a unique cyclization-fragmentation reaction sequence. *Biochemistry* **2015**, *54*, 7142–7155. [CrossRef] [PubMed]
93. Liato, V.; Aider, M. Geosmin as a source of the earthy-musty smell in fruits, vegetables and water: Origins, impact on foods and water, and review of the removing techniques. *Chemosphere* **2017**, *181*, 9–18. [CrossRef]
94. Verdoes, J.C.; Krubasik, P.; Sandmann, G.; van Ooyen, A.J.J. Isolation and functional characterisation of a novel type of carotenoid biosynthetic gene from Xanthophyllomyces dendrorhous. *Mol. Gen. Genet.* **1999**, *262*, 453–461. [CrossRef]
95. Lamb, D.C.; Ikeda, H.; Nelson, D.R.; Ishikawa, J.; Skaug, T.; Jackson, C.; Omura, S.; Waterman, M.R.; Kelly, S.L. Cytochrome P450 complement (CYPome) of the avermectin-producer Streptomyces avermitilis and comparison to that of Streptomyces coelicolor A3(2). *Biochem. Biophys. Res. Commun.* **2003**, *307*, 610–619. [CrossRef] [PubMed]
96. Whitehouse, C.J.C.; Bell, S.G.; Wong, L.L. P450(BM3) (CYP102A1): Connecting the dots. *Chem. Soc. Rev.* **2012**, *41*, 1218–1260. [CrossRef]
97. Podust, L.M.; Sherman, D.H. Diversity of P450 enzymes in the biosynthesis of natural products. *Nat. Prod. Rep.* **2012**, *29*, 1251–1266. [CrossRef]
98. Li, S.Y.; Tietz, D.R.; Rutaganira, F.U.; Kells, P.M.; Anzai, Y.; Kato, F.; Pochapsky, T.C.; Sherman, D.H.; Podust, L.M. Substrate recognition by the multifunctional cytochrome P450 MycG in mycinamicin hydroxylation and epoxidation reactions. *J. Biol. Chem.* **2012**, *287*, 37880–37890. [CrossRef]
99. Tudzynski, B.; Kawaide, H.; Kamiya, Y. Gibberellin biosynthesis in *Gibberella fujikuroi*: Cloning and characterization of the copalyl diphosphate synthase gene. *Curr. Genet.* **1998**, *34*, 234–240. [CrossRef]
100. Rojas, M.C.; Hedden, P.; Gaskin, P.; Tudzynki, B. The P450-1 gene of gibberella fujikuroi encodes a multifunctional enzyme in gibberellin biosynthesis. *Proc. Natl. Acad. Sci. USA* **2001**, *98*, 5838–5843. [CrossRef]
101. Tudzynski, B.; Rojas, M.C.; Gaskin, P.; Hedden, P. The gibberellin 20-oxidase of *Gibberella fujikuroi* is a multifunctional monooxygenase. *J. Biol. Chem.* **2002**, *277*, 21246–21253. [CrossRef] [PubMed]
102. Tudzynski, B.; Mihlan, M.; Rojas, M.C.; Linnemannstons, P.; Gaskin, P.; Hedden, P. Characterization of the final two genes of the gibberellin biosynthesis gene cluster of *Gibberella fujikuroi*-des and P450-3 encode GA(4) desaturase and the 13-hydroxylase, respectively. *J. Biol. Chem.* **2003**, *278*, 28635–28643. [CrossRef] [PubMed]
103. Bhattacharya, A.; Kourmpetli, S.; Ward, D.A.; Thomas, S.G.; Gong, F.; Powers, S.J.; Carrera, E.; Taylor, B.; Gonzalez, F.N.D.; Tudzynski, B.; et al. Characterization of the fungal gibberellin desaturase as a 2-Oxoglutarate-dependent dioxygenase and its utilization for enhancing plant growth. *Plant Physiol.* **2012**, *160*, 837–845. [CrossRef] [PubMed]
104. Inouye, M.; Takada, Y.; Muto, N.; Beppu, T.; Horinouchi, S. Characterization and expression of a P-450-like mycinamicin biosynthesis gene using a novel *Micromonospora escherichia*-coli shuttle cosmid vector. *Mol. Gen. Genet.* **1994**, *245*, 456–464. [CrossRef] [PubMed]
105. Anzai, Y.; Li, S.Y.; Chaulagain, M.R.; Kinoshita, K.; Kato, F.; Montgomery, J.; Sherman, D.H. Functional analysis of MycCI and MycG, cytochrome P450 enzymes involved in biosynthesis of mycinamicin macrolide antibiotics. *CHEM Biol.* **2008**, *15*, 950–959. [CrossRef]

106. Anzai, Y.; Tsukada, S.; Sakai, A.; Masuda, R.; Harada, C.; Domeki, A.; Li, S.Y.; Kinoshita, K.; Sherman, D.H.; Kato, F. Function of cytochrome P450 enzymes MycCI and MycG in *Micromonospora griseorubida*, a producer of the macrolide antibiotic mycinamicin. *Antimicrob. Agents Chemother.* **2012**, *56*, 3648–3656. [CrossRef] [PubMed]
107. Richter, M.E.A.; Traitcheva, N.; Knupfer, U.; Hertweck, C. Sequential asymmetric polyketide heterocyclization catalyzed by a single cytochrome P450 monooxygenase (AurH). *Angew. Chem. Int. Ed.* **2008**, *47*, 8872–8875. [CrossRef]
108. Muller, M.; He, J.; Hertweck, C. Dissection of the late steps in aureothin biosynthesis. *ChemBioChem* **2006**, *7*, 37–39. [CrossRef]
109. Zocher, G.; Richter, M.E.A.; Mueller, U.; Hertweck, C. Structural fine-tuning of a multifunctional cytochrome P450 monooxygenase. *J. Am. Chem. Soc.* **2011**, *133*, 2292–2302. [CrossRef]
110. Lin, H.C.; Tsunematsu, Y.; Dhingra, S.; Xu, W.; Fukutomi, M.; Chooi, Y.H.; Cane, D.E.; Calvo, A.M.; Watanabe, K.; Tang, Y. Generation of complexity in fungal terpene biosynthesis: Discovery of a multifunctional cytochrome P450 in the fumagillin pathway. *J. Am. Chem. Soc.* **2014**, *136*, 4426–4436. [CrossRef]
111. Iizaka, Y.; Takeda, R.; Senzaki, Y.; Fukumoto, A.; Anzai, Y. Cytochrome P450 enzyme RosC catalyzes a multistep oxidation reaction to form the non-active compound 20-carboxyrosamicin. *FEMS Microbiol. Lett.* **2017**, *364*, fnx110. [CrossRef] [PubMed]
112. Iizaka, Y.; Higashi, N.; Ishida, M.; Oiwa, R.; Ichikawa, Y.; Takeda, M.; Anzai, Y.; Kato, F. Function of cytochrome P450 enzymes RosC and RosD in the biosynthesis of rosamicin macrolide antibiotic produced by *Micromonospora rosaria*. *Antimicrob. Agents Chemother.* **2013**, *57*, 1529–1531. [CrossRef] [PubMed]
113. Qi, Z.; Kang, Q.J.; Jiang, C.Y.; Han, M.; Bai, L.Q. Engineered biosynthesis of pimaricin derivatives with improved antifungal activity and reduced cytotoxicity. *Appl. Microbiol. Biotechnol.* **2015**, *99*, 6745–6752. [CrossRef] [PubMed]
114. Proctor, R.H.; Plattner, R.D.; Desjardins, A.E.; Busman, M.; Butchko, R.A.E. Fumonisin production in the maize pathogen *Fusarium verticillioides*: Genetic basis of naturally occurring chemical variation. *J. Agric. Food Chem.* **2006**, *54*, 2424–2430. [CrossRef]
115. Matsuda, Y.; Iwabuchi, T.; Wakimoto, T.; Awakawa, T.; Abe, I. Uncovering the unusual D-ring construction in terretonin biosynthesis by collaboration of a multifunctional cytochrome P450 and a unique isomerase. *J. Am. Chem. Soc.* **2015**, *137*, 3393–3401. [CrossRef]
116. Barriuso, J.; Nguyen, D.T.; Li, J.W.H.; Roberts, J.N.; MacNevin, G.; Chaytor, J.L.; Marcus, S.L.; Vederas, J.C.; Ro, D.K. Double oxidation of the cyclic nonaketide dihydromonacolin l to Monacolin j by a single cytochrome P450 monooxygenase, LovA. *J. Am. Chem. Soc.* **2011**, *133*, 8078–8081. [CrossRef]
117. Cochrane, R.V.K.; Vederas, J.C. Highly selective but multifunctional oxygenases in secondary metabolism. *Acc. Chem. Res.* **2014**, *47*, 3148–3161. [CrossRef]
118. Udwary, D.W.; Casillas, L.K.; Townsend, C.A. Synthesis of 11-hydroxyl O-methylsterigmatocystin and the role of a cytochrome P-450 in the final step of aflatoxin biosynthesis. *J. Am. Chem. Soc.* **2002**, *124*, 5294–5303. [CrossRef]
119. Yu, J.; Chang, P.K.; Ehrlich, K.C.; Cary, J.W.; Montalbano, B.; Dyer, J.M.; Bhatnagar, D.; Cleveland, T.E. Characterization of the critical amino acids of an *Aspergillus parasiticus* cytochrome P-450 monooxygenase encoded by ordA that is involved in the biosynthesis of aflatoxins B1, G1, B2, and G2. *Appl. Environ. Microbiol.* **1998**, *64*, 4834–4841. [CrossRef]
120. Xue, Y.Q.; Wilson, D.; Zhao, L.S.; Liu, H.W.; Sherman, D.H. Hydroxylation of macrolactones YC-17 and narbomycin is mediated by the pikC-encoded cytochrome P450 in *Streptomyces venezuelae*. *CHEM Biol.* **1998**, *5*, 661–667. [CrossRef]
121. Zhao, B.; Lin, X.; Lei, L.; Lamb, D.C.; Kelly, S.L.; Waterman, M.R.; Cane, D.E. Biosynthesis of the sesquiterpene antibiotic albaflavenone in *Streptomyces coelicolor* A3(2). *J. Biol. Chem.* **2008**, *283*, 8183–8189. [CrossRef] [PubMed]
122. Zhao, B.; Lei, L.; Vassylyev, D.G.; Lin, X.; Cane, D.E.; Kelly, S.L.; Yuan, H.; Lamb, D.C.; Waterman, M.R. Crystal structure of albaflavenone monooxygenase containing a moonlighting terpene synthase active site. *J. Biol. Chem.* **2009**, *284*, 36711–36719. [CrossRef]
123. Zhao, B.; Waterman, M.R. Moonlighting cytochrome P450 monooxygenases. *IUBMB Life* **2011**, *63*, 473–477. [CrossRef]
124. Huijbers, M.M.E.; Montersino, S.; Westphal, A.H.; Tischler, D.; van Berkel, W.J.H. Flavin dependent monooxygenases. *Arch. Biochem. Biophys.* **2014**, *544*, 2–17. [CrossRef]
125. Walsh, C.T.; Wencewicz, T.A. Flavoenzymes: Versatile catalysts in biosynthetic pathways. *Nat. Prod. Rep.* **2013**, *30*, 175–200. [CrossRef] [PubMed]
126. Rebehmed, J.; Alphand, V.; de Berardinis, V.; de Brevern, A.G. Evolution study of the Baeyer-Villiger monooxygenases enzyme family: Functional importance of the highly conserved residues. *Biochimie* **2013**, *95*, 1394–1402. [CrossRef] [PubMed]
127. Kong, L.X.; Zhang, W.K.; Chooi, Y.H.; Wang, L.; Cao, B.; Deng, Z.X.; Chu, Y.W.; You, D.L. A multifunctional monooxygenase XanO4 catalyzes xanthone formation in xantholipin biosynthesis via a cryptic demethoxylation. *Cell Chem. Biol.* **2016**, *23*, 508–516. [CrossRef] [PubMed]
128. Hu, Y.C.; Dietrich, D.; Xu, W.; Patel, A.; Thuss, J.A.J.; Wang, J.J.; Yin, W.B.; Qiao, K.J.; Houk, K.N.; Vederas, J.C.; et al. A carbonate-forming Baeyer-Villiger monooxygenase. *Nat. Chem. Biol.* **2014**, *10*, 552–554. [CrossRef] [PubMed]
129. Cheng, Q.; Xiang, L.; Izumikawa, M.; Meluzzi, D.; Moore, B.S. Enzymatic total synthesis of enterocin polyketides. *Nat. Chem. Biol.* **2007**, *3*, 557–558. [CrossRef] [PubMed]
130. Xiang, L.; Kalaitzis, J.A.; Moore, B.S. EncM, a versatile enterocin biosynthetic enzyme involved in Favorskii oxidative rearrangement, aldol condensation, and heterocycle-forming reactions. *Proc. Natl. Acad. Sci. USA* **2004**, *101*, 15609–15614. [CrossRef]
131. Aik, W.; McDonough, M.A.; Thalhammer, A.; Chowdhury, R.; Schofield, C.J. Role of the jelly-roll fold in substrate binding by 2-oxoglutarate oxygenases. *Curr. Opin. Struct. Biol.* **2012**, *22*, 691–700. [CrossRef] [PubMed]

132. Clifton, I.J.; McDonough, M.A.; Ehrismann, D.; Kershaw, N.J.; Granatino, N.; Schofield, C.J. Structural studies on 2-oxoglutarate oxygenases and related double-stranded beta-helix fold proteins. *J. Inorg. Biochem.* **2006**, *100*, 644–669. [CrossRef] [PubMed]
133. Hewitson, K.S.; Granatino, N.; Welford, R.W.D.; McDonough, M.A.; Schofield, C.J. Oxidation by 2-oxoglutarate oxygenases: Non-haem iron systems in catalysis and signalling. *Philos. Trans. R. Soc. A Math. Phys. Eng. Sci.* **2005**, *363*, 807–828. [CrossRef] [PubMed]
134. Matsuda, Y.; Awakawa, T.; Wakimoto, T.; Abe, I. Spiro-ring formation is catalyzed by a multifunctional dioxygenase in austinol biosynthesis. *J. Am. Chem. Soc.* **2013**, *135*, 10962–10965. [CrossRef]
135. Lo, H.C.; Entwistle, R.; Guo, C.J.; Ahuja, M.; Szewczyk, E.; Hung, J.H.; Chiang, Y.M.; Oakley, B.R.; Wang, C.C.C. Two separate gene clusters encode the biosynthetic pathway for the meroterpenoids austinol and dehydroaustinol in *Aspergillus nidulans*. *J. Am. Chem. Soc.* **2012**, *134*, 4709–4720. [CrossRef]
136. Matsuda, Y.; Iwabuchi, T.; Fujimoto, T.; Awakawa, T.; Nakashima, Y.; Mori, T.; Zhang, H.P.; Hayash, F.; Abe, I. Discovery of key dioxygenases that diverged the paraherquonin and acetoxydehydroaustin pathways in *Penicillium brasilianum*. *J. Am. Chem. Soc.* **2016**, *138*, 12671–12677. [CrossRef] [PubMed]
137. Nakashima, Y.; Mori, T.; Nakamura, H.; Awakawa, T.; Hoshino, S.; Senda, M.; Senda, T.; Abe, I. Structure function and engineering of multifunctional non-heme iron dependent oxygenases in fungal meroterpenoid biosynthesis. *Nat. Commun.* **2018**, *9*, 104. [CrossRef]
138. Hamed, R.B.; Gomez-Castellanos, J.R.; Henry, L.; Ducho, C.; McDonough, M.A.; Schofield, C.J. The enzymes of beta-lactam biosynthesis. *Nat. Prod. Rep.* **2013**, *30*, 21–107. [CrossRef]
139. Townsend, C.A. New reactions in clavulanic acid biosynthesis. *Curr. Opin. Chem. Biol.* **2002**, *6*, 583–589. [CrossRef]
140. Lloyd, M.D.; Lipscomb, S.J.; Hewitson, K.S.; Hensgens, C.M.H.; Baldwin, J.E.; Schofield, C.J. Controlling the substrate selectivity of deacetoxycephalosporin/deacetylcephalosporin C synthase. *J. Biol. Chem.* **2004**, *279*, 15420–15426. [CrossRef]
141. Tarhonskaya, H.; Szollossi, A.; Leung, I.K.H.; Bush, J.T.; Henry, L.; Chowdhury, R.; Iqbal, A.; Claridge, T.D.W.; Schofield, C.J.; Flashman, E. Studies on deacetoxycephalosporin c synthase support a consensus mechanism for 2-oxoglutarate dependent oxygenases. *Biochemistry* **2014**, *53*, 2483–2493. [CrossRef] [PubMed]
142. Phelan, R.M.; Townsend, C.A. Mechanistic insights into the bifunctional non-heme iron oxygenase carbapenem synthase by active site saturation mutagenesis. *J. Am. Chem. Soc.* **2013**, *135*, 7496–7502. [CrossRef] [PubMed]
143. Velasquez, J.E.; van der Donk, W.A. Genome mining for ribosomally synthesized natural products. *Curr. Opin. Chem. Biol.* **2011**, *15*, 11–21. [CrossRef] [PubMed]
144. Arnison, P.G.; Bibb, M.J.; Bierbaum, G.; Bowers, A.A.; Bugni, T.S.; Bulaj, G.; Camarero, J.A.; Campopiano, D.J.; Challis, G.L.; Clardy, J.; et al. Ribosomally synthesized and post-translationally modified peptide natural products: Overview and recommendations for a universal nomenclature. *Nat. Prod. Rep.* **2013**, *30*, 108–160. [CrossRef] [PubMed]
145. Oman, T.J.; van der Donk, W.A. Follow the leader: The use of leader peptides to guide natural product biosynthesis. *Nat. Chem. Biol.* **2010**, *6*, 9–18. [CrossRef]
146. Knerr, P.J.; van der Donk, W.A. Discovery, biosynthesis, and engineering of lantipeptides. *Annu. Rev. Biochem.* **2012**, *81*, 479–505. [CrossRef]
147. Krawczyk, B.; Voller, G.H.; Voller, J.; Ensle, P.; Sussmuth, R.D. Curvopeptin: A new lanthionine-containing class III lantibiotic and its co-substrate promiscuous synthetase. *ChemBioChem* **2012**, *13*, 2065–2071. [CrossRef]
148. Zhang, Q.; Yu, Y.; Velasquez, J.E.; van der Donk, W.A. Evolution of lanthipeptide synthetases. *Proc. Natl. Acad. Sci. USA* **2012**, *109*, 18361–18366. [CrossRef]
149. Wang, H.; van der Donk, W.A. Biosynthesis of the class III lantipeptide catenulipeptin. *ACS Chem. Biol.* **2012**, *7*, 1529–1535. [CrossRef]
150. van der Donk, W.A.; Nair, S.K. Structure and mechanism of lanthipeptide biosynthetic enzymes. *Curr. Opin. Struct. Biol.* **2014**, *29*, 58–66. [CrossRef]
151. Ortiz-Lopez, F.J.; Carretero-Molina, D.; Sanchez-Hidalgo, M.; Martin, J.; Gonzalez, I.; Roman-Hurtado, F.; de la Cruz, M.; Garcia-Fernandez, S.; Reyes, F.; Deisinger, J.P.; et al. Cacaoidin, first member of the new lanthidin RiPP family. *Angew. Chem.* **2020**, *59*, 12654–12658. [CrossRef] [PubMed]
152. Xu, M.; Zhang, F.; Cheng, Z.; Bashiri, G.; Wang, J.; Hong, J.L.; Wang, Y.M.; Xu, L.J.; Chen, X.F.; Huang, S.X.; et al. Functional genome mining reveals a class v lanthipeptide containing ad-amino acid introduced by an F420H2-dependent reductase. *Angew. Chem.* **2020**, *59*, 18029–18035. [CrossRef] [PubMed]
153. Repka, L.M.; Chekan, J.R.; Nair, S.K.; van der Donka, W.A. Mechanistic understanding of lanthipeptide biosynthetic enzymes. *Chem. Rev.* **2017**, *117*, 5457–5520. [CrossRef] [PubMed]
154. Lagedroste, M.; Reiners, J.; Knospe, C.V.; Smits, S.H.J.; Schmitt, L. A structural view on the maturation of lanthipeptides. *Front. Microbiol.* **2020**, *11*, 1183. [CrossRef]
155. Kodani, S.; Hudson, M.E.; Durrant, M.C.; Buttner, M.J.; Nodwell, J.R.; Willey, J.M. The SapB morphogen is a lantibiotic-like peptide derived from the product of the developmental gene ramS in *Streptomyces coelicolor*. *Proc. Natl. Acad. Sci. USA* **2004**, *101*, 11448–11453. [CrossRef]
156. Muller, W.M.; Schmiederer, T.; Ensle, P.; Sussmuth, R.D. In vitro biosynthesis of the prepeptide of Type-III lantibiotic labyrinthopeptin A2 including formation of a c-c bond as a post-translational modification. *Angew. Chem.* **2010**, *49*, 2436–2440. [CrossRef]

157. Goto, Y.; Li, B.; Claesen, J.; Shi, Y.X.; Bibb, M.J.; van der Donk, W.A. Discovery of unique lanthionine synthetases reveals new mechanistic and evolutionary Insights. *PLoS Biol.* **2010**, *8*, e1000339. [CrossRef]
158. Ruiz-Castillo, P.; Buchwald, S.L. Applications of palladium-catalyzed c-n cross-coupling reactions. *Chem. Rev.* **2016**, *116*, 12564–12649. [CrossRef]
159. Gkotsi, D.S.; Dhaliwal, J.; McLachlan, M.M.W.; Mulholand, K.R.; Goss, R.J.M. Halogenases: Powerful tools for biocatalysis (mechanisms applications and scope). *Curr. Opin. Chem. Biol.* **2018**, *43*, 119–126. [CrossRef]
160. Gribble, G.W. Naturally occurring organohalogen compounds. *Acc. Chem. Res.* **1998**, *31*, 141–152. [CrossRef]
161. Chankhamjon, P.; Tsunematsu, Y.; Ishida-Ito, M.; Sasa, Y.; Meyer, F.; Boettger-Schmidt, D.; Urbansky, B.; Menzel, K.D.; Scherlach, K.; Watanabe, K.; et al. Regioselective dichlorination of a non-activated aliphatic carbon atom and phenolic bismethylation by a multifunctional fungal flavoenzyme. *Angew. Chem.* **2016**, *55*, 11955–11959. [CrossRef] [PubMed]
162. Liu, M.; Ohashi, M.; Hung, Y.-S.; Scherlach, K.; Watanabe, K.; Hertweck, C.; Tang, Y. AoiQ catalyzes geminal dichlorination of 1,3-diketone natural products. *J. Am. Chem. Soc.* **2021**, *143*, 7267–7271. [CrossRef] [PubMed]
163. Kong, L.X.; Wang, Q.; Deng, Z.X.; You, D.L. Flavin adenine dinucleotide-dependent halogenase XanH and engineering of multifunctional fusion halogenases. *Appl. Environ. Microbiol.* **2020**, *86*, e01225-20. [CrossRef]
164. Andorfer, M.C.; Belsare, K.D.; Girlich, A.M.; Lewis, J.C. Aromatic halogenation by using bifunctional flavin reductase-halogenase fusion enzymes. *ChemBioChem* **2017**, *18*, 2099–2103. [CrossRef] [PubMed]
165. Higashide, E.; Asai, M.; Ootsu, K.; Tanida, S.; Kozai, Y.; Hasegawa, T.; Kishi, T.; Sugino, Y.; Yoneda, M. Ansamitocin, a group of novel maytansinoid antibiotics with antitumour properties from *Nocardia*. *Nature* **1977**, *270*, 721–722. [CrossRef] [PubMed]
166. Yu, T.W.; Bai, L.Q.; Clade, D.; Hoffmann, D.; Toelzer, S.; Trinh, K.Q.; Xu, J.; Moss, S.J.; Leistner, E.; Floss, H.G. The biosynthetic gene cluster of the maytansinoid antitumor agent ansamitocin from *Actinosynnema pretiosum*. *Proc. Natl. Acad. Sci. USA* **2002**, *99*, 7968–7973. [CrossRef]
167. Moss, S.J.; Bai, L.Q.; Toelzer, S.; Carroll, B.J.; Mahmud, T.; Yu, T.W.; Floss, H.G. Identification of Asm19 as an acyltransferase attaching the biologically essential ester side chain of ansamitocins using N-desmethyl-4,5-desepoxymaytansinol, not maytansinol, as its substrate. *J. Am. Chem. Soc.* **2002**, *124*, 6544–6545. [CrossRef]
168. Spiteller, P.; Bai, L.Q.; Shang, G.D.; Carroll, B.J.; Yu, T.W.; Floss, H.G. The post-polyketide synthase modification steps in the biosynthesis of the antitumor agent ansamitocin by *Actinosynnema pretiosum*. *J. Am. Chem. Soc.* **2003**, *125*, 14236–14237. [CrossRef]
169. Zhao, P.J.; Bai, L.Q.; Ma, J.; Zeng, Y.; Li, L.; Zhang, Y.R.; Lu, C.H.; Dai, H.Q.; Wu, Z.X.; Li, Y.Y.; et al. Amide N-glycosylation by Asm25, an N-glycosyltransferase of ansamitocins. *CHEM Biol.* **2008**, *15*, 863–874. [CrossRef]
170. Sun, P.; Zhao, Q.F.; Yu, F.T.; Zhang, H.; Wu, Z.H.; Wang, Y.Y.; Wang, Y.; Zhang, Q.L.; Liu, W. Spiroketal Formation and modification in avermectin biosynthesis involves a dual activity of AveC. *J. Am. Chem. Soc.* **2013**, *135*, 1540–1548. [CrossRef]

catalysts

MDPI

Review

Diclofenac Biodegradation by Microorganisms and with Immobilised Systems—A Review

Danuta Wojcieszyńska , Karolina Łagoda and Urszula Guzik *

Institute of Biology, Biotechnology and Environmental Protection, Faculty of Natural Science, University of Silesia in Katowice, Jagiellońska 28, 40-032 Katowice, Poland
* Correspondence: urszula.guzik@us.edu.pl

Abstract: Diclofenac is one of the most popular non-steroidal anti-inflammatory drugs. Due to its over-the-counter availability and high consumption along with municipal and hospital wastewater, it enters the sewage treatment plant, where it is not completely degraded. This results in the appearance of diclofenac in the effluents from the treatment plant, and with them, it enters the surface waters. Due to its structure, it is characterised by its high resistance to degradation in the environment. At the same time, it shows documented acute and chronic toxicity to non-target organisms. For this reason, it is necessary to look for cheap solutions that enhance the degradation of diclofenac. The paper discusses both the pathways of microbiological degradation of this drug described so far, as well as modern systems of biocatalyst immobilisation, with a particular emphasis on laccases involved in the biotransformation of diclofenac.

Keywords: biodegradation; diclofenac; immobilization; toxicity; sewage treatment; laccase

Citation: Wojcieszyńska, D.; Łagoda, K.; Guzik, U. Diclofenac Biodegradation by Microorganisms and with Immobilised Systems—A Review. *Catalysts* **2023**, *13*, 412. https://doi.org/10.3390/catal13020412

Academic Editors: Zhilong Wang and Tao Pan

Received: 22 January 2023
Revised: 11 February 2023
Accepted: 13 February 2023
Published: 15 February 2023

1. Introduction

Pain is a significant public health problem worldwide, with chronic pain affecting approximately 27% of the adult population in Europe and over 100 million adults in the United States. Non-steroidal anti-inflammatory drugs (NSAIDs) were discovered over 100 years ago, and the mechanism of action is based on the inhibition of cyclooxygenase (COX) isoenzymes. To this day, they remain a crucial element in the pharmacological treatment of inflammation and acute and chronic pain. They are essential for treating acute pain in the perioperative period and the cornerstone of treating osteoarthritis and other chronic pain conditions [1]. Due to their low addictive potential, good efficacy and long history of clinical use are often preferred by physicians [2]. In recent years, there has been a steady increase in the production and consumption of these drugs. Currently, over 50 types of NSAIDs are available on the world market, and diclofenac is one of this group's best-known and popular drugs. It is difficult to calculate the exact global intake of NSAIDs as they are sold under different trade names and are often available over the counter. However, it has been estimated that worldwide consumption of diclofenac is 940 tons per year in capsules, suppositories, tablets, intravenous solutions and ointments, not including veterinary consumption [3].

Due to the constantly growing problem related to detecting active pharmaceutical compounds (PhACs) in groundwater, surface water and drinking water, modern and environmentally friendly methods of wastewater bioremediation with greater efficiency and effectiveness are sought. The threat is exacerbated by the natural processes of water circulation in nature, hydrological connections between ecosystems and the accumulation of various pollutants introduced into the environment for decades. Recently, the microbial degradation of PhACs has been the subject of many studies due to the possibility of complete or partial degradation of harmful compounds or their transformation into less toxic compounds. Due to the unfavourable environmental conditions in which the biodegradation process is carried out, its efficiency drops significantly. Therefore, appropriate immobilisation methods are sought, allowing the immobilisation of biocatalysts. It is conducive to increasing the degradability of pollutants, extending life, and increasing the catalytic activity of biocatalysts. Moreover, it increases the chances of survival and adaptation of microbial cells to the changing environment, including the concentration of toxic compounds [4,5].

The study aims to assess the toxicity of diclofenac to non-target organisms and to analyse the possibility of using immobilised preparations in the biodegradation processes of diclofenac, particularly those based on immobilised laccase and microorganisms with an increased potential for decomposition of this drug.

2. Diclofenac—Characteristics and Distribution in the Environment

Diclofenac [2- (2,6-dichloroanilino)phenylacetic acid] is one of the most widely used non-steroidal anti-inflammatory drugs (NSAIDs), acting as an inhibitor of cyclooxygenase responsible for prostanoid synthesis. After oral administration, diclofenac is rapidly and completely absorbed in the intestines and is detoxified by hydroxylation and glucuronidation. CYP2C9 and CYP3A4 (cytochrome P450 family of proteins) catalyse its oxidation to 4′- and 5′-hydroxylated derivatives, and UDP-glucuronosyltransferase-2B7 (UGT2B70) catalyses glucuronidation. The kidneys excrete 65% of the oxidised metabolites. The rest, as acyl glucuronide, is excreted in the bile. Diclofenac acyl glucuronides are chemically unstable compounds that can be epimerised by acyl migration to 2-, 3- or 4-*O*-glucuronide, especially in the alkaline environment of bile [6]. Part of diclofenac is not metabolised after ingestion, and the sewage system discharges it to the sewage treatment plant in an unchanged or slightly changed form. It is estimated that the maximum concentrations of this drug in wastewater range from 0.01 to 510 µg/L of diclofenac. Even though the efficiency of removing diclofenac by advanced oxidation processes is as high as 80%, the limitations of physic-chemical methods often preclude their use [7,8].

Alternatives to chemical methods using aggressive chemicals are ecologically safe biological methods. However, diclofenac, a hydrophobic chlorinated derivative with electron-withdrawing and donor groups having log D < 3.2 at pH 8.0, is not susceptible to biological degradation [9]. The treatment efficiency in biological treatment plants is 0–80%, but most often, it is 21–40% [8,10]. Since sewage treatment plants are characterised by a low degradation efficiency of this drug, diclofenac and its derivatives enter the waters [7,8]. Diclofenac appears in soil, surface waters, groundwater and even in drinking water in various parts of the world (Table 1) [11–14].

Table 1. Diclofenac concentration in the environment.

Sources	Concentration	References
Europe		
Soil (Jerez de la Frontera, Spain)	Nd[1] –5.06 ng/g	[11]
Sediments Ebro Delta region (Catalonia, Spain)	6.8–7.5 ng/g	[15]
Wisła river (Skoczów, Poland)	74 ng/L	[16]
Odra river (Wrocław, Poland)	0.429 µg/L	[16]
Warta river (Częstochowa, Poland)	0.277 µg/L	[16]
Danube river (Budapest, Hungary)	7–90 ng/L	[12]
Aabach river (Switzerland)	11–310 ng/L	[12]
Swiss lakes (Switzerland)	1–12 ng/L	[12]
Vltava river (Prague, The Czech Republic)	0.104 µg/L	[16]
Tejo estuary (Portugal)	51.8 ng/L	[17]
Seawater (Portugal)	30.6 ng/L	[18]
Isar River (Germany)	9–13 ng/L	[19]
Wörthsee lake (Germany)	10–15 ng/L	[19]
Asia		
Beiyun River (China)	1.8–1300 ng/L	[20]
Huangpu River (China)	13.6 ng/L	[21]

Table 1. *Cont.*

Sources	Concentration	References
Malir River (Karachi, Pakistan)	0.08–0.3 μg/L	[13]
Korang River (Rawalpindi-Islamabad, Pakistan)	28 μg/L	[22]
Sawan River (Rawalpindi-Islamabad, Pakistan)	62 μg/L	[22]
Gumrah Kas (Rawalpindi-Islamabad, Pakistan)	14 μg/L	[22]
Ling Stream (Rawalpindi-Islamabad, Pakistan)	23 μ/l	[22]
Kaveri river (India)	103 ng/L	[23]
Africa		
Mbokodweni river (KwaZulu-Natal, South Africa)	0.9–5.3 μg/L	[24]
Umgeni River (KwaZulu-Natal, South Africa)	10 μg/L	[25]
Red Sea (Saudi Arabia)	26.9 ng/L	[14]
Antarctica		
Stream (Fildes Peninsula, Antarctica)	84 ng/L	[26]
Stream (Seymour/Marambio Island, Antarctica)	77 ng/L	[26]
North America		
Groundwater survey (Montana, USA)	46 ng/L	[27]
Mississippi river (Louisiana, USA)	22–107 ng/L	[12]
South America		
Natural waters (Rio de Janeiro, Brazil)	0.01–0.06 mg/L	[12]

Nd[1]—not detected.

The average world concentration in rivers is estimated at 0.021+/−0.722 μg/L, and its concentration in fresh water in extreme cases was recorded even in the range of μg/L [28,29]. Moreover, the presence of diclofenac was observed both in sewage sludge, at concentrations up to 87 ng/g, and in soils, where the observed concentrations reach 5.6 ng/g soil, depending on changing climatic conditions [11]. Such a situation forces the development of new, efficient and, at the same time, cost-effective methods, including those based on immobilised organisms, of removing this compound from water.

3. Toxicity of Diclofenac to Non-Target Organisms

The most tragic effect of diclofenac's influence on non-target organisms was the almost complete extinction of three species of vultures in the Indian subcontinent [28,30,31]. In the 1990s, in India and Pakistan, this drug was widely used to reduce inflammation caused by trauma and infectious diseases in cattle and buffaloes. When vultures ate the corpse of diclofenac-treated animals, the drug accumulated in the bird's bodies, causing kidney failure and death [7]. The cause of nephrotoxicity in vultures was the accumulation of uric acid crystals in the visceral organs (visceral fundus). A detailed study found that uric acid and alanine aminotransferase (ALT) levels were significantly elevated, and renal architecture was disturbed [30].

However, effects such as those described above are infrequent due to the low drug concentration observed in the environment. Nowadays, research is underway to analyse the risk of adverse effects on organisms inhabiting the aquatic environment and exposed long-term to low concentrations of diclofenac [32] (Table 2).

Table 2. Toxicity of diclofenac to non-target organisms.

Organism	Exposition Time	Concentration mg/L	Effect	References
Danio rerio	96 h	0.48 ± 0.05	Mortality-LC_{50}	[33]
		0.09 ± 0.02	Teratogenicity-EC_{50}	
	90 min	0.00003	Decreased level of lipid peroxidation in zebrafish embryo	
	96 h	0.001	Reduced viability of gill cells	[34]
	48 h	0.01	Reduced viability of digestive cells	
	48 h	0.001	Reduced viability of haemocytes	
Danio magna	21 days	2.0	Mortality-LC_{50}	
		0.5	Reduction in egg production	
Gammarus pulex	24 h	216	Mortality-LC_{50}	[28]

Table 2. *Cont.*

Organism	Exposition Time	Concentration mg/L	Effect	References
Hyalella azteca	24 h	175	Mortality-LC_{50}	
Oncorhynchus mykiss	-	0.001	Cytological alterations in the liver, kidney, and gills	
Gasterosteus aculeatus	28 days	0.0046	Renal hematopoietic hyperplasia, jaw lesions	[35]
	21 days	0.271	Mortality-LOEC	
Salmo trutta f. fario	25 days	0.1	Irregularly shaped and vesiculated hepatocytes with a lack of glycogen storage and degenerating nuclei	[36]
Dreissena polymorpha	6 months	0.00382	High mortality rates, effects on immunity, and high genotoxicity	[29]
	1 h	0.25	Destabilisation of lysosomal membranes	[34]
		0.06	DNA fragmentation	
Clarias gariepinus	96 h	25.12	Mortality-LC_{50}	[37]
Lithobates catesbeianus	96 h	1	Induction malformations such as axial malformations in the tail and notochord, oedema and stunted growth	[38]
Xenopus laevis	96 h	1	Induction malformations such as axial malformations in the tail and notochord, oedema and stunted growth	[38]
Lemna minor	10 days	0.0001	Decrease in the content of photosynthetic pigments, increased amount of reactive nitrogen and oxygen species in roots, increased lipid peroxidation, disturbation in membrane integrity	[39]
Mytillus galloprovincialis	15 days	0.25	Induction of superoxide dismutase and glutathione reductase in the gills, high catalase activity and lipid peroxidation levels in the digestive gland	[40]
Oryzias latipes	4 days	0.001	Induction of *p53* gene expression	[41]
Cirrhinus mrigala	35 days	0.001	Decrease of thyroxine and triiodothyronine levels	[42]
Gyps bengalensis	36–58 h	0.25/kg	Death from renal failure and visceral gout	[43]

Diclofenac physicochemical properties, mainly the n-octanol/water partition coefficient (log Kow, 4.51), are responsible for bioaccumulation in living organisms, primarily aquatic microorganisms [9]. The toxic impact of diclofenac, even at low concentrations of µg/L, has been demonstrated in studies on such species as common carp (*Cyprinus carpio*), brown trout (*Salmo trutta fario*), rainbow trout (*Oncorhynchus mykiss*) and stickleback (*Gasterosteus aculeatus*) [28]. Exposure of brown trout embryos to diclofenac did not show any evidence of embryotoxicity of this drug up to concentrations of 100 µg/L. NOEC is administered at a level of 500 µg/L for mortality, hatching, development and teratogenicity. Similar results were obtained for the embryonic and larval stages of *Danio rerio*, *Oncorhynchus mykiss* and *Cyprinus carpio*, where significant effects were found only at concentrations higher than 1 mg/L. Juvenile brown trout responded much more sensitively to diclofenac exposure than the larvae. The increase in mortality is alarming, occurring at diclofenac concentrations in the low µg/L range. There was a concentration-dependent increase in mortality in sticklebacks exposed to diclofenac, reaching a significance at 320 µg/L. A similar effect was observed by Näslund et al. [35] for the tricuspid stickleback at 271 µg/L of diclofenac. Acute toxicity tests of diclofenac in adult fish showed that the EC_{50} for carp was 71 mg/L [36]. Exposure to increasing concentrations of diclofenac from 2 to 32 mg/L also changed the growth curves of the populations of rotifers *Plationus patulous* and *Moina macrocopa*, leading to a decrease in the density of organisms with increasing drug concentration and a reduction in the daily population growth [34]. Mortality (LC_{50} 480 $\pm$ 50 µg/L) and teratogenicity (EC_{50} 90 $\pm$ 20 µg/L) have been demonstrated in *Danio rerio* after 96–h exposure to diclofenac. Chronic toxicity bioassays on the viability of *Danio rerio* embryos exposed for ten days to diclofenac allowed for the determination of NOEC and LOEC values at the levels of 4000 and 8000 µg/L, respectively [33]. The chronic toxicity of diclofenac has also been studied at the molecular and biochemical levels in *Daphnia magna*. Mortality of individuals increased after 24 h of exposure to high concentrations of diclofenac (486 mg/L). Exposure to 2 mg/L of diclofenac resulted in 50% mortality of *D. magna* after 21-day exposure and a significant reduction in egg production at a concentration of 0.50 mg/L. 96–h exposure to 50 µg/L diclofenac induced substantial changes in the expression of some genes related to detoxification, growth, development and reproduction. Their expression was inhibited after 24 h, and overexpression was observed after 48 h of exposure [34].

Joachim et al. [29] conducted studies on the harmfulness of diclofenac concerning primary producers and consumers in a long-term freshwater mesocosm experiment. The effective concentrations were 0.041, 0.44 and 3.82 µg/L, and the experimental time was six months. In such a constructed experiment simulating natural conditions, it was shown that the toxicity of diclofenac towards non-target organisms was higher than previously observed in laboratory conditions. During the six-month exposure period, the bio-volume of macrophytes (*Nasturtium officinale* and *Callitriche platycarpa*) decreased significantly. In *Dreissena polymorpha*, high mortality, reduced immunity and high genotoxicity were observed at all examined concentrations. Moreover, the highest concentration used changed the structure of the *Gasterosteus aculeatus* population. After one month of exposure, the total fish stock and the percentage of juveniles decreased while the percentage of adults increased. It led to a general change in the F1 generation length and frequency distribution compared with the control [29].

In the context of the negative impact of diclofenac on non-target organisms, the toxicity of the intermediates of diclofenac biotransformation is very important. Fu et al. [28] studied this phenomenon in two key aquatic invertebrates: *Gammarus pulex* and *Hyalella azteca*. In both of these species, diclofenac was converted into several oxidation products and conjugates, including the taurine-diclofenac conjugate and the diclofenac methyl ester. A significant increase in the bioconcentration factor for these intermediates relative to the parent drug has been demonstrated. Moreover, diclofenac methyl ester was also characterised by higher acute toxicity than diclofenac for both species, which correlated well with the increased potential for bioconcentration. The LC_{50} of diclofenac for *H. azteca* was 216 mg/L, while the LC_{50} diclofenac methyl ester was only 0.53 mg/L, which is a 430-fold increase in acute toxicity compared to diclofenac. The diclofenac-taurine conjugate was less toxic to *H. azteca* than its parent compound, which may be due to its slightly lower hydrophobicity [28]. In addition, it was observed that the two most frequently detected hydroxylated derivatives of diclofenac: 4′-OH-diclofenac and 5-OH-diclofenac can be further oxidised to reactive benzoquinone imines that interact with the protein nucleophilic groups, resulting in the formation of adducts [9]. These studies clearly show that when researching the toxicity of drugs toward non-target organisms, it is also crucial to look at the toxicity of their biotransformation products more broadly [28].

Histological evaluation helps to better understand the mechanism of diclofenac toxicity. In fish exposed to this drug, severe tissue reactions and lesions, especially in the liver, were more frequently observed. Näslund et al. [35] showed renal hematopoietic hyperplasia, and jaw lesions of the tricuspid stickleback at the lowest concentration tested at 4.6 µg/L. Importantly, it is an identified concentration in the environment. Ultrastructural studies of rainbow trout liver revealed glycogen level reduction and macrophage infiltration [36]. Moreover, it has been shown that even low concentrations of diclofenac found in the environment (1 µg/L) lead to cellular changes in the liver, kidneys and gills of rainbow trout, which reduces the functionality of the kidneys and gills [41]. In another in vitro experiment, the toxicity of diclofenac was tested in a range of concentrations of 0.001, 0.01, 0.1, 1 and 10 mg/L on three different cell types of zebra mussels: haemocytes, gill cells and digestive gland cells. After 96 h of exposure, a significantly reduced viability of diclofenac-treated gill cells was observed already in the presence of the lowest concentration applied. Moreover, the viability of diclofenac-treated digestive cells was significantly reduced after 48 h of exposure to 0.01 mg/L, while the haematocyte viability was decreased at a concentration of 0.001 mg/L [34].

The apparent effect of diclofenac on inter-individual relationships in exposed fish is surprising. The percentage of individuals showing signs of aggressive behaviour increased significantly with increasing diclofenac concentration, with a LOEC of 10 µg/L. Behavioural changes of African catfish (*Rhamdia quelen*) at 25 mg/L of diclofenac were also observed. These included respiratory failure, loss of balance, and irregular swimming, but no signs of aggression were seen. It may be the effect of reduced ability or propensity to act in defence, resulting from the weakened condition of the animals. For example, stickleback was leaving food behind and having a higher percentage of skin ulceration after exposure to diclofenac. Open wounds can cause infections with pathogens, making animals vulnerable to potentially lethal consequences. On the other hand, diclofenac-induced mortality may increase aggressive behaviour through changes in fish density. Another possible explanation for the increased aggressiveness observed in juvenile fish relates to the finding that diclofenac leads to corneal perforation in rainbow trout. The behavioural abnormalities of juvenile brown trout can be attributed to panic reactions due to visual impairment [36,37].

Studies on carp showed a significant increase in hydroperoxide and lipid peroxidation content at a diclofenac concentration of 7.1 mg/L [36]. Reduced lipid peroxidation (LPO) levels were observed in zebrafish embryos after exposure to 30 ng/L diclofenac for 90 min. At the environmental concentration (ng/L), diclofenac showed toxicity to *Perna perna* mussels, leading to decreased lysosomal membrane stability and increased cyclooxygenase activity, higher levels of oxidative stress and DNA damage. An increased level of lipid peroxidation was also observed in zebra mussels treated with diclofenac at a concentration of 1 μg/L. However, oxidative stress was induced in the mussel *Mytilus galloprovincialis* even after exposition to 0.25 μg/L of diclofenac [32]. Moreover, in zebrafish, in response to oxidative stress induced by diclofenac and its photolysis products, elevated levels of enzymes such as catalase, superoxide dismutase, and glutathione transferase and lipid peroxidation were observed [44]. The presence of diclofenac results in oxidative stress also in the cultures of microorganisms. Among other things, increased activity has been shown in superoxide dismutase and catalase and the formation of lipid peroxidation products. Changes accompanied this phenomenon on the cell surface and within the biological membrane. Multivariate analysis showed that exposure of the *Pseudomonas moorei* KB4 strain to diclofenac caused a decrease in the zeta potential with a simultaneous increase in the hydrophobicity of the cell wall. In addition, significant stiffening of the membrane was observed as a result of changes in the fatty acid composition of the membrane (including the appearance of a branched fatty acid-19:0 *anteiso* and cyclopropane-17:0 *cyclo*) [9,45]. Not only the parent form of diclofenac but also its intermediates can cause oxidative stress in cells of non-target organisms. Aissaoui et al. [46] assessed the toxicity of diclofenac intermediates obtained during the degradation of this drug by *Enterobacter cloacae* isolated from compost toward mouse liver cells. It was shown that diclofenac at therapeutic concentrations and its metabolites affected oxidative stress parameters, including a decrease in glutathione reserve, lipid peroxidation and disorders of the liver detoxifying enzymes, including superoxide dismutase, catalase, and glutathione S-transferase. However, these researchers emphasise the lack of a negative influence of diclofenac and its metabolites on the oxidative stress parameters in mice cells after applying environmental concentrations of this drug [46].

Lymnaea stagnalis snails were used to assess the immunotoxicity of diclofenac. They were exposed to diclofenac for three days at environmental concentrations (1–10 μg/L) and therapeutic concentrations (100–1000 μg/L). Diclofenac significantly influenced the immune capacity and the performance of the cochlear haemocytes. This effect is typical of the inflammatory response, confirmed by an increase in NADPH oxidase activity, mainly after using the drug at a concentration of 1000 μg/L [34]. The expression of hepatic *c7* genes was also shown to be dependent on the concentration of diclofenac. The c7 protein is a complement of the system part of the innate immune system. It forms a membrane attack complex with other complement component proteins that lead to the lysis of foreign cells. Complement components are linked via the arachidonic acid pathway, which explains the effect of NSAID exposure on c7 [35]. In turn, the cytogenotoxicity of diclofenac was tested in vitro through the 1-h exposition of haemocytes collected from *Dreissena polymorpha* to 60, 126 and 250 μg/L of the drug. A significant cytotoxic effect in the destabilisation of lysosomal membranes was noted only after exposure to 250 μg/L of diclofenac, while both primary genetic changes (e.g., DNA fragmentation) and permanent DNA damage occurred after exposure to all tested concentrations [34]. Increased DNA fragmentation was also observed in the mussel *Mytilus galloprovincialis* after exposure to the environmental concentration (2.5 μg/L) [32]. On the other hand, *Oryzias latipes* showed induction of *p53* gene expression after 4-day exposure to 1 μg/L of diclofenac. The *p53* gene is an important biomarker in analysing environmental toxin carcinogenicity, and DNA damage, as its product plays a crucial role in cell cycle arrest, apoptosis and DNA repair [41]. The studies on the toxicity of diclofenac do not give conclusive results, and some reports show that in environmental concentrations, diclofenac does not threaten aquatic animals. Memmert et al. [47] indicated very low bioconcentration of diclofenac in fish, corresponding to the chemical properties of diclofenac, including the dissociation constant of pKa 3.99–4.16. Moreover, these authors indicate that although the NOEC is estimated at 10 μg/L, only a concentration above 320 μg/L causes a reduction in growth in zebrafish [46]. However, considering the number of reports of adverse effects of diclofenac on non-target organisms, the conclusions of Näslund et al. [35] seem to be justified. These authors postulate the reduction in the environmental risk associated with diclofenac toxicity through the substitution of diclofenac, where possible from a therapeutic point of view, with naproxen-a drug with a similar effect in the treatment of pain. Although naproxen and diclofenac have similar effects in fish, environmental hazards and risks are lower with naproxen than diclofenac because the toxic effects appear at higher naproxen concentrations than diclofenac [35].

Due to the increasing concentration of NSAIDs in soils, plants are also exposed to the toxic effects of diclofenac. Studies of the stress response to diclofenac of two crops, maize and tomato, showed that the sensitivity to diclofenac is species-dependent. The tomato was more sensitive, with growth inhibition, a decrease in the content of photosynthetic pigments and a decrease in the maximum PSII quantum efficiency and PSII activity. However, no effect of diclofenac in maize was observed in the content of photosynthetic pigments or growth. However, an effect of diclofenac on the quantum efficiency of the PSII photosystem was observed. In both plants, oxidative stress was also observed, manifested by an increased concentration of hydrogen peroxide. In response, the plants triggered a defence mechanism in the form of the synthesis of phenolic compounds [48]. Similarly, Copolovici et al. [49] showed in beans the effect of diclofenac on the reduction in assimilation coefficients and stomatal conductivity for water vapor. In addition, an increase in the concentration of monoterpenes was also found: a-pinene, camphene and 3-carene). The authors postulate that diclofenac may also interfere with the methylerythritol phosphate pathway in plastids [49].

4. Diclofenac Biodegradation by Bacteria and Fungi

Few fungi and bacteria capable of degrading diclofenac have been described so far. Moreover, only a partial decomposition of this drug is most often described in the literature due to the emergence of difficult-to-degradable intermediates. The fungi that can degrade diclofenac include *Trametes trogii, T. polyzona, Yarrowia lipolytica, Aspergillus niger, Phanerochaete chrysosporium, Mucor circinelloides, Trichoderma longibrachiatum, Rhizopus microspore* [50,51]. Among the bacteria that degrade diclofenac, strains of *Raoultella* sp. DD4, *Bacillus subtilis, Brevibacillus laterosporus, Rhodococcus ruber, Labrys portucalensis* F11, *Alcaligenes faecalis, Staphylococcus aureus, Staphylococcus haemolyticus, Pseudomonas moorei* KB4, *Klebsiella* sp. KSC and *Proteus mirabilis* have been described [9,10,45,52–56].

The most frequently observed transformation of diclofenac by fungi is hydroxylation by laccase, manganese peroxidase and the cytochrome P450 enzyme system to intermediates such as 4-hydroxydiclofenac, 5-hydroxydiclofenac, 3-hydroxydiclofenac and 4,5-dihydroxydiclofenac [50,51] (Figure 1).

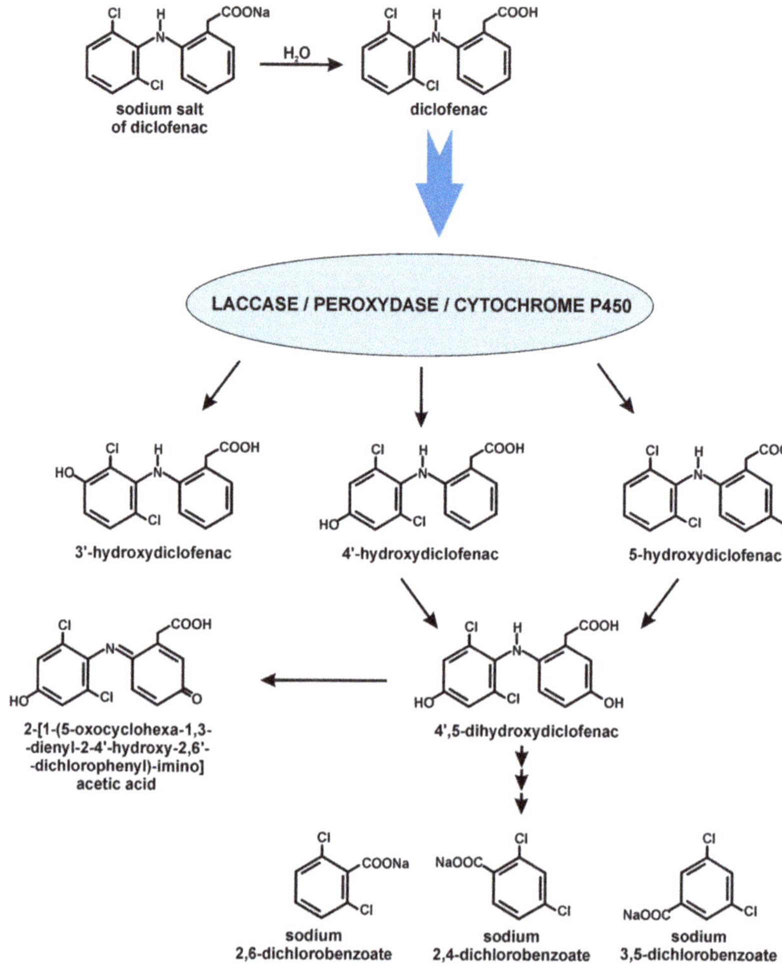

Figure 1. Diclofenac biotransformation by fungi [50,51].

Moreover, ligninolytic fungi: *T. polyzona*, *M. circinelloides*, and *T. longibrachiatum* degraded diclofenac into the intermediates: adduct of 2,6-dicholorobenzoic, 2,4-dichlorobenzoic and 3,5-dichlorobenzoic acid, which proved the breaking of the C-N bond in the drug structure. After ten days of incubation, these intermediates disappeared from the culture, indicating their further degradation with ring cleavage. A positive correlation was observed between the activity of manganese peroxidase and the drug tolerance of the fungi. In contrast, no such correlation was observed concerning the drug degradation efficiency, which shows that these strains' mechanism of diclofenac degradation is more complex [51].

Metabolites of diclofenac, commonly observed in fungi, were also shown during bacterial degradation. Among others, strains of *Bacillus* and *Brevibacillus* decomposed diclofenac into 4'-hydroxydiclofenac [53].

The first described bacterial strains capable of diclofenac degradation were *Raoultella* sp. DD4 and *Rhodococcus ruber* IEGM 346 [52,54]. *Raoultella* sp. DD4 degraded 0.6 mg/L of diclofenac within 28 days and was highly resistant to the toxic effects of this drug [52]. *Rhodococcus ruber* strain IEGM 346 can degrade high diclofenac concentrations (50 mg/L). It was confirmed that the C-N bond is broken during degradation, and the aromatic ring opens in the structure of diclofenac. 16 intermediates of the decomposition of this drug by the IEGM 346 strain have been identified. The described pathway leads through a series of oxidation reactions to homogentisic acid. The further oxidation of this acid through a quinone derivative leads to the end products: acetoacetic acid, fumaric acid and 4,6,7-trioxooct-2-enedioic acid (Figure 2).

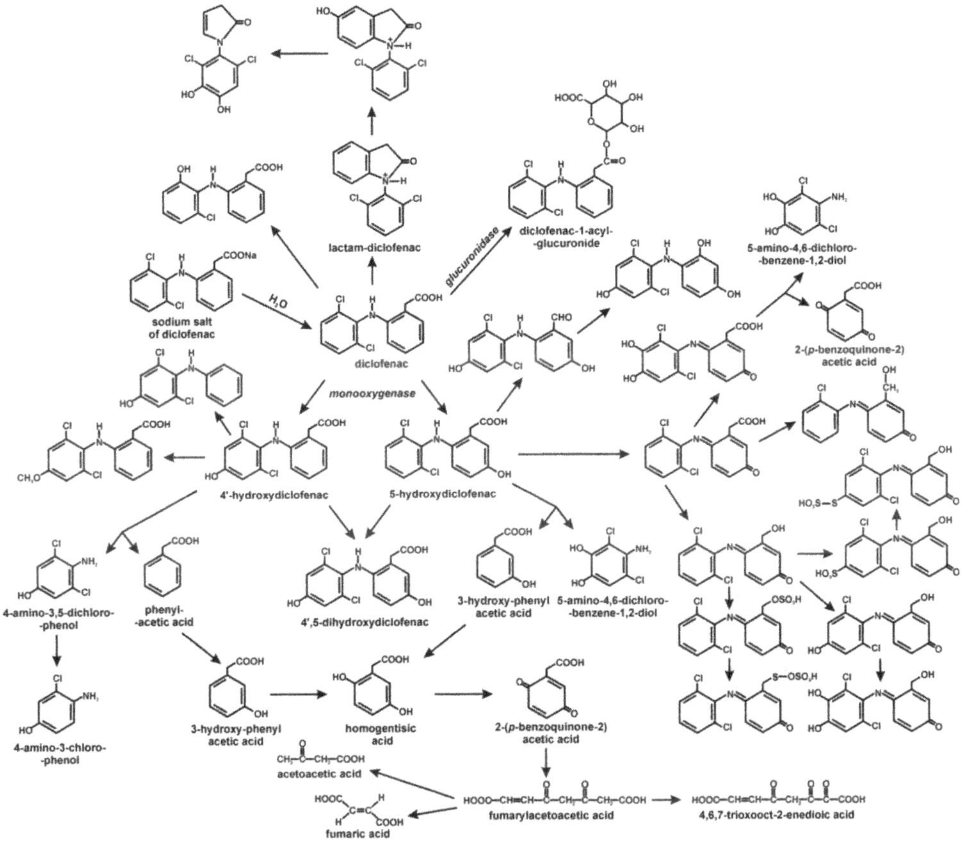

Figure 2. Diclofenac degradation by bacteria [9,10,55,56].

The adaptation mechanisms of *Rhodococcus ruber* IEGM 346 to high concentrations of this drug are altered ζ potential of bacterial cells, increased cell hydrophobicity and total cell lipid content, formation of multicellular conglomerates, and altered surface-to-volume ratio [54]. In addition, Moreira et al. [10] showed that the degradation of diclofenac by the *Labrys portucalensis* F11 strain took place through hydroxylation and the formation of benzoquinone imine as a key metabolite. The resulting product was further decarboxylated and hydroxylated. The stoichiometric release of chlorine and the lack of detected metabolites at the end of the experiments indicated complete degradation of the drug by the F11 strain. It was also the first time that a sulphation reaction was described during bacterial diclofenac decomposition, indicating the similarity of metabolites during bacterial diclofenac degradation to the conjugates which appeared during the Phase II detoxification diclofenac in mammals [8]. For another example, the similarities between the detoxification pathways of drugs in mammals and bacterial metabolism have also been shown by Murshid et al. [55]. They described *Staphylococcus* sp. and *Alcaligenes* sp. with the glucuronidase activity responsible for the conjugation of diclofenac with glucuronic acid to diclofenac 1-acyl-glucuronide [55].

Żur et al. [9,45] described the *Pseudomonas moorei* KB4 strain as capable of degrading 0.5 mg/L of diclofenac in a mono-substrate culture. In contrast, in a culture supplemented with glucose and sodium acetate, this strain degraded 1 mg/L of diclofenac within 12 days. 4′-OH-diclofenac and diclofenac-lactam were identified as intermediates. Gene expression analysis revealed up-regulation of selected genes encoding biotransformation enzymes in the presence of diclofenac such as monooxygenase, dihydroxylating and aromatic ring cleaving dioxygenases, and cytochrome p450 system [9,45]. In turn, the bacterial strain *Klebsiella* sp. KSC, isolated from the livestock soil, has been described as a strain capable of biodegradation of diclofenac high concentrations. *Klebsiella* sp. KSC exposition to 70 mg/L of diclofenac caused the mineralisation of diclofenac after 72 h. In this case, 12 biotransformation products of diclofenac have been identified, indicating that hydroxylation, dehydroxylation, decarboxylation and dechlorination are critical steps in the degradation of this compound. As a result of these mechanisms, alcohols and ketones compounds are formed. Both mono-, di-, tri- and tetrahydroxylated derivatives were observed. The generation of such products resulted from removing the carboxyl group and two hydrogens from diclofenac with the simultaneous addition of hydroxyl groups to the parent compound. In addition, hydroxylation products formed after the cleavage of the acetate group from the structure of the parent substance were observed. In addition, the cyclisation product between the carboxyl group and the nitrogen atom was also identified [56].

5. Diclofenac Biodegradation in Immobilised Systems

Conventional diclofenac wastewater treatment methods, such as physical and chemical procedures, have severe limitations, such as poor treatment, low efficiency, high cost, generation of hazardous by-products and application to a narrow range of concentrations of organic compounds in the wastewater. Therefore, the challenge for environmental engineers and biotechnologists is to develop an efficient, economical and environmentally safe bioremediation technique to provide outstanding remediation solutions instead of current treatment technologies [57]. Due to the confirmed toxicity of diclofenac towards microorganisms, including microorganisms capable of biodegradation, more and more attention is paid to immobilised biopreparations usage (Table 3).

Table 3. Immobilisation matrix/technologies in diclofenac biodegradation.

Immobilisation Matrix/Technology	Pros and Cons of Matrix/Technology	Microorganism/Enzyme	References
Sodium alginate-silicon dioxide-polyvinyl alcohol	Highly effective in subsequent cycles with an electron mediator	Laccase (*Sphingobacterium* ksn-11)	[58]
Electrospun nanofibers poly(L-lactic acid)-co-poly(ε-caprolactone)	Thin structure, porosity, biocompatibility, a high number of functional groups	Laccase (*Trametes versicolor*)	[59]
Porcine manure biocarbon	High adsorption capacity, effectiveness, high storage stability	Laccase	[60,61]
Polyvinylidene chloride membrane modified with multi-wall carbon nanotubes	Resistance to contaminants, specific surface area, mechanical strength, water permeability, selectivity, thermal resistance	Laccase (*Trametes hirsuta*)	[62]
Granulated activated carbon	Large specific surface, high adsorption capacity, porous structure, availability	Laccase	[63]
Palladium nanoparticles	Resistance to aggregation	Microorganisms	[64]

Currently, the most commonly used methods in immobilising live microorganisms are the so-called self-immobilisation in pellets and adhesion to a fixed or porous surface. The first method is based on the natural tendency of some species of microorganisms in submerged cultures for pellet development. The second method is based on the adhesion of cells to the support material with a secreted exopolysaccharide acting as an adhesive material. Immobilisation in the support material can also occur by encapsulating in pores or through physical or chemical traps in porous solids or matrices. Recently, the use of mushroom pellets has attracted attention due to the appropriate ability to self-immobilise and the possibility of connecting another microorganism to such an aggregate or material, resulting in the formation of self-immobilised biomixes [65].

Both whole cells of microorganisms and selected oxidising enzymes have been used in biodegradation and biotransformation processes [45,52,59]. An example of the use of whole immobilised microorganisms in the degradation process of diclofenac was the designed bioreactor based on a biofilter with immobilised activated sludge microorganisms. After a two-month adaptation period, the removal efficiency of diclofenac in the designed system reached 97.63%. Based on Illumina sequencing, the major bacterial taxa in the biofilter were identified, which included *Granulicella pectinivorans, Rhodanobacter terrae, Castellaniella denitrificans, Parvibaculum lavamentivorans, Bordetella petrii, Bryocella elongata* and *Rhodopseudomonas palustris. Wickerhamiella* was the dominant fungal taxa in the immobilised cell biofilter, indicating its leading share in diclofenac degradation in activated sludge systems. Such enormous biodiversity of microorganisms allowed for the efficient operation of the reactor. However, a decrease in pH was observed during its process, which did not affect the degradation efficiency. Moreover, this arrangement proved successful during the implementation of a wide range of carbon sources [8]. Pereira et al. [65] indicate the potential of using the so-called biomixes to enhance microorganisms' ability to degrade organochlorine derivatives, including diclofenac. In the construction of biopreparation, the ability of fungi and bacteria to autoaggregate is used. These authors, however, point out that this requires research to understand interspecies interactions between fungal granules and bacteria, especially as co-immobilisation affects pellet fixation and the possibility of their use in continuous operation. Such research can contribute to developing cost-effective and efficient biodegradation techniques for diclofenac and other substances resistant to degradation [65].

The literature data show that not only microorganisms but also enzymes isolated from them can be used to degrade diclofenac. However, there are few reports on the involvement of enzymes in the degradation/biotransformation of diclofenac. The process of enzymatic degradation of diclofenac was most fully described *by* Żur et al. [9], indicating the involvement of enzymes such as hydroxylating mono- and dioxygenases, aromatic ring-cleaving enzymes (catechol 1,2-dioxygenase, homogentisate 1,2-dioxygenase and salicylate 1,2-dioxygenase) in the decomposition of the drug structure. In addition, it was shown that deaminase had a significant impact on the decomposition of the dicyclic structure of diclofenac. However, Żur et al. [9] have not proposed any purification system based on immobilized enzymes based on the enzymes shown. Other enzymes involved in the degradation of diclofenac include the cytochrome P-450 system and laccase responsible for hydroxylation, as well as the enzymes responsible for the transformation to glucuronide derivatives [5,55,66,67] However, so far, only laccases have found broader applications in systems based on immobilized enzymes used to transform diclofenac due to its low specificity and high oxidising capacity towards diclofenac (Figure 3). Both simple and popular carriers, such as alginate and new-generation synthetic carriers, were used for this enzyme [58,59,62,64]. An example of laccase usage in the biodegradation of diclofenac is the enzyme derived from *Sphingobacterium* ksn-11 immobilised in sodium alginate-silicon dioxide-polyvinyl alcohol beads. It was shown that the immobilised laccase oxidised diclofenac to 4-OH of diclofenac after 4 h of incubation, and the preparation itself was highly effective in subsequent cycles. Moreover, the application of an electron mediator (2,2-azino-bis-(3-ethylbenzthiozoline-6-sulfonic aid) shortened the transformation time to 90 min [58].

Zdarta et al. [59] immobilised laccase from *Trametes versicolor* on electrospun nanofibers poly(L-lactic acid)-co-poly(ε-caprolactone) (PLCL) by adsorption, encapsulation and covalent bonding. Thin structures characterise materials produced by the electrospinning method with a diameter ranging from 100 nm to 1 μm and a length of up to several thousand meters. Due to features such as porosity, biocompatibility and the high number of functional groups on nanofibers' surfaces, these materials are favourable in immobilising enzymes. As a result, over 90% biodegradation of 1 mg/L diclofenac was achieved under optimal conditions. The immobilised enzyme was also active in the following cycles, maintaining 40% efficiency after the fifth cycle. In addition, a thorough toxicity analysis of the biodegradation products was carried out. It was also shown that the solution obtained after this process was about 65% less toxic than the initial diclofenac solution [59].

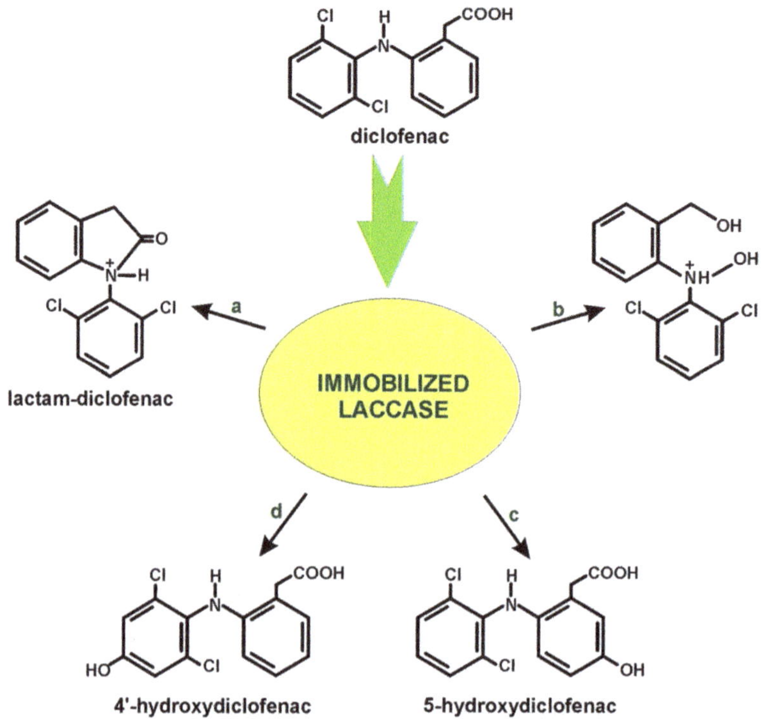

Figure 3. Products of diclofenac biodegradation in the immobilised system. Small letters indicate the type of carrier, where (**a**,**b**) electrospun nanofibers, (**c**) carbon nanotubes, and (**d**) sodium alginate-silicon dioxide-polyvinyl alcohol beads [58,59,63].

In turn, the complete biodegradation of diclofenac (500 μg/L) within five hours was observed after applying laccase immobilised by adsorption on porcine manure biocarbon (BC-PM) at the initial concentration of 500 μM diclofenac. Among other biocarbon carriers, such as those obtained from pine wood (BC-PW) or almond shell (BC-AS), BC-PM showed the highest adsorption capacity of laccase. It was observed that after using biocarbon carriers, along with an increase in the initial concentration of laccase in the solution, the enzyme binding capacity and, consequently, the effectiveness of immobilisation also increased. The pre-treatment of the biocarbon carrier with citric acid increased its ability to bind the enzyme. It was also established that the primary mechanism of enzyme adsorption on biochar is homogeneous monolayer adsorption. Immobilised laccase showed higher storage stability. Three times longer life was observed than free laccase [57,60,61].

Another example of enzymatic decomposition of diclofenac is the use of laccase derived from *Trametes hirsuta*, immobilised by a covalent bond on PVDF/MWCNT membranes ((polyvinylidene chloride (PVDF) membrane modified with multi-wall carbon nanotubes (MWCNTs)). Polymer materials increase the resistance to contamination of the resulting nanocomposite membrane. Nanomaterials can be used in the membrane matrix or on its surface. They exhibit unique properties, including increasing their specific surface area and mechanical strength, physicochemical properties such as water permeability, resistance to contaminants, selectivity and thermal resistance. Carbon-based nanomaterials, including MWCNTs, are popular membrane reinforcement materials used in wastewater treatment. Polymer membranes are compatible with sewage treatment plant devices and do not interfere with their functioning. During the immobilisation of laccase on MWCNTs, there was a problem with nanoparticle separation at the purification stage. Mixing MWCNT with PVDF membranes made it possible to avoid this problem. The addition of MWCNTs to PVDF improves the physical properties of the membrane. It increases the rate of electron transfer between the laccase and the substrate, thus, increasing the effectiveness of laccases. Covalently immobilised laccase showed a high activity of 4.47 U/cm^2 and an activity recovery of 38.31%. Using chemically immobilised laccase in the mini-membrane reactor by covalent bonding on PVDF/MWCNT membranes allowed for 95% of diclofenac degradation within 4 h [62].

As a carrier for the immobilisation of laccase by physical adsorption, granulated activated carbon (GAC) is also used. GAC characterises by a large specific surface, high adsorption capacity, porous structure and wide availability on the market. These features give GAC an excellent potential for enzyme immobilisation. It was shown that the adsorption process on GAC did not change the structure of the laccase, which allowed it to maintain its activity. GAC can efficiently adsorb micropollutants such as diclofenac. However, as with all adsorbents, GAC adsorption of the micropollutants decreases with exposure time due to surface saturation. Due to this issue, GAC regeneration is required to maintain system performance. Hence, a regeneration strategy was developed by pre-adsorbing laccase to the GAC. Immobilised laccase degrades the adsorbed micropollutants, thus releasing the adsorption sites. In addition, the co-adsorption of laccase and micropollutants on GAC improves biodegradation efficiency due to the increased electron transfer between laccase and micropollutants. The immobilisation of laccase on GAC does not influence micropollutants' adsorption because, after the enzyme's immobilisation, about 65% of the carbon surface is still available for the adsorbates. Moreover, sorption sites on the GAC surface are released after biodegradation, and the sorption-degradation cycle can start anew. As a result, the efficiency of diclofenac removal increases with subsequent cycles. Laccase prevents the complete saturation of GAC, which is a must for continuous operation. Higher laccase load, i.e., "full saturation", allowed obtaining the highest diclofenac degradation result in all cycles. The use of GAC-bound laccase overcomes some problems associated with using free laccase for the catalytic degradation of micropollutants. The immobilisation of laccase improved its reusability and stability over a wide range of pH and temperature, and the enzyme removed micropollutants more effectively [63].

Recently, catalysts based on palladium (Pd) have aroused increasing interest. They can catalyse many reactions, such as denitrification and hydrodechlorination. Palladium catalysts are usually prepared by chemical methods and then immobilised on supports such as silica. This avoids their aggregation and facilitates recycling. The production of palladium nanoparticles using microbial reduction is a promising solution. This process allows for using fewer toxic chemicals and does not require stabilisers or carriers, which makes it environmentally friendly. The biomass-supported nanoparticles show more excellent resistance to aggregation than those supported by conventional supports. Several microorganisms, both pure strains and mixed bacterial cultures, can produce biogenic nanopalladium (Bio-Pd) in their cell membranes and cytoplasm. One example of Bio-Pd producers is the anaerobic granular sludge (AGS). AGS is a particular form of microbial aggregates consisting of mixed cultures of microorganisms with a three-dimensional, heterogeneous structure. Biocatalizator Pd-AGS combines AGS's microbial metabolic role with palladium's catalytic function. As a heterogeneous catalyst consisting of Pd nanoparticles and microbial granules, it has the advantage of initiating Pd autocatalysis using hydrogen donors or electrons generated from organic compounds by microbial transformation and fermentation. It has been shown that many factors, such as hydrogen and electron donors, remediation medium, immobilisation vehicle, and pH, can influence the catalytic activity of Pd during diclofenac bioremediation. Hydrogen is the most effective electron donor for Pd-AGS, which is then more resistant to inactivation by chloride or sulphide than free Pd nanoparticles. Using the Pd-AGS system allowed the decomposition of 96% of diclofenac during four iterations of reduction, and the purification with water quickly regenerated its catalytic activity. Pd-AGS appears to be a viable and economical alternative to homogeneous Pd complexes or a conventional supported heterogeneous Pd catalyst [64].

6. Conclusions

The literature review clearly shows that diclofenac has a multidirectional effect on non-target organisms living in waters contaminated with this drug, causing several negative changes. At the same time, data analysis shows that there are microorganisms capable of degrading low concentrations of this drug. However, most of the described fungi used in the biodegradation processes of diclofenac, in fact, only transform to hydroxylated derivatives without disturbing the aromatic structure of this compound. Such products may be more toxic than the parent compound. In addition, using immobilised systems using laccase as enzymes involved in the oxidation of diclofenac does not solve this problem because the enzymes only carry out the initial hydroxylation. To solve the problem of environmental pollution with diclofenac, it is necessary to look for microorganisms capable of completely degrading diclofenac. It may be promising to use immobilised fungal and bacterial systems, which would make it possible to use the high potential of non-specific fungal enzymes for the activation of diclofenac. The activated substrate could then be degraded sequentially into primary metabolism intermediates with the participation of bacteria. However, research related to the analysis of fungal-bacterial interactions is necessary for such systems to be used as an application.

Author Contributions: Conceptualization, D.W., K.Ł. and U.G.; writing—original draft preparation, D.W. and K.Ł.; writing—review and editing, D.W. and U.G.; visualization, D.W. and U.G.; supervision, D.W. and U.G.; project administration, U.G.; funding acquisition, U.G. All authors have read and agreed to the published version of the manuscript.

Funding: This research was funded by the National Science Centre, Poland (grant number 2018/29/B/NZ9/00424).

Data Availability Statement: No new data were created or analysed in this study. Data sharing is not applicable to this article.

Conflicts of Interest: The authors declare no conflict of interest.

References

1. Monteiro-Steagall, B.P.; Steagall, P.V.M.; Lascelles, B.D.X. Systematic review of non-steroidal anti-inflammatory drug-induced adverse effects in dogs. *J. Vet. Intern. Med.* **2013**, *27*, 1011–1019. [CrossRef]
2. Brune, K.; Patrignani, P. New insights into the use of currently available non-steroidal anti-inflammatory drugs. *J. Pain Res.* **2015**, *8*, 105–118. [CrossRef] [PubMed]
3. Guzik, U.; Wojcieszyńska, D. Biodegradation of non-steroidal anti-inflammatory drugs and their Influence on soil microorganisms. In *Microbes and Enzymes in Soil Health and Bioremediation*; Kumar, A., Sharma, S., Eds.; Springer: Singapore, 2019; pp. 379–401. Available online: https://link.springer.com/content/pdf/10.1007%2F978-981-13-9117-0_16.pdf (accessed on 10 February 2023).
4. Du, Y.; Jia, X.; Zhong, L.; Jiao, Y.; Zhang, Z.; Wang, Z.; Feng, Y.; Bilal, M.; Cui, J.; Jia, S. Metal-organic frameworks with different dimensionalities: An ideal host platform for enzyme@MOF composites. *Coord. Chem. Rev.* **2022**, *454*, 214327. [CrossRef]
5. Zhou, W.; Zhang, W.; Cai, Y. Laccase immobilization for water purification: A comprehensive review. *Chem. Eng. J.* **2021**, *403*, 126272. [CrossRef]
6. Krasniqi, V.; Dimovski, A.; Domjanović, I.K.; Bilić, I.; Božina, N. How polymorphisms of the cytochrome P450 genes affect ibuprofen and diclofenac metabolism and toxicity. *Arch. Ind. Hyg. Toxicol.* **2016**, *67*, 1–8. [CrossRef]
7. Chen, J.; Gao, H.; Zhang, Y.; Zhang, Y.; Zhou, X.; Li, C.; Gao, H. Developmental Toxicity of diclofenac and elucidation of gene regulation in zebrafish (*Danio rerio*). *Sci. Rep.* **2014**, *4*, 4841. [CrossRef]
8. Navrozidou, E.; Remmas, N.; Melidis, P.; Karpouzas, D.G.; Tsiamis, G.; Ntougias, S. Biodegradation potential and diversity of diclofenac-degrading microbiota in an immobilised cell biofilter. *Processes* **2019**, *7*, 554. [CrossRef]
9. Żur, J.; Piński, A.; Wojcieszyńska, D.; Smułek, W.; Guzik, U. Diclofenac degradation-enzymes, genetic background and cellular alterations triggered in diclofenac-metabolising strain *Pseudomonas moorei* KB4. *Int. J. Mol. Sci.* **2020**, *21*, 6786. [CrossRef]
10. Moreira, I.S.; Bessa, V.S.; Murgolo, S.; Piccirillo, C.; Mascolo, G.; Castro, P.M.L. Biodegradation of diclofenac by the bacterial strain *Labrys portucalensis* F11. *Ecotoxicol. Environ. Saf.* **2018**, *152*, 104–113. [CrossRef]

11. Biel-Maeso, M.; Corada-Fernandez, C.; Lara-Martin, P.A. Monitoring the occurrence of pharmaceuticals in soil irrigated with reclaimed wastewater. *Environ. Pollut.* **2018**, *235*, 312–321. [CrossRef]
12. Ebele, A.J.; Abdallah, M.A.-E.; Harrad, S. Pharmaceuticals and personal care products (PPCPs) in the freshwater aquatic environment. *Emerg. Contam.* **2017**, *3*, 1–16. [CrossRef]
13. Scheurell, M.; Franke, S.; Shah, R.M.; Hühnerfuss, H. Occurrence of diclofenac and its metabolites in surface water and effluent samples from Karachi, Pakistan. *Chemosphere* **2009**, *77*, 870–876. [CrossRef] [PubMed]
14. Ali, A.M.; Sydnes, L.K.; Alarif, W.M.; Al-Lihaibi, S.S.; Aly, M.M.; Aanrud, S.G.; Kallenborn, R. Diclofenac and two of its photooxidation products in the marine environment: Their toxicology and occurence in Red Sea coastal waters. *Environ. Chem. Ecotoxicol.* **2019**, *1*, 19–25. [CrossRef]
15. Čelić, M.; Gros, M.; Farre, M.; Barcelo, D.; Petrovic, M. Pharmaceuticals as chemical markers of wastewater contamination in the vulnerable area of the Ebro Delta (Spain). *Sci. Total. Environ.* **2019**, *652*, 952–963. [CrossRef] [PubMed]
16. Baranowska, I.; Kowalski, B. Using HPLC method with DAD detection for the simultaneous determination of 15 drugs in srface water and wastewater. *Polish J. Environ. Stud.* **2011**, *20*, 21–28.
17. Reis-Santos, P.; Paisa, M.; Duarte, B.; Caçador, I.; Freitas, A.; Vila Pouca, A.S.; Barbosa, J.; Leston, S.; Rosa, J.; Ramos, F.; et al. Screening of human and veterinary pharmaceuticals in estuarine waters: A baseline assessment for the Tejo estuary. *Mar. Pollut. Bull.* **2018**, *135*, 1079–1084. [CrossRef]
18. Lolić, A.; Paĕga, P.; Santos, L.H.M.L.M.; Ramos, S.; Correia, M.; Delerue-Matos, C. Assessment of non-steroidal anti-inflammatory and analgesic pharmaceuticals in seawaters of North of Portugal: Occurrence and environmental risk. *Sci. Total. Environ.* **2015**, *508*, 240–250. [CrossRef]
19. Huebner, M.; Weber, E.; Niessner, R.; Boujday, S.; Knopp, D. Rapid analysis of diclofenac in freshwater and wastewater by a monoclonal antibody-based highly sensitive ELISA. *Anal. Bioanal. Chem.* **2015**, *407*, 8873–8882. [CrossRef]
20. Yang, L.; He, J.T.; Su, S.H.; Cui, Y.F.; Huang, D.L.; Wang, G.C. Occurrence, distribution, and attenuation of pharmaceuticals and personal care products in the riverside groundwater of the Beiyun River of Beijing, China. *Environ. Sci. Pollut. Res.* **2017**, *24*, 15838–15851. [CrossRef]
21. Cherik, D.; Benali, M.; Louhab, K. Occurrence, ecotoxicology, removal of diclofenac by adsorption on activated carbon and biodegradation and its effect on bacterial community: A review. *World Sci. News* **2015**, *10*, 116–144.
22. Hanif, H.; Waseem, A.; Kali, S.; Qureshi, N.A.; Majid, M.; Iqbal, M.; Ur-Rehman, T.; Tahir, M.; Yousaf, S.; Iqbal, M.M.; et al. Environmental risk assessment of diclofenac residues in surface waters and wastewaters: A hidden global threat to aquatic ecosystem. *Environ. Monit. Assess.* **2020**, *192*, 204. [CrossRef] [PubMed]
23. Shanmugam, G.; Sampath, S.; Selvaraj, K.K.; Larsson, D.G.J.; Ramaswamy, B.R. Non-steroidal ant-inflammatory drugs in Indian rivers. *Environ. Sci. Pollut. Res.* **2014**, *21*, 921–931. [CrossRef] [PubMed]
24. Madikizela, L.M.; Chimuka, L. Occurrence of naproxen, ibuprofen, and diclofenac residues in wastewater and river water of KwaZulu-Natal Province in South Africa. *Environ. Monit. Assess.* **2017**, *189*, 348. [CrossRef]
25. Di Lorenzo, T.; Cifoni, M.; Baratti, M.; Pieraccini, G.; Di Marzio, W.D.; Galassi, D.M.P. Four scenarios of environmental risk of diclofenac in European groundwater ecosystems. *Environ. Pollut.* **2021**, *287*, 117315. [CrossRef] [PubMed]
26. Gonzales-Alonso, S.; Merino, L.M.; Esteban, S.; de Alda, M.L.; Barcelo, D.; Duran, J.J.; Lopez-Martinez, J.; Acena, J.; Perez, S.; Mastoianni, N.; et al. Occurrence of pharmaceutical, recreational and psychotropic drug residues in surface water on the northern Antarctic Penisula region. *Environ. Pollut.* **2017**, *229*, 241–254. [CrossRef]
27. Lapworth, D.J.; Baran, N.; Stuart, M.E.; Ward, R.S. Emerging organic contaminants in grounwater: A review of sources, fate and occurrence. *Environ. Pollut.* **2012**, *163*, 287–303. [CrossRef] [PubMed]
28. Fu, Q.; Fedrizzi, D.; Kosfeld, V.; Schlechtriem, C.; Ganz, V.; Derrer, S.; Rentsch, D.; Hollender, J. Biotransformation changes bioaccumulation and toxicity of diclofenac in aquatic organisms. *Environ. Sci. Technol.* **2020**, *54*, 4400–4408. [CrossRef]
29. Joachim, S.; Beaudouin, R.; Daniele, G.; Geffard, A.; Bado-Nilles, A.; Tebby, C.; Palluel, O.; Dedourge-Geffard, O.; Fieu, M.; Bonnard, M.; et al. Effects of diclofenac on sentinel species and aquatic communities in semi-natural conditions. *Ecotoxicol. Environ. Saf.* **2021**, *211*, 111812. [CrossRef]
30. Swan, G.E.; Cuthbert, R.; Quevedo, M.; Green, R.E.; Pain, D.J.; Bartels, P.; Cunningham, A.A.; Duncan, N.; Meharg, A.A.; Oaks, J.L.; et al. Toxicity of diclofenac to *Gyps* vultures. *Biol. Lett.* **2006**, *2*, 279–282. [CrossRef]
31. Naidoo, V.; Wolter, K.; Cuthbert, R.; Duncan, N. Veterinary diclofenac threatens Africa's endangered vulture species. *Regul. Toxicol. Pharmacol.* **2009**, *53*, 205–208. [CrossRef]
32. Sathishkumar, P.; Meena, R.A.A.; Palanisami, T.; Ashokkumar, V.; Palvannan, T.F.L. Occurrence, interactive effects and ecological risk of diclofenac in environmental compartments and biota-a review. *Sci. Total Environ.* **2020**, *698*, 134057. [CrossRef] [PubMed]
33. Feito, R.; Valcárcel, Y.; Catalá, M. Biomarker assessment of toxicity with miniaturised bioassays: Diclofenac as a case study. *Ecotoxicology* **2012**, *21*, 289–296. [CrossRef] [PubMed]
34. Parolini, M. Toxicity of the non-steroidal anti-inflammatory drugs (NSAIDs) acetylsalicylic acid, paracetamol, diclofenac, ibuprofen and naproxen towards freshwater invertebrates: A review. *Sci. Total Environ.* **2020**, *740*, 140043. [CrossRef] [PubMed]
35. Näslund, J.; Asker, N.; Fick, J.; Larsson, D.G.J.; Norrgren, L. Naproxen affects multiple organs in fish but is still an environmentally better alternative to diclofenac. *Aquat. Toxicol.* **2020**, *227*, 105583. [CrossRef] [PubMed]

36. Schwarz, S.; Schmieg, H.; Scheurer, M.; Köhler, H.R.; Triebskorn, R. Impact of the NSAID diclofenac on survival, development, behaviour and health of embryonic and juvenile stages of brown trout, *Salmo trutta f. fario*. *Sci. Total Environ.* **2017**, *607*, 1026–1036. [CrossRef]
37. Ajima, M.N.O.; Ogo, O.A.; Audu, B.S.; Ugwoegbu, K.C. Chronic diclofenac (DCF) exposure alters both enzymatic and haematological profile of *African catfish*, *Clarias gariepinus*. *Drug Chem. Toxicol.* **2015**, *38*, 383–390. [CrossRef]
38. Cardoso-Vera, J.D.; Islas-Flores, H.; SanJuan-Reyes, N.; Montero-Castro, E.I.; Galar-Martínez, M.; García-Medina, S.; Elizalde-Velázquez, A.; Dublán-García, O.; Gómez-Oliván, L.B. Comparative study of diclofenac-induced embryotoxicity and teratogenesis in *Xenopus laevis* and *Lithobates catesbeianus*, using the frog embryo teratogenesis assay: *Xenopus* (FETAX). *Sci. Total Environ.* **2017**, *574*, 467–475. [CrossRef]
39. Kummerová, M.; Zezulka, S.; Babula, P.; Tříska, J. Possible ecological risk of two pharmaceuticals diclofenac and paracetamol demonstrated on a model plant *Lemna minor*. *J. Hazard. Mater.* **2016**, *302*, 351–361. [CrossRef]
40. Gonzalez-Rey, M.; Bebianno, M.J. Effects of non-steroidal anti-inflammatory drug (NSAID) diclofenac exposure in mussel Mytilus galloprovincialis. *Aquat. Toxicol.* **2014**, *148*, 221–230. [CrossRef]
41. Parolini, M.; Binelli, A.; Provini, A. Assessment of the potential cyto–genotoxicity of the non-steroidal anti-inflammatory drug (NSAID) diclofenac on the zebra mussel (*Dreissena polymorpha*). *Water Air Soil Pollut.* **2011**, *217*, 589–601. [CrossRef]
42. Saravanan, M.; Hur, J.-H.; Arul, N.; Ramesh, M. Toxicological effects of clofibric acid and diclofenac on plasma thyroid hormones of an Indian major carp, *Cirrhinus mrigala* during short and long-term exposures. *Environ. Toxicol. Pharmacol.* **2014**, *38*, 948–958. [CrossRef] [PubMed]
43. Peters, A.; Crane, M.; Merrington, G.; Ryan, J. Environmental quality standards for diclofenac derived under the European water framework directive: 2. Avian secondary poisoning. *Environ. Sci. Eur.* **2022**, *34*, 28. [CrossRef]
44. Diniz, M.S.; Salgado, R.; Pereira, V.J.; Carvalho, G.; Oehmen, A.; Reis, M.A.M.; Noronha, J.P. Ecotoxicity of ketoprofen, diclofenac, atenolol and their photolysis byproducts un zebrafish (*Danio rerio*). *Sci. Total Environ.* **2015**, *505*, 282–289. [CrossRef] [PubMed]
45. Żur, J.; Marchlewicz, A.; Piński, A.; Guzik, U.; Wojcieszyńska, D. Degradation of diclofenac by new bacterial strains and its influence on the physiological status of cells. *J. Hazard. Mater.* **2021**, *403*, 124000. [CrossRef] [PubMed]
46. Aissaoui, S.; Sifou, M.; Ouled-Haddar, H.; Benguedouar, L.; Lahouel, M. Toxicity assessment of diclofenac and its biodegradation metabolites toward mice. *Toxicol. Environ. Health Sci.* **2017**, *9*, 284–290. [CrossRef]
47. Memmert, U.; Peither, A.; Burri, R.; Weber, K.; Schmidt, T.; Sumpter, J.P.; Hartmann, A. Diclofenac: New data on chronic toxicity and bioconcentration in fish. *Environ. Toxicol. Chem.* **2013**, *32*, 442–452. [CrossRef]
48. Siemieniuk, A.; Ludynia, M.; Rudnicka, M. Response of two crop plants, *Zea mays* L. and *Solanum lycopersicum* L., to diclofenac and naproxen. *Int. J. Mol. Sci.* **2021**, *22*, 8856. [CrossRef]
49. Copolovici, L.; Timis, D.; Taschina, M.; Copolovici, D.; Cioca, G.; Bungau, S. Diclofenac influence on photosynthethic parameters and volatile organic compounds emission from *Phaseolus vulgaris* L. plants. *Rev. Chem.* **2017**, *68*, 2076–2078.
50. Aracagök, D.Y.; Göker, H.; Cihangir, N. Biodegradation of diclofenac with fungal strains. *Arch. Environ. Prot.* **2018**, *44*, 55–62. Available online: https://journals.pan.pl/dlibra/publication/118181/edition/102790 (accessed on 10 February 2023).
51. Kasonga, T.K.; Coetzee, M.A.A.; Kamika, I.; Momba, M.N.B. Assessing the fungal simultaneous removal efficiency of carbamazepine, diclofenac and ibuprofen in aquatic environment. *Front. Microbiol.* **2021**, *12*, 755972. [CrossRef]
52. Domaradzka, D.; Guzik, U.; Hupert-Kocurek, K.; Wojcieszyńska, D. Toxicity of diclofenac and its biotransformation by *Raoultella* sp. DD4. *Pol. J. Environ. Stud.* **2016**, *25*, 2211–2216. [CrossRef] [PubMed]
53. Grandclément, C.; Piram, A.; Petit, M.-E.; Seyssiecq, I.; Laffont-Schwob, I.; Vanot, G.; Tiliacos, N.; Roche, N.; Doumenq, P. Biological removal and fate assessment of diclofenac using *Bacillus subtilis* and *Brevibacillus laterosporus* strains and ecotoxicological effects of diclofenac and 4′-hydroxy-diclofenac. *J. Chem.* **2020**, *2020*, 9789420. [CrossRef]
54. Ivshina, I.B.; Tyumina, E.A.; Kuzmina, M.V.; Vikhareva, E.V. Features of diclofenac biodegradation by *Rhodococcus ruber* IEGM 346. *Sci. Rep.* **2019**, *9*, 9159. [CrossRef] [PubMed]
55. Murshid, S.; Dhakshinamoorthy, G.P. Biodegradation of sodium diclofenac and mefenamic acid: Kinetic studies, identification of metabolites and analysis of enzyme activity. *Int. Biodeterior. Biodegrad.* **2019**, *144*, 104756. [CrossRef]
56. Stylianou, K.; Hapeshi, E.; Vasquez, M.I.; Fatta-Kassinos, D.; Vyrides, I. Diclofenac biodegradation by newly isolated *Klebsiella* sp. KSC: Microbial intermediates and ecotoxicological assessment. *J. Environ. Chem. Eng.* **2018**, *6*, 3242–3248. [CrossRef]
57. Bilal, M.; Rasheedb, T.; Nabeelb, F.; Iqbalc, H.M.N.; Zhao, Y. Hazardous contaminants in the environment and their laccase-assisted degradation—A review. *J. Environ. Manag.* **2019**, *234*, 253–264. [CrossRef]
58. Neelkant, K.S.; Shankar, K.; Jayalakshmi, S.K.; Sreeramulu, K. Purification, biochemical characterisation, and facile immobilisation of laccase from *Sphingobacterium* ksn-11 and its application in transformation of diclofenac. *Appl. Biochem. Biotechnol.* **2020**, *192*, 831–844. [CrossRef]
59. Zdarta, J.; Jankowska, K.; Wyszowska, M.; Kijeńska-Gawrońska, E.; Zgoła-Grześkowiak, A.; Pinelo, M.; Meyer, A.S.; Moszyński, D.; Jesionowski, T. Robust biodegradation of naproxen and diclofenac by laccase immobilised using electrospun nanofibers with enhanced stability and reusability. *Mater. Sci. Eng. C* **2019**, *103*, 109789. [CrossRef]
60. Lonappan, L.; Liu, Y.; Rouissi, T.; Pourcela, F.; Brar, S.K.; Verma, M.; Surampalli, R.Y. Covalent immobilisation of laccase on citric acid functionalised micro- biochars derived from different feedstock and removal of diclofenac. *Chem. Eng.* **2018**, *351*, 985–994. [CrossRef]

61. Lonappan, L.; Liu, Y.; Rouissi, T.; Brar, S.K.; Verma, M.; Surampalli, R.Y. Adsorptive immobilisation of agro-industrially produced crude laccase on various micro-biochars and degradation of diclofenac. *Sci. Total Environ.* **2018**, *640–641*, 1251–1258. [CrossRef]
62. Masjoudi, M.; Golgoli, M.; Nejad, Z.G.; Sadeghzadeh, S.; Borghei, S.M. Pharmaceuticals removal by immobilised laccase on polyvinylidene fluoride nanocomposite with multi-walled carbon nanotubes. *Chemosphere* **2021**, *263*, 128043. [CrossRef] [PubMed]
63. Nguyen, L.N.; Hai, F.I.; Dosseto, A.; Richardson, C.; Price, W.E.; Nghiem, L.D. Continuous adsorption and biotransformation of micropollutants by granular activated carbon-bound laccase in a packed-bed enzyme reactor. *Biores. Technol.* **2016**, *210*, 108–116. [CrossRef] [PubMed]
64. Quan, X.; Wang, X.; Sun, Y.; Li, W.; Chen, L.; Zhao, J. Degradation of diclofenac using palladized anaerobic granular sludge: Effects of electron donor, reaction medium and deactivation factors. *J. Hazard. Mater.* **2018**, *365*, 155–163. [CrossRef] [PubMed]
65. Pereira, J.C.V.; Serbent, M.P.; Skoronski, E. Application of immobilised mycelium-based pellets for the removal of organochlorine compounds: A review. *Water Sci. Technol.* **2021**, *83*, 1781–1796. [CrossRef] [PubMed]
66. Chen, L.; Li, Y.; Lin, L.; Tian, X.; Ciu, H.; Zhao, F. Degradation of diclofenac by *B. subtilis* through a cytochrome P450-dependent pathway. *Environ. Technol. Innov.* **2020**, *20*, 101160. [CrossRef]
67. Fernandez-Fernandez, M.; Sanroman, M.A.; Moldes, D. Recent development and applications of immobilized laccase. *Biotechnol. Adv.* **2013**, *31*, 1808–1825. [CrossRef]

catalysts

Article

Highly Efficient Biosynthesis of Nicotinic Acid by Immobilized Whole Cells of *E. coli* Expressing Nitrilase in Semi-Continuous Packed-Bed Bioreactor

Xue-Jiao Liu, Bao-Di Ma, Xiao-Mei Wu and Yi Xu *

School of Chemical and Environmental Engineering, Shanghai Institute of Technology, Shanghai 201418, China
* Correspondence: xuyi@sit.edu.cn; Tel.: +86-21-60873024

Abstract: A recombinant *E. coli*, expressing nitrilase from *Acidovorax facilis* 72W with dual-site expression plasmid pRSFduet (*E. coli* pRSF-AfNit2), was constructed. It showed higher soluble expression of nitrilase than that in the pET21a plasmid. The recombinant nitrilase can efficiently catalyze the hydrolysis of 3-cyanopyridine to nicotinic acid. The whole cells of *E. coli* pRSF-AfNit2 were immobilized by using sodium alginate/glutaraldehyde/polyethylene imine as the best immobilized reagents. The immobilized cells showed 95% activity recovery and excellent mechanical strength, with improved thermal stability and pH stability. They also retained 82% of initial activity after nearly two months of storage at 4 °C. A semi-continuous packed-bed bioreactor (sPBR) filled with the immobilized cells was studied for efficient production of nicotinic acid. After optimization, the highest space–time yield of 1576 g/(L·d) was obtained on 0.8 M substrate concentration at 2 mL/min of flow rate. The sPBR was repeatedly operated for 41 batches, keeping 100% conversion in the presence of 30 mM $CaCl_2$. Finally, 95 g of nicotinic acid were obtained at 90% yield after separation and purification. The developed technology has potential application value.

Keywords: nitrilase; cell immobilization; nicotinic acid; reaction optimization; semi-continuous packed-bed bioreactor

Citation: Liu, X.-J.; Ma, B.-D.; Wu, X.-M.; Xu, Y. Highly Efficient Biosynthesis of Nicotinic Acid by Immobilized Whole Cells of *E. coli* Expressing Nitrilase in Semi-Continuous Packed-Bed Bioreactor. *Catalysts* **2023**, *13*, 371. https://doi.org/10.3390/catal13020371

Academic Editors: Zhilong Wang and Tao Pan

Received: 27 December 2022
Revised: 4 February 2023
Accepted: 6 February 2023
Published: 8 February 2023

1. Introduction

Nicotinic acid (NA) and nicotinamide are two main forms of vitamin B3, and they are collectively referred to as vitamin PP, and they are also known as anti-sculpturial factors. Named 3-pyridinic acid in the chemistry of NA, commonly known as Nicotinic acid or vitamin B3, is an indispensable nutrient component in the human body [1,2]. NA, usually in the form of coenzymes (coenzyme I and coenzyme II) in the human body, participate in the oxidative reactions in the body, thereby promoting the metabolic activity of the human body, maintaining the normal function of the human organs [3].

Traditional chemical synthetic methods of NA typically require harsh conditions, such as strong alkali, high temperatures, etc., which not only increase the process costs, but also produce certain pollution to the environment. Therefore, biocatalytic methods have been increasingly applied to NA synthesis [4–9]. There are two main routes (Scheme 1) for the biocatalytic synthesis of NA. One is the direct hydrolysis of 3-cyanopyridine to NA by nitrilase, releasing the ammonia molecule at the same time. So far, there were many reports on the synthesis of NA by wild nitrile hydrolase from different sources [10–13]. Mathew, C.D. et al. reported the production of NA from 3-cyanopyridine by resting whole cells of *Rhodococcus rhodochrous* J1, and the highest yield was achieved, being 172 mg of NA per 1.0 mL of reaction mixture containing 2.89 mg (dry weight) of cells in 26 h according to feed batch addition of substrate [8]. Another route is that the 3-cyanopyridine is hydrolyzed by nitrile hydratase to nicotinamide, which is then hydrolyzed by amidase to NA and ammonia. Cantarella, L. et al. reported a cascade bioconversion of 3-cyanopyridine to NA using a nitrile hydratase–amidase containing resting cells of *Microbacterium imperiale*

CBS 498-74 in an ultrafiltration-membrane reactor, operated in either batch or continuous mode [9]. It is worth mentioning that the Swiss company Lonza commecialized the bioconversion of 3-cyanopyridine to nicotinamide by immobilized cells of *Rhodococcus rhodochrous* J1 at >10,000 t/year scale [14].

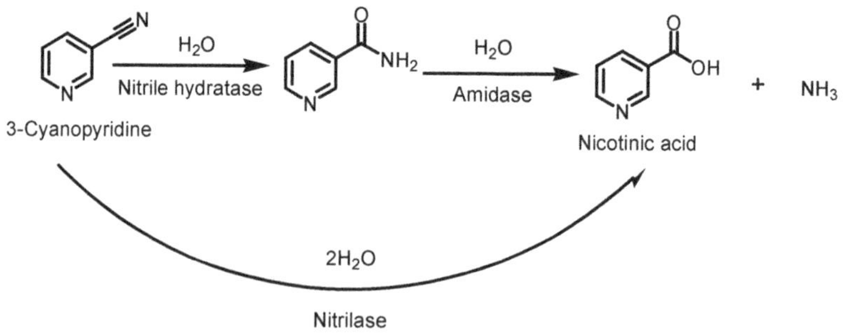

Scheme 1. Enzymatic synthesis of nicotinic acid via nitrilase or nitrile hydratase–amidase route.

Although the production of NA by recombinant nitrilases were reported, there were several shortcomings, which restricted the practical application of nitrilase, such as low substrate concentration, low catalytic efficiency, and low stability of the enzyme [15,16]. In addition, most of such reaction was carried out in a batch reactor [17–19]. There were few reports on using packed-bed bioreactor for the production of NA by recombinant nitrilase. Therefore, we constructed the recombinant nitrile with dual-site expression. In order to achieve the industrial production of niacin, cell immobilization of recombinant bacteria was studied and constructed a semi-continuous packed-bed bioreactor (sPBR).

The vector pRSFDuet-1 contains two multiple cloning sites, and each of them is preceded by a T7 promoter and ribosome binding site. There were several reports on the successful co-expression of two enzymes [20–22]. The use of pRSFDuet as recombinant plasmid may improve the expression level of nitrilase in *E. coli*.

Therefore, the development of a highly efficient bioprocess by high active and stable nitrilase under high substrate concentration in a suitable kind of bioreactor will facilitate the practical application process. The current research was mainly focused on the construction of recombinant *E. coli* pRSF-AfNit2 capable of two-site expression of nitrilase for the hydrolysis of 3-cyanopyridine to nicotinic acid, cell immobilization, and the development of an efficient process in a semi-continuous packed-bed bioreactor (sPBR). Finally, the preparation of NA by an immobilized whole cell of nitrilase at hectogram scale in sPBR was firstly reported.

2. Results

2.1. Expression of Recombinant Bacteria

After the strains of *E. coli* pRSF-AfNit2 were constructed, the strains of *E. coli* pRSF-AfNit2 and *E. coli* pET21a-AfNit were cultured [23,24]. SDS-PAGE was conducted in order to verify the expression level of nitrile hydrolase from different recombinant strain *E. coli*. The size of the band was similar to the target band and the theoretical values were consistent. It can be seen, from Figure S1, that the strains of *E. coli* pRSF-AfNit2 had higher expression of soluble nitrilase. Under the condition of optimal flask level, the cell amount and enzyme activitiy of *E. coli* pRSF-AfNit2 reached 4.8 g_{dcw}/L and 3218 U/L, respectively (the enzyme activity was determined according to the hydroysis of 3-cyanopyridine at 30 °C and pH 7.0, see Section 4.4 for details).

2.2. Immobilization of Whole Cells of E. coli pRSF-AfNit2 by Different Entrapment Method

This experiment was mainly based on the sodium alginate (SA) gel carrier, selecting other different complexes to further combine with calcium alginate to immobilize the cells [25]. The beads prepared using different immobilization methods were between

3~5 mm in size. Immobilized cells were prepared by different methods, and their enzyme activity and mechanical strength were compared to judge which immobilization method was better (Table 1).

Table 1. Effect of different cellular immobilization methods on stability of nitrile hydrolase.

Immobilization Method	Bead Dimeter (mm)	Color	Mechanical Strength [a]	Relative Activity (%)
SA gel beads	3	milky	*	87.3 ± 0.1
Fe_3O_4-SA	5	dark brown	*	70.3 ± 0.3
SA-GA	3	light pink	**	91.7 ± 0.2
SA-GA/PEI	3	pink	***	95.1 ± 0.1
Activated carbon-SA	4	black	*	78.8 ± 0.4
PVA-SA	3	white	**	87.7 ± 0.5

[a] The mechanical strength is expressed by the time (t) required for the breakage rate of the immobilized pellets to reach 25%. One star (*): $t < 30$ min; two stars (**): 30 min $< t <$ 60 min; three stars (***): $t > 60$ min.

It can be seen that, because the polyvinyl alcohol (PVA) solution was too viscous and the solubility was too poor, it was not easy to prepare the polyvinyl alcohol–calcium alginate (PVA-SA) immobilized cells, and other methods were relatively easy to prepare; the mechanical strength test results showed that the calcium alginate, Fe_3O_4-calcium alginate pellets, and activated carbon–calcium alginate pellets had poor mechanical strength and were easy to swell and break. Calcium alginate–glutaraldehyde (SA–GA) and calcium alginate–glutaraldehyde-polyethyleneimine (SA–GA/PEI) had moderate to high mechnical strength, with both having over 90% activity recovery. Therefore, the best method was the SA–GA/PEI method [25].

2.3. *Enzymatic Properties of SA-GA/PEI Immobilized Cells and Free Cells*

Temperature has always been an important factor affecting an enzymatic reaction. Too high of a temperature will cause an enzyme to denature and inactivate; if the temperature is too low, the catalytic reaction rate will be slower, and the enzyme activity will not be high. This experiment compared the enzyme activity of free cells and immobilized cells at different temperatures to determine the optimal temperature for free cells and immobilized cells. It can be seen from Figure 1a that the optimum temperature for free cells was 55 °C, and the optimum temperature for immobilized cells was 60 °C. When the temperature was greater than the optimum temperature, as the temperature rised, the enzyme activity of both free cells and immobilized cells all have decreased, but the enzyme activity of immobilized cells was higher than that of free cells at the same temperature. For example, the relative enzyme activity of free cells was only 18% at 70 °C, while the relative enzyme activity of immobilized cells was 50%. This showed that its heat resistance was improved, and the temperature range has been broadened [26].

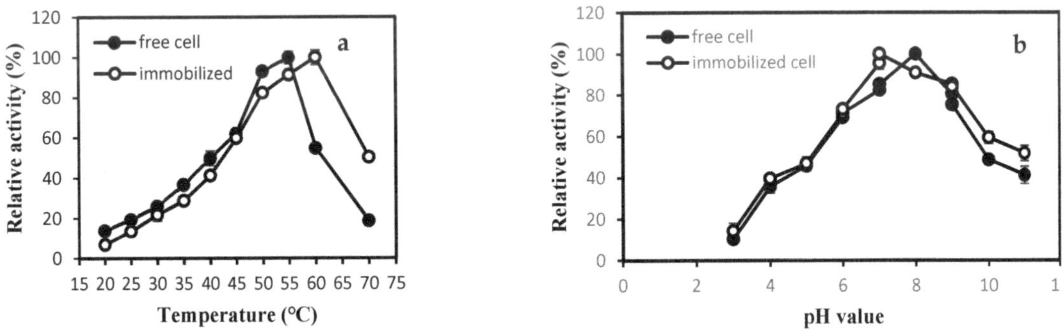

Figure 1. Effect of reaction temperature (**a**) and pH (**b**) on the activity of free and SA–GA/PEI immobilized cells.

In the enzymatic reaction, only when the catalyst exists in an appropriate pH buffer system can the enzyme achieve higher catalytic activity. As shown in Figure 1b, the optimal pH for free cells was 8.0, while it was 7.0 for immobilized cells. The activity of immoilized cells decreased less than that of free cells with the increase in pH value under alkaline conditions, which indicated that immobilized cells had a higher pH tolerance than free cells in the alkaline range [26].

2.4. Storage Stability of Immobilized Cells

In practical industrial applications, an important evaluation criterion for the application value of a catalyst was its storage stability. In this experiment, the storage stability of free cells and immobilized cells was studied, and the results were shown in Figure 2. The residual enzyme activity of free cells was only about 15% after being stored in a refrigerator at 4 °C after 55 days. However, it was still more than 80% for immobilized cells after nearly two months, which showed a much better storage stability of immobilized cells than free cells [27].

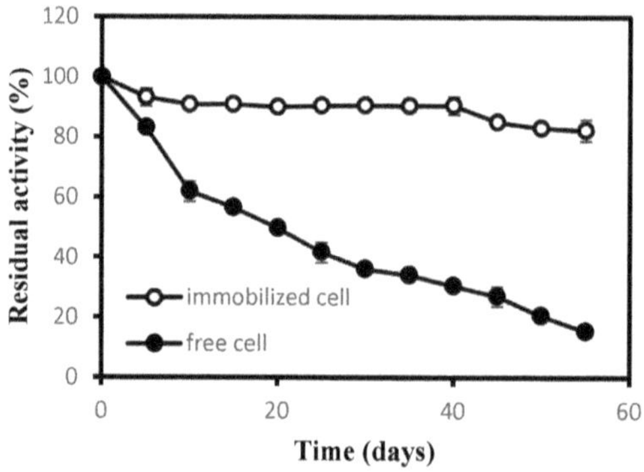

Figure 2. Storage stability of free and immobilized cells.

2.5. Research on Semi-Continuous Synthesis of NA in a Packed-Bed Reactor with Immobilized Cells

The packed-bed bioreactor can also be called a fixed-bed bioreactor, because the biocatalyst in the reactor was always in a static state, and the reaction happened when substrate solution flowed through the biocatalyst [28]. Since different methods of immobilizing cells produced different cell shapes, the catalyst in the reactor can be filled in different forms. When there existed product inhibition, the use of this type of reactor had certain advantages. Since the catalyst in the reactor is always static, the fluid flow pattern can be regarded as a plug flow. Compared with the continuous stirred tank bioreactor (CSTR), the shear force was smaller, which facilitated keeping the stability of the biocatalyst. Up to now, only scarce research was focused on the biosynthesis of NA in packed-bed bioreactors. In the current research, we studied the effect of flow rate of substrate solution and substrate concentration on the efficiency of catalytic reaction.

It can be seen, from Figure 3a, that the reaction rate increased with the increase in the flow rate when it was increased from 0.5 to 2.0 mL/min. Further increasing the flow rate from 2.0 to 3.0 mL/min, it was maintained in a stable state, which means the influence of external diffusion could be negligible. Therefore, the optimal flow rate was 2.0 mL/min.

After determining the optimal flow rate, the substrate concentration was investigated. It can be seen from Figure 3b that as the substrate concentration was increased, and the conversion rate was decreased at the same time. When the substrate concentration was greater than 0.8 M, the time for full conversion of the substrate was gradually increased. The space–time yield at different substrate concentrations was shown in Table 2. When the substrate concentration was 0.8 M, the space–time yield reached the highest value of 1576 g/(L·d). Therefore, the optimal substrate concentration was 0.8 M.

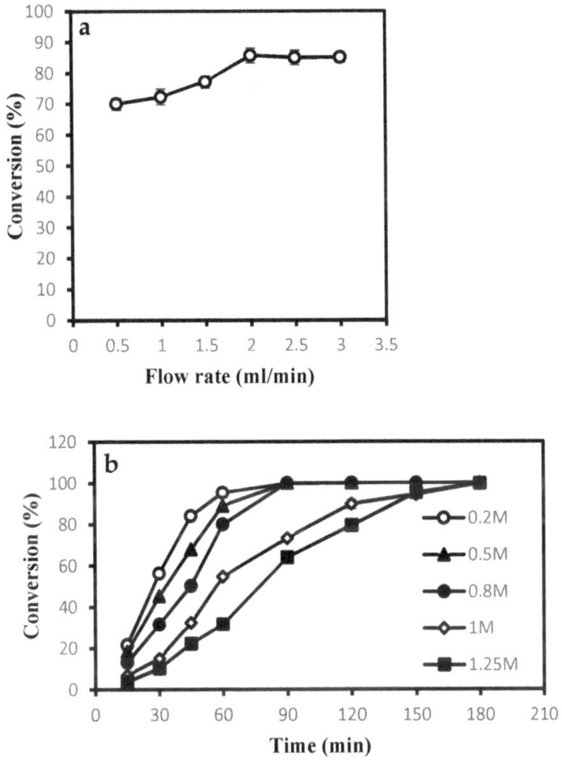

Figure 3. Study on different parameters on the conversion of NA in semi-continuous packed-bed bioreactor. (**a**) Flow rate (after operation for 90 min); (**b**) substrate concentration.

Table 2. The space–time yield at different substrate concentrations.

Substrate Concentration (M)	0.2	0.5	0.8	1.0	1.25
Full conversion time (h)	1.5	1.5	1.5	3.0	3.0
Space–time yield (g/(L·d))	393 ± 1.0	984 ± 0.5	1576 ± 0.7	981 ± 0.5	1227 ± 0.5

2.6. Reusability of Immobilized Cells in the Bioreactor

According to the optimal substrate flow rate and substrate concentration, repeated batch experiments were performed in the reactor to verify the operational stability of the immobilized cells. The effect of calcium ions on the stability of immobilized cells was also investigated (Figure 4). When no calcium ions were added to the substrate solution, the immobilized cells began to swell and broke after only five batches, resulting in a sharp decrease in the conversion. However, the immobilized cells with $CaCl_2$ (30 mM) in solution can be reused for at least 41 batches, keeping 100% of conversion. After 41 batches, the immobilized cells started to swell and partially break, leading to a decrease in conversion.

When the reaction finished, the immobilized cells in the bioreactor were washed. The washing liquid was combined with all the product solutions for each batch, and NA was furthur seperated and purified. Finally, 95 g of pure NA were obtained from 43 batches, corresponding to 90% of isolated yield [28]. Since the immobilized cells packed in sPBR contained around 0.6 g dry cell weight of cells, it can be estimated that preparing 1 ton of NA requires an immobilized biocatalyst containing only 6 kg of dry weight cells.

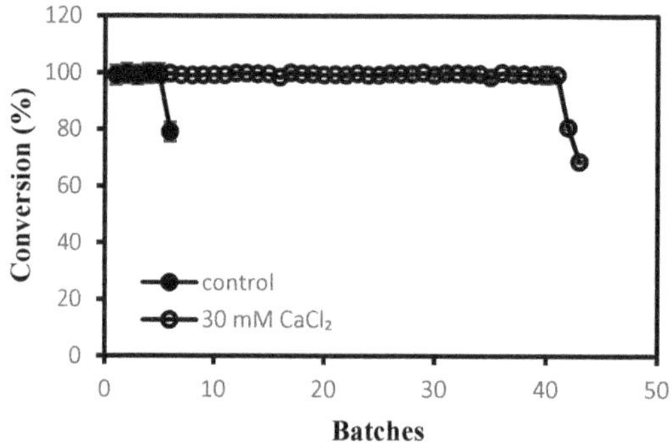

Figure 4. Repeated use of immobilized cells in sPBR for the production of nicotinic acid.

3. Discussion

In this paper, by molecular cloning, the nitrilase gene derived from *Acidovorax facilis* was ligated to a vector with two independent multiple cloning sites for two-site expression, and the nitrilase-producing recombinant strain *E. coli* pRSF-AfNit2 showed higher soluble expression of nitrilase than that of *E. coli* pET21a-AfNit2.

In order to further improve the stability of nitrilase, the whole cell immobilization of nitrilase was systematically studied, and the immobilized cells were used to catalyze the semi-continuous hydrolysis reaction of 3-cyanopyridine in a packed bed column bioreactor. The results of the study were the following: the best immobilization method was the calcium alginate–glutaraldehyde–polyethyleneimine cross-linking method, and the enzyme activity of immobilized cells was 95% of free cells, and the residual enzyme activity of immobilized cells was still more than 80% after nearly two months storage at 4 °C. In the semi-continuous packed-bed column reactor, it was found that both the substrate solution flow rate and the substrate concentration affected the catalytic efficiency. Regarding this, the catalytic efficiency reached its highest when the substrate flow rate was 2.0 mL/min and the substrate concentration was 0.8 M. The space–time yield of the reactor during stable operation can reach 1576 g/(L·d). When the substrate solution contains 30 mM $CaCl_2$, the immobilized cells can be reused for at least 41 batches, and the conversion was still kept at 100%. Finally, 95 g of NA were obtained in 90% isolated yield through seperation and purification. It can be estimated that producing 1 ton of NA would require immobilized biocatalyst containing only 6 kg of dry weight cells, which fufill the needs of pactical applications.

4. Materials and Methods

4.1. Chemicals, Plasmids and Strains

3-Cyanopridine, nicotinic acid, protein marker, and soluble starch were from the Aladdin reagent company (Shanghai, China); NaCl and yeast extract powder were from sinopharm chemical reagent Co., Ltd. (Shanghai, China); anhydrous ethanol, $CaCl_2$, K_2HPO_4, KH_2PO_4 and $MgSO_4 \cdot 7H_2O$ were purchased from Shanghai Taitan Technology Co., Ltd. (Shanghai, China); kanamycin was from Shanghai Yuanye Biotechnology Co., Ltd. (Shanghai, China).

Plasmids pET-21a(+) (Novagen, Merck, Darmstadt, Germany) and pRSFDuet-1 (Novagen, Merck, Darmstadt, Germany) were preserved in the Biocatalysis and Biopharmaceutical Laboratory (Shanghai Institute of Technology, Shanghai, China). Restriction endonucleases (*Nde*I, *Xho*I, *Pst*I, *Nco*I) were purchased from Takara Biotechnology (Dalian, China). *E. coli* BL21 (DE3) and *E. coli* DH5α were purchased from Friction Biological Engineering (Shanghai) Co., Ltd. (Shanghai, China). The recombinant plasmid pET21a-AfNit was constructed in our previous work [23].

4.2. Construction of Recombinant E. coli Strains Expressing Nitrilase

The nitrilase gene (FJ851547) [29] from *Acidovorax facilis* 72W (AfNit) was synthesized by Shanghai Generay Biotech Co., Ltd. (Shanghai, China). AfNit was ligated into pET21a via *Nde*I/*Xho*I restriction sites to generate expression plasmid pET21a-AfNit. The pET21a-AfNit was transformed into *E. coli* BL21 (DE3) to construct the recombinant strain *E. coli* pET21a-AfNit. For the construction of *E. coli* pRSF-AfNit2, the AfNit gene was ligated into the first multi clone site (MCS1) via *Pst*I/*Nco*I restriction sites, and the second MCS (MCS2) was ligated via the *Nde*I/*Xho*I restriction sites to generate the expression plasmid pRSF-AfNit2. The pRSF-AfNit2 was transformed into *E. coli* BL21 (DE3) to construct the recombinant strain *E. coli* pRSF-AfNit2. The construction of recombinant pRSFDuet-1 plasmids and their transformation into *E. coli* were performed according to standard protocols.

4.3. Cultivation of Recombinant E. coli Expressing Nitrilase

A pure colony (*E. coli* pRSF-AfNit2 or *E. coli* pET21a-AfNit) was picked and cultivated overnight in the 50 mL LB medium with 50 μg/mL kanamycin at 37 °C and 200 rpm. Then, the 2 mL culture was incubated in 50 mL of culture medium (Soluble starch: 20 g/L, yeast extract powder: 15 g/L, NaCl: 5 g/L, K_2HPO_4: 4 g/L, $MgSO_4 \cdot 7H_2O$ 1 g/L) with 50 μg/mL kanamycin, and the cells were cultured at 30 °C and 200 rpm. When the OD_{600} value reached 0.4 to 0.6, IPTG (0.4 mM) was added to the culture to induce enzyme expression at 30 °C for 4 h [23]. Harvested cells were suspended in 50 mM PBS buffer (pH 7.0) and then ultrasonicated on ice. Centrifugation was used to remove cell debris for 10 min at 12,000× *g* and 4 °C. The protein molecular mass was evaluated by SDS-PAGE [30,31].

4.4. Enzyme Assay

To assay the enzyme activity of AfNit, 0.5 mg of dry weight resting cells were suspended in 475 μL of phosphate buffer (pH 7.0, 100 mM). The cell suspension was preheated for 5 min on thermo-mixer compact at 30 °C and 1000 rpm, then 25 μL of 3-cyanopyridine solution (1 M) in ethanol were added to the cell suspension, corresponding to 50 mM of final substrate concentration. The reaction was performed at 30 °C for 15 min, and the reaction solution was immediately added to 100 μL of HCl (2 M) to terminate the reaction. Then, 1900 μL of anhydrous ethanol was added to the reaction mixture for dilution. After centrifugation at 12,000× *g* for 5 min, the supernatant was filtered with organic filter membrane (25 mm × 0.45 μm, sinopharm chemical reagent Co., Ltd., Shanghai, China.), and the filtrate was assessed by HPLC. One unit of the enzyme activity was defined as the amount of enzyme releasing 1 μmol of NA per minute under the described conditions [25].

4.5. Analytical Methods

The NA and 3-cyanopridine were quantitatively determined by HPLC (Shimadzu LC-20AT, Kyoto, Japan), equipped with UV detector and a reversed phase column (Diamonsil® Plus C18 5 μm 250 × 4.6 mm, Cat. No.: 99409, Dikma Co., Shanghai, China), and monitored under UV 217 nm at a column temperature of 25 °C. The mobile phase was acetonitrile: phosphate buffer (10 mM sodium dihydrogen phosphate, pH 2.5) = 15:85 (*v*:*v*). The sample was eluted at a flow rate of 1.0 mL/min. The retention time of the product and substrate are 3.2 min and 6.8 min, respectively.

The molar response factor of NA to 3-caynopyridine was 0.5345. It was determined by the analysis of equal molar amounts of NA and 3-cyanopyridine standards with HPLC, and the molar response factor was obtained by mapping the peak area of NA and 3-caynopyridine.

4.6. Preparation of Immobilized Cells and Enzyme Assay

General procedue for the immobilization of *E. coli* pRSF-AfNit2 whole cells (exampled as SA-GA/PEI): the *E. coli* fermentation broth was washed twice with 0.1 M phosphate buffer (pH 7.0). It was resuspended in certain amount of 0.8% NaCl solutions to 10 g/L. The cell suspension was then mixed with equivalent volume of alginate solution (30 g/L). The mixture was added to 2% (*w*/*v*) $CaCl_2$ solution with a syringe or a peristaltic pump, and the curing time was 2 h, then a small ball having an approximately diameter of 3 mm can be obtained, and it is finally washed with deionized water. The prepared alginate immobilized pellets were crosslinked in 2 wt% glutaraldehyde aqueous solution for 1 h. The additional glutaraldehyde was washed away, and 3% of the polyethyleneimine solution was added for 1 h, and the surface residual polyethyleneimine was washed away. After curing for 12 h, it was finally washed three times with deionized water, and the enzyme activity of the immobilized cells was measured [25,26].

Immobilized cells containing 0.5 mg dry weight of free cells were placed in 475 μL of Tris-HCl (0.1 M, pH 7.0) buffer solution, and they were preheated in a constant temperature mixing instrument at 30 °C and 1000 rpm for 5 min, and then 25 μL of 3-cyanopyridine solution (1 M) in ethanol was added to the mixture (final substrate concentration: 50 mM). After reaction for 15 min, the product concentration was determined in the sample by HPLC, as described in Section 4.4, and the enzyme activity was calculated from the degree of conversion. One unit of the immobilized enzyme activity was defined as the amount of enzyme releasing 1 μmol of NA per minute under the described conditions.

4.7. Determination of Mechanical Strength of Immobilized Cells

The immobilized cells studied in this experiment were spherical gel particles. Therefore, when measuring the mechanical strength of immobilized cells, 100 balls of immobilized cells were directly counted and placed in Tris-HCl (0.1 M, pH 7.0) buffer at 30 °C and 180 rpm while shaking, and the gel breakage was observed at different times.

4.8. Effect of Different Reaction Temperature and pH on the Activity of Free and Immobilized Cells

To determine the effect of temperature, the activity of free cells (1 g_{dcw}/L) or immobilized cells (containing 1 g_{dcw}/L free cells) was assayed in Tris-HCl buffers (100 mM, pH 7.0) at various temperatures (20–70 °C). The relative activity referred to the ratio of the enzyme activity at different temperatures to their respective highest activity (free cell: 3076 U/g_{dcw}, immobilized cell: 2462 U/g_{dcw}). Experiments were independently performed in triplicate.

To determine the effect of pH value, the activity of free cells (1 g_{dcw}/L) or immobilized cells (containing 1 g_{dcw}/L free cells) was assayed at 30 °C in various buffers over a pH range of 3.0~11.0 (pH 3.0~6.0, citrate–citric acid buffer; pH 6.0~7.0, PBS buffer; pH 7.0~9.0, Tris-HCl buffer; pH 9.0~11.0, Gly-NaOH buffer). The relative activity was expressed as a percentage of their respective maximum activity of free or immobilized cells. Experiments were independently performed in triplicate.

4.9. Semi-Continuous Packed-Bed Bioreactor

The column bioreactor used in this experiment (Figure 5) was made by Chengxin Glass Technology Co., Ltd. (Shanghai, China). It was a jacketed glass column reactor, and the temperature of the bioreactor was controlled by an external circulating water bath. The height of column was 100 mm, and the inner diameter was 17 mm. The effective volume of the column was approximately 20 mL. There was a round sand core at the bottom of column to prevent the beads from leaking. The wet weight of immobilized cells packed into the bioreactor column was 7.8 g (~10 mL volume, containing 600 mg dry weight of cells). When the reaction begins, the substrate solution entered from the bottom at a constant flow rate, made contact with the immobilized cells, and reacted. The reaction solution flowed out from the top of the column and was pumped back into the substrate storage tank.

To investigate the effect of flow rate, 25 mL of 0.8 M substrate solution in (Tris-HCl buffer, 0.1 M, pH 7.0) was pumped into the column with different flow rate (from 0.5 mL/min to 3.0 mL/min). Substrate conversion was determined at different times to determine the best flow rate.

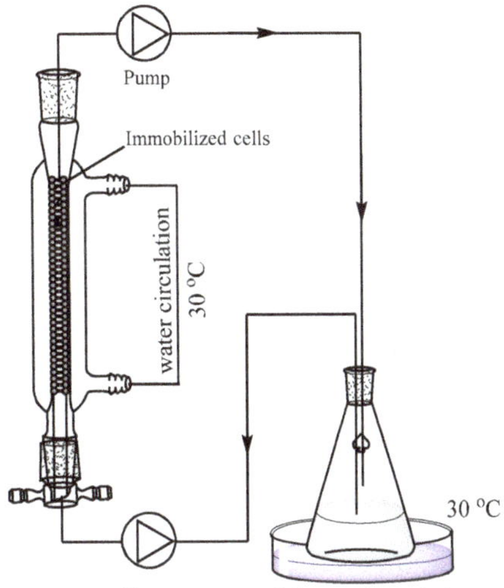

Figure 5. Enzymatic synthesis of NA by immobilized cells in semi-continuous packed-bed bioreactor.

To investigate the effect of substrate concentration, 25 mL of different concentrations of substrate (0.2 M, 0.5 M, 0.8 M, 1.0 M, 1.25 M) solutions were pumped into the column at 2.0 mL/min, and the conversion of the substrate was monitored at different time to determine the optimal substrate concentration.

To study the operational stability of immobilized cells in sPBR, 25 mL of 0.8 M substrate solution (Tris-HCl buffer, 0.1 M, pH 7.0, with or without 30 mM $CaCl_2$) were pumped into the column with a flow rate of 2 mL/min, and the conversion of the substrate was monitored by HPLC. When the reaction for each batch finished, the reaction solution was stored in a collection tank, and 25 mL of fresh substrate solution (0.8 M) was added into substrate storage tank to operate a new batch [28]. The batch reaction was repeated until the conversion at the same time decreased significantly. When the batch reaction finished, the immobilized cells were washed with 0.1 M Tris-HCl buffer (25 mL × 3, pH 7.0). The collected washing liquid was centrifuged (10,000 rpm, 5 min) to obtain the clear supernatant. The supernatant and the accumulated reaction solutions for each batch were combined together for the preparation of NA.

4.10. Preparation of NA

The combined solutions mentioned in Section 4.9 were heated at 70 °C for 30 min, and then they were centrifuged for 30 min at 10,000 rpm. The precipitate (mainly broken immobilized cells) was washed with normal saline (20 mL × 2) to dissolve NA absorbed in the precipitate. Finally, the supernatant and washing solution were collected, and concentrated hydrochloric acid was dropped into the collected solution at pH 3.7 (NA isoelectric points). After all the crystals were precipitated, they were heated to 70 °C, and the obtained solution was filtered while it was hot by a Büchner funnel to remove insoluble impurities. The filtrate was placed at room temperature for natural cooling for 30 min, and then it was placed in a refrigerator at 4 °C for low temperature recrystallization. The crystals were obtained by air pump filtration with a Büchner funnel. The filtrated liquid was concentrated by rotary evaporation, and the above operation of recrystallization was repeated to further obtain more NA and to increase the total yield of NA [32].

The NA product (see Figure S2a for its morphology) was analyzed qualitatively and quantitatively. Its melting point was determined by melting point analyzer (mp. 238.7–238.9 °C, ref. 236–239 °C), and its purity was analyzed by HPLC (Figure S2b). The chemical structure was determined by nuclear magnetic resonance (NMR) ^{1}H NMR

(501 MHz, DMSO-d6): δ 13.49 (s, 1H), 9.27–9.06 (m, 1H), 8.83 (dd, J = 4.8, 1.7 Hz, 1H), 8.32 (dt, J = 7.9, 2.0 Hz, 1H), 7.58 (ddd, J = 7.9, 4.8, 0.9 Hz, 1H) (Figure S2c) and Fourier infrared spectroscopy (Figure S2d).

5. Conclusions

A highly efficient biocatalytic process for NA production from 3-cyanopyridine by immobilized whole cells of recombinant strain *E. coli* pRSF-AfNit2 in sPBR was developed. The immobilized cells prepared by SA-GA/PEI method showed excellent activity recovery, improved stability, and the highest mechanical strength among the tested ones. The immobilized cells were used to catalyze the hydrolysis reaction of 3-cyanopyridine in sPBR. After optimization, the space–time yield of the reaction reached 1576 g/(L·d) during stable operation. The sPBR was repeatedly operated for 41 batches, keeping 100% conversion in the presence of 30 mM $CaCl_2$, resulting in 95 g of NA (90% yield). It can be estimated that producing 1 ton of NA would require an immobilized biocatalyst containing only 6 kg of dry weight cells, which fufill the needs of practical applications. In summary, the developed method exhibits good application potential for the biosynthesis of NA.

Supplementary Materials: The following supporting information can be downloaded at: https://www.mdpi.com/article/10.3390/catal13020371/s1, Figure S1: SDS-PAGE of recombinant AfNit from *E. coli* pET21a-AfNit and *E. coli* pRSF-AfNit2; Figure S2: Characterization spectrum and picture of prepared nicotinic acid.

Author Contributions: Conceptualization, Y.X.; methodology, B.-D.M. and X.-J.L.; validation, X.-J.L. and Y.X.; formal analysis, X.-J.L.; data curation, X.-J.L. and X.-M.W.; writing—original draft preparation, X.-J.L.; writing—review and editing, X.-M.W., B.-D.M. and Y.X.; funding acquisition, Y.X. All authors have read and agreed to the published version of the manuscript.

Funding: This research was funded by the Project of Leading Talents in Shandong Taishan Industry (Grant No. LJNY202019).

Data Availability Statement: Data are contained within the article.

Conflicts of Interest: The authors declare no conflict of interest.

References

1. Chuck, R. Technology development in nicotinate production. *Appl. Catal. A Gen.* **2005**, *280*, 75–82.
2. Chuck, R. A catalytic green process for the production of NA. *Chim. Int. J. Chem.* **2000**, *54*, 508–513.
3. Velankar, H.; Clarke, K.G.; Preez, R.; Cowan, D.A.; Burton, S.G. Developments in nitrile and amide biotransformation processes. *Trends Biotechnol.* **2010**, *28*, 561–569. [PubMed]
4. Gong, J.S.; Lu, Z.M.; Li, H.; Shi, J.S.; Zhou, Z.M.; Xu, Z.H. Nitrilases in nitrile biocatalysis: Recent progress and forthcoming research. *Microb. Cell Factories* **2012**, *11*, 142–144.
5. Banerjee, A.; Sharma, R.; Banerjee, U. The nitrile-degrading enzymes: Current status and future prospects. *Appl. Microbiol. Biotechnol.* **2002**, *60*, 33–44.
6. Chen, J.; Zheng, R.C.; Zheng, Y.G.; Shen, Y.C. Microbial transformation of nitriles to high-value acids or amides. *Adv. Biochem. Eng.-Biotechnol.* **2009**, *113*, 33–77. [PubMed]
7. Park, W.J.; Kriechbaumer, V.; Müller, A.; Piotrowski, M.; Meeley, R.B.; Gierl, A.; Glawischnig, E. The nitrilase ZmNIT2 converts indole-3-acetonitrile to indole-3-acetic acid. *Plant Physiol.* **2003**, *133*, 794–802. [CrossRef]
8. Mathew, C.D.; Nagasawa, T.; Kobayashi, M.; Yamada, H. Nitrilase-Catalyzed Production of Nicotinic Acid from 3-Cyanopyridine in *Rhodococcus rhodochrous* J1. *Appl. Environ. Microbiol.* **1988**, *54*, 1030–1032.
9. Cantarella, L.; Gallifuoco, A.; Malandra, A.; Martínková, L.; Pasquarelli, F.; Spera, A.; Cantarella, M. Application of continuous stirred membrane reactor to 3-cyanopyridine bioconversion using the nitrile hydratase–amidase cascade system of *Microbacterium imperiale* CBS 498-74. *Enzym. Microb. Technol.* **2010**, *47*, 64–70.
10. Straathof, A.J.J.; Panke, S.; Schmid, A. The production of fine chemicals by biotransformations. *Curr. Opin. Biotechnol.* **2003**, *13*, 548–556.
11. Zhou, Z.M.; Hashimoto, Y.; Kobayashi, M. Nitrile degradation by rhodococcus: Useful microbial metabolism for industrial productiions. *Actinomycetologica* **2005**, *19*, 18–26. [CrossRef]
12. Gupta, N.; Balomajumder, C.; Agarwal, V.K. Enzymatic mechanism and biochemistry for cyanide degradation: A review. *J. Hazard. Mater.* **2010**, *176*, 1–13. [PubMed]

13. Fischer-Colbrie, G.; Matama, T.; Heumann, S.; Martinkova, L.; Cavaco Paulo, A.; Guebitz, G. Surface hydrolysis of polyacrylonitrile with nitrile hydrolysing enzymes from *Micrococcus luteus* BST20. *J. Biotechnol.* **2007**, *129*, 62–68.
14. Robins, K.; Gordon, J. Whole Cell Production of Fine Chemicals and Intermediates. In *Biocatalysis for Green Chemistry and Chemical Process Development*; John Wiley & Sons, Inc.: Hoboken, NJ, USA, 2011; pp. 299–326.
15. Pace, H.C.; Brenner, C. The nitrilase superfamily: Classification, structure and function. *Genome Biol.* **2001**, *2*, reviews0001.1. [CrossRef]
16. Desantis, G.; Zhu, Z.; Greenberg, W.A.; Wong, K.; Chaplin, J.; Hanson, S.R.; Farwell, B.; Nicholson, L.W.; Rand, C.L.; Weiner, D.P.; et al. An enzyme library approach to biocatalysis: Development of nitrilases for enantioselective production of carboxylic acid derivatives. *J. Am. Chem. Soc.* **2002**, *124*, 9024–9025. [CrossRef]
17. Obanijesu, E.O.; Bello, O.O.; Osinowo, F.A.O.; Macaulay, S.R.A. Development of a packed-bed reactor for the recovery of metals from industrial wastewaters. *Int. J. Environ. Pollut.* **2004**, *22*, 701–702.
18. Jorge, R.M.F.; Livingston, A.G. Microbial dynamics in a continuous stirred tank bioreactor exposed to alternating organic compounds. *Biotechnol. Bioeng.* **2000**, *69*, 409–417. [CrossRef]
19. England, E.; Fitch, M.W.; Mormile, M.; Roberts, M. Toluene removal in membrane bioreactors under recirculating and non-recirculating liquid conditions. *Clean Technol. Environ. Policy* **2005**, *7*, 259–269.
20. Pham, S.Q.; Gao, P.; Li, Z. Engineering of recombinant *E. coli* cells coexpressing P450pyrTM monooxygenase and glucose dehydrogenase for highly region and stereoselective hydroxylation of alicycles with cofactor recycling. *Biotechnol. Bioeng.* **2013**, *110*, 363–373.
21. Chen, Y.T.; Ma, B.D.; Cao, S.S.; Wu, X.M.; Xu, Y. Efficient synthesis of Ibrutinib chiral intermediate in high space-time yield by recombinant *E. coli* coexpressing alcohol dehydrogenase and glucose dehydrogenase. *RSC Adv.* **2019**, *9*, 2325–2331. [CrossRef]
22. Wu, J.; Ma, B.D.; Xu, Y. One–pot synthesis of β–alanine from maleic acid via three–enzyme cascade biotransformation. *Catalysts* **2023**, *13*, 267.
23. Qiu, J.W.; Zheng, X.S.; Ma, B.D.; Xu, Y. Efficient Improvement the Production of Recombinant Nitrilase by Optimizing Culture Conditions Using Response Surface Methodology (RSM). *J. Comput. Theor. Nanosci.* **2016**, *13*, 2269–2276. [CrossRef]
24. Li, H.; Dong, W.L.; Zhang, Y.; Liu, K.; Zhang, W.M.; Zhang, M.; Ma, J.F.; Jiang, M. Enhanced catalytic efficiency of nitrilase from *Acidovorax facilis* 72W and application in bioconversion of 3-cyanopyridine to nicotinic acid. *J. Mol. Catal. B Enzym.* **2017**, *133*, 459–467. [CrossRef]
25. He, L.; Chen, Y.; Wu, X.M.; Xu, Y. Magnetic Fe_3O_4/Alginate Cell Beads: Application in Enzymatic Synthesis of Pharmaceutical Intermediate. *J. Comput. Theor. Nanosci.* **2016**, *13*, 2264–2268.
26. Kaul, P.; Banerjee, A.; Banerjee, U.C. Stereoselective Nitrile Hydrolysis by Immobilized Whole-Cell Biocatalyst. *Biomacromolecules* **2006**, *7*, 1536–1541.
27. Zhong, X.; Yang, S.M.; Su, X.Y.; Shen, X.X.; Zhao, W.; Chan, Z. Production of Cyanocarboxylic Acid by Acidovorax facilis 72W Nitrilase Displayed on the Spore Surface of *Bacillus subtilis*. *J. Microbiol. Biotechnol.* **2019**, *29*, 749–757.
28. Liu, Q.G.; Zhao, N.; Zou, Y.N.; Ying, H.J.; Liu, D.; Chen, Y. Feasibility study on long-term continuous ethanol production from Cassava supernatant by immobilized yeast cells in packed bed reactor. *J. Microbiol. Biotechnol.* **2020**, *30*, 1227–1234. [CrossRef]
29. Chauhan, S.; Wu, S.; Blumerman, S.; Fallon, R.D.; Gavagan, J.E.; Dicosimo, R.; Payne, M.S. Purification, cloning, sequencing and over-expression in *Escherichia coli* of a regioselective aliphatic nitrilase from *Acidovorax facilis* 72W. *Appl. Microbiol. Biotechnol.* **2003**, *61*, 118–122. [CrossRef]
30. Gong, J.S.; Dong, T.T.; Gu, B.C.; Li, H.; Dou, W.F.; Lu, Z.M.; Zhou, Z.M.; Shi, J.S.; Xu, Z.H. Semi-rational engineering accelerates the laboratory evolution of nitrilase catalytic efficiency for nicotinic acid biosynthesis. *ChemCatChem* **2017**, *9*, 3395–3401.
31. Gong, J.S.; Li, H.; Lu, Z.M.; Zhang, X.J.; Zhang, Q.; Yu, J.H.; Zhou, Z.M.; Shi, J.S.; Xu, Z.H. Engineering of a fungal nitrilase for improving catalytic activity and reducing by-product formation in the absence of structural information. *Catal. Sci. Technol.* **2016**, *6*, 4134–4141. [CrossRef]
32. Roy, R.B.; Merten, J.J. Evaluation of urea-acid system as medium of extraction for the B-group vitamins. Part II. Simplified semi-automated chemical analysis for niacin and niacinamide in cereal products. *J. Assoc. Off. Anal. Chem.* **1983**, *66*, 291–296. [CrossRef] [PubMed]

catalysts

Review

Biodegradation of Different Types of Bioplastics through Composting—A Recent Trend in Green Recycling

Wazir Aitizaz Ahsan [1], Adnan Hussain [1], Chitsan Lin [2,3,*] and Minh Ky Nguyen [2,4]

1 Ph.D. Program of Aquatic Science and Technology, National Kaohsiung University of Science and Technology, Kaohsiung 81157, Taiwan
2 Ph.D. Program in Maritime Science and Technology, National Kaohsiung University of Science and Technology, Kaohsiung 81157, Taiwan
3 Department of Marine Environmental Engineering, National Kaohsiung University of Science and Technology, Kaohsiung 81157, Taiwan
4 Faculty of Environment and Natural Resources, Nong Lam University, Ho Chi Minh City 700000, Vietnam
* Correspondence: ctlin@nkust.edu.tw

Abstract: In recent years, the adoption of sustainable alternatives has become a powerful tool for replacing petroleum-based polymers. As a biodegradable alternative to petroleum-derived plastics, bioplastics are becoming more and more prevalent and have the potential to make a significant contribution to reducing plastic pollution in the environment. Meanwhile, their biodegradation is highly dependent on their environment. The leakage of bioplastics into the environment and their long degradation time frame during waste management processes are becoming major concerns that need further investigation. This review highlights the extent and rate of the biodegradation of bioplastic in composting, soil, and aquatic environments, and examines the biological and environmental factors involved in the process. Furthermore, the review highlights the need for further research on the long-term fate of bioplastics in natural and industrial environments. The roles played by enzymes as biocatalysts and metal compounds as catalysts through composting can help to achieve a sustainable approach to the biodegradation of biopolymers. The knowledge gained in this study will also contribute to the development of policies and assessments for bioplastic waste, as well as provide direction for future bioplastics research and development.

Keywords: biopolymers; aerobic composting; sustainability; depolymerization; enzymes

Citation: Ahsan, W.A.; Hussain, A.; Lin, C.; Nguyen, M.K. Biodegradation of Different Types of Bioplastics through Composting—A Recent Trend in Green Recycling. *Catalysts* **2023**, *13*, 294. https://doi.org/10.3390/catal13020294

Academic Editors: Zhilong Wang and Tao Pan

Received: 24 December 2022
Revised: 22 January 2023
Accepted: 25 January 2023
Published: 28 January 2023

1. Introduction

Single-use plastic consumption has been increasing for years due to its durability, light weight, and low cost [1]. The use of plastic has led to many technological advances, including high strength-to-weight ratio construction, automotive materials, and highly resistant packaging materials for food [2]. Approximately 9.2 billion tons of plastic have been produced worldwide, and the annual global production of plastic increased to 368 million tons in 2019 [3,4]. As estimated, the annual production of plastic waste is 34 million tons, and 93% of it is disposed of in landfills and oceans [5]. In 2015, 322 million tons of petroleum-based plastic were produced globally, compared with 1.7 million tons in 1950 [6]. Synthetic petroleum-based plastic leads to an increase in plastic waste, which contributes to adverse effects on the environment, such as ozone depletion, eco-toxicity, the release of carcinogens, global warming, and eutrophication [7]. Approximately 2.8 kg of CO_2 is released into the environment when 1 kg of plastic is burned [8]. Bioplastics emerged in response to environmental concerns about non-biodegradable plastics.

In the circular economy, bioplastics are expected to play an important role in achieving sustainable development goals, such as avoiding fossil fuels, introducing new degradation or recycling approaches, and reducing toxic chemicals during the manufacturing process. Biodegradable plastics derived from renewable biomass have become increasingly popular and bioplastics are currently produced on a scale of 4 million tons annually [9]. Globally, bioplastic production is expected to increase from 2.11 million tons in 2019 to 2.42 million tons by 2024. A major market for bioplastics is the packaging industry, which accounts for nearly 40% of global production [10]. Although many reviews discuss bioplastics, few address the positive and negative impacts of bioplastics on the environment comprehensively and simultaneously [11]. Nonetheless, not all polymers that are derived from bio-based sources are biodegradable, and not all polymers that are derived from fossil sources are non-biodegradable [12]. In nature, bioplastics are primarily composed of renewable resources, such as cellulose, starch, sugar, etc. [13]. In fact, biodegradation rates differ among bioplastics, and biopolymer properties depend on external environmental factors, intrinsic biopolymer properties, and filler properties in blends and composites [14]. In addition to their original source, production processes also have a great deal to do with degradation [15]. Moreover, many reports show that bioplastic composites and films degrade slowly in normal water and soil environments [16]. Due to this, there are concerns about their disposal in landfills and in soils at the end of their useful lives. Thus, composting bioplastics becomes an important tool for their effective environmental management at end-of-life.

Composting is considered more environmentally friendly and cost-effective than recycling or incineration. Specific microorganisms, such as Pseudomonaceae, Comamonadaceae, Erythrobacteraceae, Streptomycetaceae, Caulobacteraceae families, and Enterobacteriaceae, and enzymes, such as N-acetyl-β-glucosaminidase, esterase, β-glucosidase, acid phosphatase, alkaline, and phosphohydrolase, are involved in the degradation and microbial decomposition of bioplastics [17,18]. Specifically, enzymatic decomposition has been regarded as a means of minimizing environmental pollution. Microbiological degradation of bioplastics, particularly microbial enzymatic catalysis, has drawn attention as a means of reducing the amount of pollution in the environment.

The process of composting involves decomposing organic matter and turning it into humus, which can be used to strengthen soil structure and its fertility rate [19,20]. Bioplastic waste is typically disposed of in landfills, followed by recycling, incineration, and composting [16,21]. In contrast, landfilling produces greenhouse gases and creates environmental concerns. Landfilling not only produces greenhouse gases but also occupies and contaminates future agricultural land [22,23]. Therefore, composting would be a more profitable and desirable method for disposing of bioplastic waste. As a cost-effective and safe waste management solution, composting technology is being adopted by several industries [24]. In the literature, industrial composting of bioplastics has been demonstrated to be one of the most desirable methods for managing the material's end of life [25].

Compostable polymers are being developed as environmentally friendly alternatives, especially if they can be recycled organically and derived from renewable resources. Using lifecycle assessment techniques, ASTM D7075 and ISO 14000 have developed standards to evaluate biobased products and their environmental performance [26,27]. However, only some of the biopolymers are listed as compostable materials by ASTM. In order for a polymer to be considered compostable, it must convert 90% of its carbon content to carbon dioxide in accordance with ASTM International (D5338). An ASTM International (D5338) polymer can only be considered compostable if 90% of its carbon content is converted into carbon dioxide. This prepared polymer undergoes three primary steps in order to become biodegradable: biodeterioration, fragmentation, and assimilation [28]. In addition, plant-based polymers, thermoplastic starch, polyhydroxyalkanoates (PHAs), and polylactic acid or polylactide (PLA) are commonly reported as biopolymers [29]. It is important to know that a number of factors affect the biodegradation rate of biopolymers in nature, such as their chemical structures, functional groups, crystallinities, and polymer chains [8]. Furthermore, temperature, oxygen, and pH content play a significant role in polymer biodegradation [30]. It has been reported that PLA degradation in the soil is much slower than in compost medium because compost has a higher moisture content and temperature range encouraging PLA hydrolysis and the assimilation of PLA by thermophilic microorganisms. According to [31], Zn was used as catalyst in PLA depolymerization, but the problem with these catalysts was that they could not be recycled or re-used. However, ref. [32] reported that the degradation of PLA in soil takes much longer than in compost medium due to thermophilic bacteria which are able to hydrolyze and assimilate PLA with a higher temperature range and moisture content in the compost. After 47 days of composting, it was determined that the average rate of biodegradation for cellulose was $96.8 \pm 6.7\%$ [33] according to standard composting methods [34]. In addition, the composting of biobased polymers and the use of the compost in agriculture can result in significant emission and energy credits. Biobased polymers can be made even more sustainable through composting, which is an integral part of sustainable agriculture practices.

This review aims to gather information about the biodegradation of bioplastics in diverse environments and to discuss it to examine the compostability rate of different types of bioplastics through composting. Finally, this review concludes by discussing the composting technology in the biodegradation of bioplastics as well as classifications of different bioplastics according to the degradation rate through home and industrial composting.

2. Types of Bioplastics

2.1. Starch-Based Bioplastic

Biopolymers made from starch are becoming increasingly popular due to their abundant availability, renewability, low-cost, and biodegradability. In addition, starch is regarded as a promising raw material for biopolymer production. After polylactic acid (PLA), starch-based plastics accounted for the second-highest share of the total bioplastics production [35]. There are two types of polymers involved in its composition: linear amylose and branched amylopectin [36]. An important feature of bioplastics is their elasticity, which is provided by linear amylose, while amylopectin has a branched structure that controls tensile strength and elongation [37]. Among the most promising biopolymers for producing edible films, starch is particularly popular because of its affordability.

2.2. PLA-Based Bioplastic

Polylactic acid is a commercial biodegradable thermoplastic based on lactic acid also called polylactide or PLA (also known as polylactic acid, lactic acid polymer). The most widely used biodegradable aliphatic polyester, PLA is a thermoplastic that is aliphatic non-cyclic, non-aromatic, derived from lactic acid and lactide, and formed by polymerizing sugars obtained from various agricultural biomass sources [38]. Polylactides are developed for degradable packaging materials, and polylactide decomposes within three weeks in industrial composting processes. Polylactide is the first synthetic polymer to be synthesized from renewable resources [39]. Moreover, polylactic acid exhibits a number of desirable characteristics, including being easy to fabricate, biocompatible, biodegradable, non-toxic, and having better thermal properties [40]. When polylactic acid biodegrades, it releases water, CO_2, and decomposed organic matter that green plants are able to utilize, which reduces greenhouse gas emissions. Additionally, when oxygen is added to polylactic acid, no toxic intermediates or byproducts are produced. In comparison with other synthetic polymers, polylactic acid emits relatively fewer greenhouse gases [41].

2.3. PHAs-Based Bioplastic

Several types of microalgae produce PHAs, which are biodegradable biopolymers [42,43]. In nutrient-limited environments, diverse prokaryotic microbes produce PHAs for carbon storage [44]. In PHAs, the carboxylate group of one monomer forms ester bonds with the hydroxyl group of the adjoining monomer to form polymers of 3 hydroxy-acid, sometimes called hydroxy alkanoic acids [45]. In terms of physical properties, PHAs can be compared to petro-chemical polymers, which makes them viable alternatives for the growing global bioplastic market [46]. In bioplastics, PHAs have not been widely applied, and this may be due to their high production and recovery costs [47]. Scientists are searching for cost-effective feedstocks to replace PHA. Approximately 90 percent of the microbes that degrade PHAs also breakdown starch as the biodegradation pathways are similar [48].

2.4. Cellulose-Based Bioplastic

A variety of biomass can be used to produce cellulose, including wood, seed fibers, bast fibers, grass, marine animals (tunicates), algae, fungi, invertebrates, and bacteria [49]. Additionally, acetic acid bacteria can synthesize cellulose in addition to higher plants [50]. As with starch, cellulose consists of linear chains with glycosidic bonds that join a few hundred to more than ten thousand glucose units. Although starch and cellulose have the same monomer unit, they differ in how their polymeric chains are oriented [51]. In recent years, cellulose-based biopolymers have gained attention due to their strength, stiffness, high durability, and biodegradability [52]. In addition to being low-density, low-price, and nonabrasive, cellulose-based reinforced composites are also non-abrasive [1]. As cellulose-based bioplastics contain distant tenuous molecules with weak hydrogen bonds, they degrade rapidly. Conversely, bioplastics made from cellulose have weaker hydrogen bonds, and therefore have lower mechanical properties, such as strength and flexibility.

3. Biodegradation of Biopolymers in Soil and Aquatic Environments

In biodegradation, naturally occurring microorganisms, such as bacteria and fungi, mineralize materials through their action [53]. The degradation of bioplastics varies in three different surroundings (soil, aquatic system, and compost). In contrast, bioplastics derived from biological sources take significantly less time to degrade than petroleum-based plastics. Because plastics have a high molecular weight, chemical structure, low water solubility, and contain xenobiotics, their biodegradation is limited [54]. In previous studies in the literature, many scholars investigated the biodegradation of bioplastics which are listed in Table 1.

Bioplastic Degradation in Soil

Soil contains a wide diversity of microorganisms, making plastic biodegradation more feasible than in other environments such as water and air [55]. A number of microorganisms isolated from soil media utilized bioplastic as a carbon source. Actinobacteria species, such as Nonomuraea, Amycolatopsis, Streptomyces, Laceyella, Actinomadura, and Thermomactimyces species, were obtained from soil. However, among these the Streptomyces and Amycolatopsis were the most common species that play a crucial role in bioplastic degradation in soil environments. Bulkholderia, Pseudomonas, Paenibacillus, and Bacillus species were mainly isolated from different soil environments, and they were capable of degrading the bioplastics. Most commonly, Aspergillus, Fusarium, and Penicillium were identified as soil-isolated fungi responsible for bioplastic degradation [56]. In spite of the fact that cellulose, which was used as a positive control, was fully degraded, the biodegradation process was slow. Possibly, this is due to the lower temperature of the system under real conditions and the longer time span of the experiment. Consequently, these bioplastics required higher temperatures and longer degradation times to degrade effectively [57]. The biodegradation of polymers depends on the chemical nature of the polymer as well as on environmental factors, such as moisture, temperature, acidic nature, etc. [58]. Including these factors, bioplastics biodegrade differently in different soil compositions. Figure 1 depicts the biodegradation mechanism of biopolymers in soil environments. Starch-based plastics are found to reduce in weight and faster degradation were observed in field soil than PHAs and PLA, while PLA sustains its weight for a long period of time, about 12 weeks [59]. The highest biodegradability was found with cellulose-based bioplastics (80 to 100%) after 100 days [60,61]. Based on the kinetic constants of degradation of the three blends studied in soil, PHAs, blends showed the highest kinetic constant, followed by PLA blends [62]. Overall, the bioplastic-composted soil increases the soil fertility and increase the yield of crops. It is generally observed that microbiological content increases after biodegradable films are buried, as the organic mulch increases bacterial populations because of the different chemical compositions and decomposition rates of these materials [63]. However, in composting processes, the PHAs' films enrich the soil more than PLA since they increase the microbial population present in the soil [64]. In addition to an increase in Clostridia species and mesophilic aerobic bacteria, there was also a significant increase in fungi. There is no doubt that these changes were caused by the swift degradation of the protein-based bioplastic, which resulted in the release of carbon and nitrogen compounds, which served as food and increased the microbial population.

Table 1. Biodegradation of different types of bioplastics in soil and aquatic environments.

Bioplastics	Environment	Temperature/Moisture/pH	Biodegradability	Days Taken for Biodegradation	References
Starch-based	Soil	20 °C, 60%	14.2%	110	[54]
Starch-based blends	Sea water	25 °C	1.5%	90	[65]
Starch/chitosan (35/65)	Soil	Soil burial test method	96%	28	[66]
Starch-based	Sea water	Room temperature	1.5%	90	[67]
PLA	Soil	30%	10%	98	[68]
PLA	Soil	25 °C, 60%	13.8%	28	[69]
PLA (powdered)	Soil	25 °C, 60%	13.8%	28	[66]
PLA	Sea water	25 °C	8.4%	365	[70]
PHA	Soil	20 °C, 60%	48.5%	280	[54]
PHA	Compost/Soil (10/90%)	25 °C, 65%	50%	15	[71]
PHA	Soil	39% pH 6.8	75%	80	[72]
PHAs	Sea water	25 °C	8.5%	365	[73]
Cellulose	Soil	Undefined	100%	103	[60]
Sponge fibers	Compost containing synthetic soil	Aerobic, 58 °C	>80%	154	[74]
Cellulose	Municipal solid waste	Room temperature	44%	14	[75]

There is no doubt that the aquatic environment is the most susceptible to plastic contamination. However, bioplastic degradation in both seawater and fresh water generally appears to be slower than biodegradation in composting, anaerobic digestion, and soil environments. Specifically, this was related to the characteristics of aquatic environments that play a critical role in bioplastic degradation. In addition to bioplastics' properties, some environmental parameters, such as nutrients content, temperature, pH, microbial diversity, and microbial population density, have an important impact on bioplastic degradation in aquatic environments. As a result of the study in [76], the PHAs degraded in seawater, and temperature played a significant role in the degradation process. According to the authors, seasonal changes in water temperature led to the difference in degradation rates. There are a number of factors that could contribute to the slow biodegradation of bioplastics under aquatic environments, including low temperatures, nutrient levels, and microbe population density. Several bacteria species were capable of degrading bioplastics in aquatic environments, such as river water and marine environments; Bacillus, Lepthotrix, Tenacibaculum, Pseudomonas, Entrobacter, Variovorax Gracilibacillus, and Avanivorax were isolated from these environments as reported in several studies [55]. Figure 2 depicts the biodegradation mechanism of biopolymers in aquatic environments.

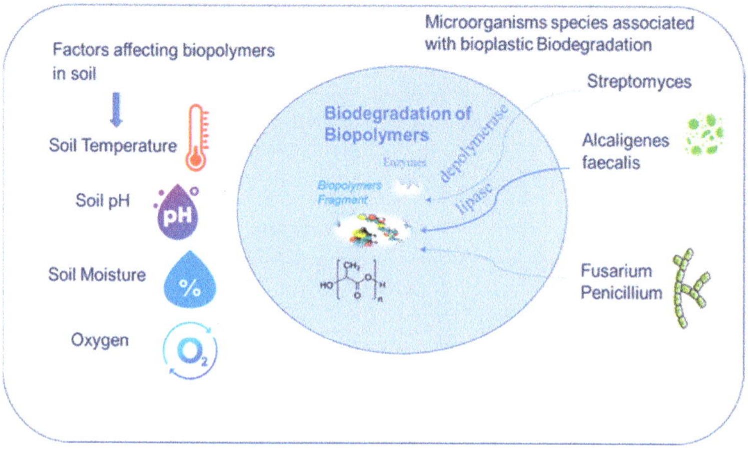

Figure 1. Interactions of microbes and environmental factors in microplastic-contaminated soils.

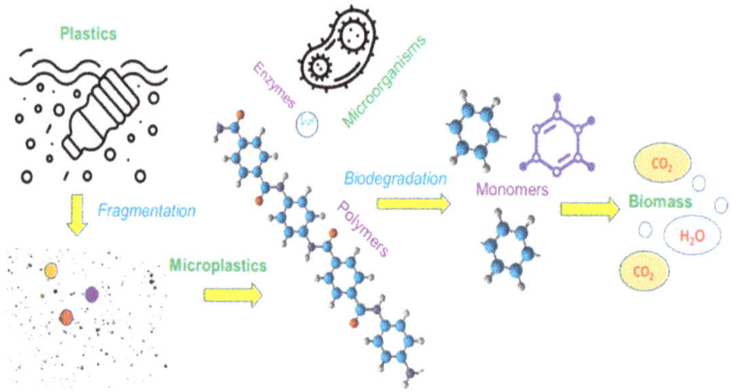

Figure 2. Bioplastic degradation in the aquatic environment.

In aquatic environments, PLA blends showed the slowest rate of degradation. There was an average estimated time of more than ten years for complete degradation. The degradation of starch-based blends in fresh water and seawater shows high variability. The authors of the study [65] concluded that the starch-based bioplastic obtain only a 1.5% degradation under marine and freshwater environments (25 °C, 90 days), while other studies have reported significantly higher degradation rates. The results of [67] showed that starch-based shoppers degraded by 69% (weight basis) within 236 days, probably due to both the material characteristics and the environmental conditions (sea water and sediment). In addition, aquatic environments are less likely to degrade bioplastics than soil environments due to the lack of microbial diversity.

Recently, microscale plastics have entered the marine environment through wastewater discharges, which have caught the attention of researchers. In the current literature, however, no information was found about wastewater discharges releasing and/or shedding bioplastics.

4. Biodegradation of Bioplastics in Compost

There is a significant amount of plastic waste disposed in landfills, which eventually generates greenhouse gases and leachate. Recycling or composting are generally regarded as more suitable ways of recovering plastic from solid waste. Composting occurs when microorganisms convert organic matter into CO_2 and humus by consuming it [58,77]. A number of modern techniques have been introduced to detect the presence of microorganisms, among them, polymerase chain reaction (PCR) and next-generation sequencing (NGS) are two well-known techniques that can be used to analyze and sequence in depth the specific microbial communities involved in the degradation of plastics [78,79]. In accordance with the ASTM's definition of compostable plastics [80], the decomposition of such polymers produces CO_2, water, inorganic compounds, and biomass without leaving behind any visible or toxic residues. The composting process is particularly appropriate for dealing with food-contaminated packaging, as recycling facilities cannot deal with food-contaminated plastics, and the compost formed can be used for soil improvement [81]. A biodegradable plastic is not necessarily a bio-based plastic, as degradability depends on the structure and polymer chemical composition, as well as its interaction with its surroundings [82]. In addition to reducing our global ecological footprint, composting is an excellent end-of-life option. It has been extensively studied over the past decade how compost can be used to biodegrade different types of bioplastics (Table 2).

4.1. *Degradation of PLA through Composting*

PLA is one of the latest materials to be commercialized for use in organic food packaging, such as bags, containers, and films, and it has been proven to decompose under composting conditions [32,83]. PLA degradation in compost occurs only in high-temperature, humid environments containing relevant microorganisms [84]. In the process of biodegradation of PLA, it undergoes two stages: first, hydrolysis or oxidation into monomers and oligomers, and then finally metabolization by microorganisms that produce CO_2 and H_2O [85]. During degradation, PLA is chemically hydrolyzed in thermophilic conditions to reduce its molecular weight, and then microorganisms assimilate lactic acid oligomers as an energy source. A number of enzymes play an important role in the depolymerization of PLA, including carboxylesterase, cutinase, lipase, and serine protease [86]. Serine protease has been identified as the most important enzyme involved in PLA degradation by actinobacteria of the genus amycolatopsis [87]. Moreover, enzymes encoded by a multitude of bacteria and fungi can partially degrade plastics; enzymes are crucial to the depolymerization of polymers, even those that are considered resistant [88]. The enzymes in this group mostly include carboxylesterases, lipases, cutinases, and proteases as well as several other enzyme groups (i.e., laccases, oxidoreductases, manganese peroxidases, and alkane hydroxylases monooxygenases) [89] involved in the degradation of plastics. Despite this, little information exists on the characterization of PLA-degrading enzymes in previous studies. Figure 3 shows the biodegradation mechanism of biopolymers in compost. In terms of chemical and biological degradation, temperature is considered the important restraining parameter, as the increased flexibility of the chains occurs only above the PLA glass transition temperature of 55 °C [32]. Another relevant parameter is the PLA amount in the composting pile: in a mixture of 70:30 wt% garden waste/PLA, the chemical hydrolysis of PLA lowers the pH due to the large amounts of lactic acid that are produced, reducing the degrading action of compost microorganisms [90]. According to a study, ref. [91], poly lactic acid (PLA) bioplastics degrade completely through industrial composting within four to six weeks. This makes PLA incompatible with home composting as the moisture-rich environments favor chemical hydrolysis [92].

4.2. *Degradation of PHAs through Composting*

Although PHA is not as well-known as PLA, its easy disposal makes it more popular among the environmental community [93]. In low-temperature or low-pH home composting conditions, PHAs show minimal or no biodegradation. Biodegradation is improved by higher temperatures in industrial composting [94]. Ref. [95] studied the biodegradation of PHAs when exposed to temperatures between 8 and 30 °C for 152 days, and PHB, PHBV (10% HV), and PHBV (20% HV) degraded at 4%, 6–17%, and 67%, respectively. Ref. [96] reported that PHBV (HV26%) had fifty-nine-percent mass loss in 186 days, suggesting that it is biodegrading more slowly even though the temperature ranged from 40 to 63 degrees Celsius throughout the study. Probably, the differences in inoculum and temperature profiles resulted in PHB degrading by 50% in 84 days at 34–66 °C and 74–89% humidity in organic waste home compost [97]. It was observed that degradation rates for PHBV were increased more rapidly than those for PHBA, PHBHHx, and PHB when a medium with PHA depolymerase was used. According to their hypothesis, PHAs with longer side chains, including PHBA and PHBHHx, have a lower degradation rate than PHAs with shorter side chains because their side chains slow down depolymerase degradation (Wang et al., 2018). According to Danimer Scientific, PHAS produced by the company are compostable in backyard and industrial composting systems. They have obtained third-party certification from Vinçotte that demonstrates that their products are compostable both in home and industrial composting systems [98].

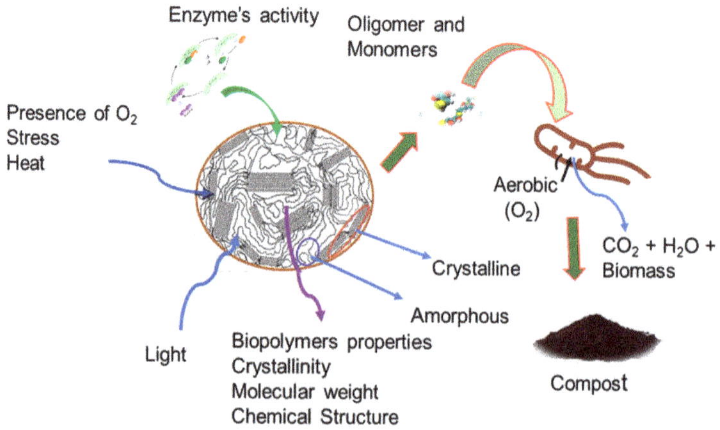

Figure 3. Mechanism of biodegradation of bioplastics through composting.

4.3. Degradation of Starch-Based Bioplastic through Composting

In bioplastic production, starch is used frequently because of its abundance in nature (especially agricultural products) and low cost [99]. Additionally, starch has been reported in the literature to be a good compostable material for plastic films, bags, and agricultural mulching films [100]. Microorganisms can directly attack starch and cellulose molecules since they are capable of producing enzymes to depolymerize or cleave the polymer physical structure. This can result in molecular weight abatement outside the microbial cells [53]. For example, the mineralization of corn starch at 58 °C took 44 days under aerobic conditions [34]. The pH range of 7.0–8.0 and 50% moisture facilitate the biodegradation of starch films in organic compost obtained from different crops. In the first stage of degradation, mainly caused by plasticizer leaching, around 30% of the weight was lost within 24 h. In the second stage, primarily due to biological activity and glycosidic bond scission, weight slowly decreased until 90% of the original weight in approximately 20 days [101]. Interestingly, several papers reported on starch-based blends' decomposition during mesophilic composting (23–25 °C) and found that under aerobic conditions, temperature also played a key role in bioplastic degradation. Ref. [65] reported starch-based bioplastic degradation in non-industrial composting conditions after about 9 weeks of composting [102].

4.4. Degradation of Cellulose-Based Bioplastic through Composting

Just like starch, cellulose is a polysaccharide arising from glucose monomers. However, in cellulose structures, these monomers are bound with stronger glycosidic bonds, making them more resistant to decomposition. Plant cell walls, which are made of cellulose, contain high levels of natural cellulose [103]. Among biopolymers, cellulose is considered relatively fast-degrading in compost environments [54]. A 47-day composting experiment found that 97 + 7% of cellulose mineralized after standard composting methods were applied [34]. Ref. [104] studied the degradation of cellulose powder through a composter bin at lab scale. During the composting process, the maximum temperature was above 60 °C for at least one week. This shows that cellulose powder biodegrades at a rate of 69% after 65 days. Furthermore, the degradation of cellulose-based products may be dependent on a favorable environment for microorganisms in terms of temperature, moisture, and oxygen. In some cases, however, adding other substances (salts, pigments) can inhibit cellulose decomposition [54].

Bioplastic degradation in composting environments is influenced by a number of factors, with temperature and the chemical composition of bioplastics playing major roles [27]. In composting, maximum temperatures (above 55 °C) allow the most common bioplastics to reach their glass transition temperature, resulting in amorphous polymers that are more hydrophilic [105], increased hydrolyzation, and enhanced bioplastic degradation kinetics. The conditions for industrial composting are adequately standardized and controlled. Comparatively, home composting conditions tend to be much more variable, and temperatures tend to be lower. Thus, composting at home is less effective and slower than composting in industrial settings. In composting, biodegradable mulch by itself does not provide sufficient nutrients to compost. However, polymer carbon must be accounted for in determining the appropriate ratio of carbon to nitrogen.

In general, PLA-, PAHs-, starch-, and cellulose-based bioplastics which are easy to hydrolyze are considered an end-of-life option by biodegradation, but it should only be performed under controlled industrial conditions to ensure complete digestion and prevent side effects that are uncontrollable, such as the formation of microplastics or the leakage of contaminants on the site. Moreover, composting requires chemical compounds to fully degrade the bioplastics, but on the other hand these chemical compounds or approaches have a heavy load on the environment. For a sustainable approach, some studies [106] reported that some metal compounds and enzymes are commonly used as catalysts to degrade bioplastics, such as PLA-, PHAs-, starch-, and cellulose-based bioplastics, from the environment.

Table 2. Biodegradation of different types of bioplastics in compost.

Bioplastics	Feedstock	Temperature/Moisture Contents	Biodegradability in Percentage	Composting Time Frame (Days)	References
Starch-based (potato almidon)	Compost	Aerobic, 58 °C	85%	90	[107]
Plastarch	Compost	Aerobic, 55 °C, 60%	50%	85	[54]
Starch-based blends	Compost/Food waste	45–65 °C	60%	90	[108]
PLA	Compost	58 °C, 60%	60%	30	[109]
PLA +Clay film	Compost	Aerobic, 58 °C, 55%	34%	130	[110]
PLA	Compost	65 °C, pH = 8.5, 63%	84%	58	[111]
PHA-based	Compost	55 °C, 70%	80%	28	[112]
PHAs blends	Compost/Cow manure	50 °C	30%	60	[113]
Cellulose-based	Compost containing synthetic material	Aerobic, 58 °C	>80%	154	[74]
Sponge cloth (Cellulose-based)	Compost	Aerobic, 58 °C	80%	154	[74]
Nylon4 (polyamides, bio-based)	Composted soil	25 °C, pH 7.5–7.6, 80%	100%	120	[61]

5. Current Gaps and Future Research Directions

There is a need for further research on the degradation of bioplastics in backyard compost piles as well as in industrial compost piles. In spite of the fact that there are many composability standards for bioplastics, data from the literature showed good performance in industrial composting when proper conditions were followed. In addition, compost is the most suitable environment for biodegradation, followed by soil and aquatic environments. Compost or anaerobic digestion can easily degrade some biodegradable plastics, but soil may not. It is therefore important that biodegradable plastics have clearly defined end-of-life targets. Aquatic environments may degrade some biodegradable polymers, but they should never be used as end-of-life disposal. Biodegradation is less feasible in aquatic environments due to the lower temperatures and less microbial activity than in compost and soil.

The information here can help industrial companies to categorize the current limits of bioplastic degradation and identify potential growth areas. In a way, it will boost the food packaging industry's sustainable progress toward producing cleaner, environmentally friendly packaging, meeting consumer and industry expectations for the future of this important sector. In terms of research, it is also important to look at the relationship between the biopolymer's chemical structure and its composability in industrial plants. Therefore, the understanding of biodegradation processes is progressing and the advancement from a technological point of view makes this approach an actual opportunity within a certain maturity level. There are some problems that need to be addressed, such as pollution from non-compostable plastics, the accumulation of plastics with long degradation times, and confusion about how additives affect biodegradation rates. A compostable material is the perfect solution for some applications, such as food waste bags, where organic matter cannot be separated from plastics. In addition, there is a need to introduce biocatalysts, such as enzymes and microorganisms, to selectively depolymerize bioplastic waste into its constituent monomers or other value-added products.

6. Conclusions

Bioplastics are emerging as a sustainable alternative to traditional plastics. The identification and biodegradation of bioplastics have been developed through various methods in recent years. Bioplastics have been reported to biodegrade in various studies. In the studies, bioplastics were studied in their production as well as their environmental persistence. A variety of standard biodegradation test methods were described in aerobic biodegradation. The current knowledge about the degradation of bioplastics through composting, soil, and aquatic environments is summarized in this review paper. A composting process can only degrade compostable polymers, and mineralization can begin within the composting period for other biodegradable materials. With the help of industrial composting, the volume of bioplastic waste can be reduced in a sustainable way. Therefore, it is important to identify the conditions that result in safe compost production.

Author Contributions: Writing—original draft, visualization and conceptualization, W.A.A.; supervision, validation, C.L.; writing—reviewing and editing, A.H.; software and editing, M.K.N. All authors have read and agreed to the published version of the manuscript.

Funding: This research received no external funding.

Data Availability Statement: Not applicable.

Conflicts of Interest: The authors declare no conflict of interest.

References

1. Polman, E.M.; Gruter, G.-J.M.; Parsons, J.R.; Tietema, A. Comparison of the aerobic biodegradation of biopolymers and the corresponding bioplastics: A review. *Sci. Total Environ.* **2021**, *753*, 141953. [CrossRef] [PubMed]
2. Koch, D.; Mihalyi, B. Assessing the change in environmental impact categories when replacing conventional plastic with bioplastic in chosen application fields. *Chem. Eng. Trans.* **2018**, *70*, 853–858.

3. Huang, J.; Veksha, A.; Chan, W.P.; Giannis, A.; Lisak, G. Chemical recycling of plastic waste for sustainable material management: A prospective review on catalysts and processes. *Renew. Sustain. Energy Rev.* **2022**, *154*, 111866. [CrossRef]
4. Mellinas, C.; Valdés, A.; Ramos, M.; Burgos, N.; Garrigos, M.d.C.; Jiménez, A. Active edible films: Current state and future trends. *J. Appl. Polym. Sci.* **2016**, *133*. [CrossRef]
5. Mekonnen, T.; Mussone, P.; Khalil, H.; Bressler, D. Progress in bio-based plastics and plasticizing modifications. *J. Mater. Chem. A* **2013**, *1*, 13379–13398. [CrossRef]
6. Suman, T.Y.; Li, W.-G.; Alif, S.; Faris, V.R.P.; Amarnath, D.J.; Ma, J.-G.; Pei, D.-S. Characterization of petroleum-based plastics and their absorbed trace metals from the sediments of the Marina Beach in Chennai, India. *Environ. Sci. Eur.* **2020**, *32*, 110. [CrossRef]
7. Chae, Y.; An, Y.-J. Current research trends on plastic pollution and ecological impacts on the soil ecosystem: A review. *Environ. Pollut.* **2018**, *240*, 387–395. [CrossRef]
8. Burgos, N.; Valdés, A.; Jiménez, A. Valorization of agricultural wastes for the production of protein-based biopolymers. *J. Renew. Mater.* **2016**, *4*, 165–177. [CrossRef]
9. Ali, S.S.; Elsamahy, T.; Abdelkarim, E.A.; Al-Tohamy, R.; Kornaros, M.; Ruiz, H.A.; Zhao, T.; Li, F.; Sun, J. Biowastes for biodegradable bioplastics production and end-of-life scenarios in circular bioeconomy and biorefinery concept. *Bioresour. Technol.* **2022**, *363*, 127869. [CrossRef]
10. Yasin, N.M.; Akkermans, S.; Van Impe, J.F. Enhancing the biodegradation of (bio) plastic through pretreatments: A critical review. *Waste Manag.* **2022**, *150*, 1–12. [CrossRef]
11. Atiwesh, G.; Mikhael, A.; Parrish, C.C.; Banoub, J.; Le, T.-A.T. Environmental impact of bioplastic use: A review. *Heliyon* **2021**, *7*, e07918. [CrossRef] [PubMed]
12. Garrison, T.F.; Murawski, A.; Quirino, R.L. Bio-based polymers with potential for biodegradability. *Polymers* **2016**, *8*, 262. [CrossRef] [PubMed]
13. Wicaksono, J.A.; Purwadaria, T.; Yulandi, A.; Tan, W.A. Bacterial dynamics during the burial of starch-based bioplastic and oxo-low-density-polyethylene in compost soil. *BMC Microbiol.* **2022**, *22*, 309. [CrossRef] [PubMed]
14. Thakur, S.; Chaudhary, J.; Sharma, B.; Verma, A.; Tamulevicius, S.; Thakur, V.K. Sustainability of bioplastics: Opportunities and challenges. *Curr. Opin. Green Sustain. Chem.* **2018**, *13*, 68–75. [CrossRef]
15. Endres, H.-J. Bioplastics. In *Biorefineries*; Springer: Cham, Switzerland, 2017; pp. 427–468.
16. Cucina, M.; de Nisi, P.; Tambone, F.; Adani, F. The role of waste management in reducing bioplastics' leakage into the environment: A review. *Bioresour. Technol.* **2021**, *337*, 125459. [CrossRef]
17. Sánchez, Ó.J.; Ospina, D.A.; Montoya, S. Compost supplementation with nutrients and microorganisms in composting process. *Waste Manag.* **2017**, *69*, 136–153. [CrossRef]
18. Tiquia, S. Evolution of extracellular enzyme activities during manure composting. *J. Appl. Microbiol.* **2002**, *92*, 764–775. [CrossRef]
19. Báreková, A.; Demovičová, M.; Tátošová, L.; Danišová, L.; Medlenová, E.; Hlaváčiková, S. Decomposition of Single-Use Products Made of Bioplastic under Real Conditions of Urban Composting Facility. *J. Ecol. Eng.* **2021**, *22*, 265–272. [CrossRef]
20. Tran, H.-T.; Lin, C.; Bui, X.-T.; Ngo, H.-H.; Cheruiyot, N.K.; Hoang, H.-G.; Vu, C.-T. Aerobic composting remediation of petroleum hydrocarbon-contaminated soil. Current and future perspectives. *Sci. Total Environ.* **2021**, *753*, 142250. [CrossRef]
21. Megías-Sayago, C.; Navarro-Jaén, S.; Drault, F.; Ivanova, S. Recent Advances in the Brønsted/Lewis Acid Catalyzed Conversion of Glucose to HMF and Lactic Acid: Pathways toward Bio-Based Plastics. *Catalysts* **2021**, *11*, 1395. [CrossRef]
22. de Freitas Batista, G.; Roman, F.F.; de Tuesta, J.L.D.; Mambrini, R.V.; Praça, P.; Gomes, H.T. Assessment of pretreatments for highly concentrated leachate waters to enhance the performance of catalytic wet peroxide oxidation with sustainable low-cost catalysts. *Catalysts* **2022**, *12*, 238. [CrossRef]
23. Diaz de Tuesta, J.L.; Pantuzza, G.F.; Silva, A.M.T.; Praça, P.; Faria, J.L.; Gomes, H.T. Catalysts prepared with matured compost derived from mechanical-biological treatment plants for the wet peroxide oxidation of pollutants with different lipophilicity. *Catalysts* **2020**, *10*, 1243. [CrossRef]
24. Zainudin, M.H.M.; Zulkarnain, A.; Azmi, A.S.; Muniandy, S.; Sakai, K.; Shirai, Y.; Hassan, M.A. Enhancement of agro-industrial waste composting process via the microbial inoculation: A brief review. *Agronomy* **2022**, *12*, 198. [CrossRef]
25. Sikorska, W.; Musiol, M.; Nowak, B.; Pajak, J.; Labuzek, S.; Kowalczuk, M.; Adamus, G. Degradability of polylactide and its blend with poly [(R, S)-3-hydroxybutyrate] in industrial composting and compost extract. *Int. Biodeterior. Biodegrad.* **2015**, *101*, 32–41. [CrossRef]
26. Bishop, G.; Styles, D.; Lens, P.N. Environmental performance comparison of bioplastics and petrochemical plastics: A review of life cycle assessment (LCA) methodological decisions. *Resour. Conserv. Recycl.* **2021**, *168*, 105451. [CrossRef]
27. Emadian, S.M.; Onay, T.T.; Demirel, B. Biodegradation of bioplastics in natural environments. *Waste Manag.* **2017**, *59*, 526–536. [CrossRef]
28. Lucas, N.; Bienaime, C.; Belloy, C.; Queneudec, M.; Silvestre, F.; Nava-Saucedo, J.-E. Polymer biodegradation: Mechanisms and estimation techniques—A review. *Chemosphere* **2008**, *73*, 429–442. [CrossRef]
29. Naser, A.Z.; Deiab, I.; Darras, B.M. Poly (lactic acid)(PLA) and polyhydroxyalkanoates (PHAs), green alternatives to petroleum-based plastics: A review. *RSC Adv.* **2021**, *11*, 17151–17196. [CrossRef]
30. Kaur, H.; Banipal, T.S.; Thakur, S.; Bakshi, M.S.; Kaur, G.; Singh, N. Novel biodegradable films with extraordinary tensile strength and flexibility provided by nanoparticles. *ACS Sustain. Chem. Eng.* **2013**, *1*, 127–136. [CrossRef]

31. Song, X.; Bian, Z.; Hui, Y.; Wang, H.; Liu, F.; Yu, S. Zn-Acetate-Containing ionic liquid as highly active catalyst for fast and mild methanolysis of Poly (lactic acid). *Polym. Degrad. Stab.* **2019**, *168*, 108937. [CrossRef]
32. Itävaara, M.; Karjomaa, S.; Selin, J.-F. Biodegradation of polylactide in aerobic and anaerobic thermophilic conditions. *Chemosphere* **2002**, *46*, 879–885. [CrossRef] [PubMed]
33. Devahasdin, S.; Fan Jr, C.; Li, K.; Chen, D.H. TiO_2 photocatalytic oxidation of nitric oxide: Transient behavior and reaction kinetics. *J. Photochem. Photobiol. A Chem.* **2003**, *156*, 161–170. [CrossRef]
34. Degli-Innocenti, F.; Tosin, M.; Bastioli, C. Evaluation of the biodegradation of starch and cellulose under controlled composting conditions. *J. Environ. Polym. Degrad.* **1998**, *6*, 197–202. [CrossRef]
35. Agarwal, S.; Singhal, S.; Godiya, C.B.; Kumar, S. Prospects and Applications of Starch based Biopolymers. *Int. J. Environ. Anal. Chem.* **2021**. [CrossRef]
36. Hernandez-Carmona, F.; Morales-Matos, Y.; Lambis-Miranda, H.; Pasqualino, J. Starch extraction potential from plantain peel wastes. *J. Environ. Chem. Eng.* **2017**, *5*, 4980–4985. [CrossRef]
37. Dang, K.M.; Yoksan, R. Development of thermoplastic starch blown film by incorporating plasticized chitosan. *Carbohydr. Polym.* **2015**, *115*, 575–581. [CrossRef]
38. Helanto, K.E.; Matikainen, L.; Talja, R.; Rojas, O.J. Bio-based polymers for sustainable packaging and biobarriers: A critical review. *BioResources* **2019**, *14*, 4902–4951.
39. Giordano, G. Making Packaging Pop: Film Packaging Gets Personal: Film packaging must now do double duty, protecting perishables and catching the attention of busy consumers. *Plast. Eng.* **2018**, *74*, 38–42. [CrossRef]
40. Jiménez-Rosado, M.; Zarate-Ramírez, L.; Romero, A.; Bengoechea, C.; Partal, P.; Guerrero, A. Bioplastics based on wheat gluten processed by extrusion. *J. Clean. Prod.* **2019**, *239*, 117994. [CrossRef]
41. Karamanlioglu, M.; Preziosi, R.; Robson, G.D. Abiotic and biotic environmental degradation of the bioplastic polymer poly (lactic acid): A review. *Polym. Degrad. Stab.* **2017**, *137*, 122–130. [CrossRef]
42. Geueke, B. *Dossier—Bioplastics as Food Contact Materials*; Food Packaging Forum: Zurich, Switzerland, 2014.
43. Kim, G.M.; Chang, W.-S.; Kim, Y.-K. Biocomposites Using Whole or Valuable Component-Extracted Microalgae Blended with Polymers: A Review. *Catalysts* **2021**, *12*, 25. [CrossRef]
44. Dietrich, K.; Dumont, M.-J.; Del Rio, L.F.; Orsat, V. Sustainable PHA production in integrated lignocellulose biorefineries. *New Biotechnol.* **2019**, *49*, 161–168. [CrossRef]
45. Mal, N.; Satpati, G.; Raghunathan, S.; Davoodbasha, M. Current strategies on algae-based biopolymer production and scale-up. *Chemosphere* **2022**, *289*, 133178. [CrossRef] [PubMed]
46. Peelman, N.; Ragaert, P.; De Meulenaer, B.; Adons, D.; Peeters, R.; Cardon, L.; Van Impe, F.; Devlieghere, F. Application of bioplastics for food packaging. *Trends Food Sci. Technol.* **2013**, *32*, 128–141. [CrossRef]
47. Asgher, M.; Arshad, S.; Qamar, S.A.; Khalid, N. Improved biosurfactant production from *Aspergillus niger* through chemical mutagenesis: Characterization and RSM optimization. *SN Appl. Sci.* **2020**, *2*, 1–11. [CrossRef]
48. Imam, S.; Gordon, S.; Shogren, R.; Tosteson, T.; Govind, N.; Greene, R. Degradation of starch–poly (β-hydroxybutyrate-co-β-hydroxyvalerate) bioplastic in tropical coastal waters. *Appl. Environ. Microbiol.* **1999**, *65*, 431–437. [CrossRef]
49. Nechyporchuk, O.; Belgacem, M.N.; Bras, J. Production of cellulose nanofibrils: A review of recent advances. *Ind. Crops Prod.* **2016**, *93*, 2–25. [CrossRef]
50. Tajeddin, B. Cellulose-based polymers for packaging applications. *Lignocellul. Polym. Compos.* **2014**, 477–498. [CrossRef]
51. Qasim, U.; Osman, A.I.; Al-Muhtaseb, A.a.H.; Farrell, C.; Al-Abri, M.; Ali, M.; Vo, D.-V.N.; Jamil, F.; Rooney, D.W. Renewable cellulosic nanocomposites for food packaging to avoid fossil fuel plastic pollution: A review. *Environ. Chem. Lett.* **2021**, *19*, 613–641. [CrossRef]
52. Okolie, J.A.; Nanda, S.; Dalai, A.K.; Kozinski, J.A. Chemistry and specialty industrial applications of lignocellulosic biomass. *Waste Biomass Valorization* **2021**, *12*, 2145–2169. [CrossRef]
53. Vroman, I.; Tighzert, L. Biodegradable polymers. *Materials* **2009**, *2*, 307–344. [CrossRef]
54. Gómez, E.F.; Michel, F.C., Jr. Biodegradability of conventional and bio-based plastics and natural fiber composites during composting, anaerobic digestion and long-term soil incubation. *Polym. Degrad. Stab.* **2013**, *98*, 2583–2591. [CrossRef]
55. Volova, T.; Boyandin, A.; Vasiliev, A.; Karpov, V.; Prudnikova, S.; Mishukova, O.; Boyarskikh, U.; Filipenko, M.; Rudnev, V.; Xuân, B.B. Biodegradation of polyhydroxyalkanoates (PHAs) in tropical coastal waters and identification of PHA-degrading bacteria. *Polym. Degrad. Stab.* **2010**, *95*, 2350–2359. [CrossRef]
56. Maran, J.P.; Sivakumar, V.; Thirugnanasambandham, K.; Sridhar, R. Degradation behavior of biocomposites based on cassava starch buried under indoor soil conditions. *Carbohydr. Polym.* **2014**, *101*, 20–28. [CrossRef] [PubMed]
57. Rudnik, E.; Briassoulis, D. Degradation behaviour of poly (lactic acid) films and fibres in soil under Mediterranean field conditions and laboratory simulations testing. *Ind. Crops Prod.* **2011**, *33*, 648–658. [CrossRef]
58. Kale, G.; Kijchavengkul, T.; Auras, R.; Rubino, M.; Selke, S.E.; Singh, S.P. Compostability of bioplastic packaging materials: An overview. *Macromol. Biosci.* **2007**, *7*, 255–277. [CrossRef]
59. Kawashima, N.; Yagi, T.; Kojima, K. How Do Bioplastics and Fossil-Based Plastics Play in a Circular Economy? *Macromol. Mater. Eng.* **2019**, *304*, 1900383. [CrossRef]
60. Bilo, F.; Pandini, S.; Sartore, L.; Depero, L.E.; Gargiulo, G.; Bonassi, A.; Federici, S.; Bontempi, E. A sustainable bioplastic obtained from rice straw. *J. Clean. Prod.* **2018**, *200*, 357–368. [CrossRef]

61. Hashimoto, K.; Sudo, M.; Ohta, K.; Sugimura, T.; Yamada, H.; Aoki, T. Biodegradation of nylon4 and its blend with nylon6. *J. Appl. Polym. Sci.* **2002**, *86*, 2307–2311. [CrossRef]
62. Kalita, N.K.; Hazarika, D.; Kalamdhad, A.; Katiyar, V. Biodegradation of biopolymeric composites and blends under different environmental conditions: Approach towards end-of-life panacea for crop sustainability. *Bioresour. Technol. Rep.* **2021**, *15*, 100705. [CrossRef]
63. Setti, L.; Francia, E.; Pulvirenti, A.; De Leo, R.; Martinelli, S.; Maistrello, L.; Macavei, L.I.; Montorsi, M.; Barbi, S.; Ronga, D. Bioplastic film from black soldier fly prepupae proteins used as mulch: Preliminary results. *Agronomy* **2020**, *10*, 933. [CrossRef]
64. Ong, S.Y.; Sudesh, K. Effects of polyhydroxyalkanoate degradation on soil microbial community. *Polym. Degrad. Stab.* **2016**, *131*, 9–19. [CrossRef]
65. Accinelli, C.; Saccà, M.L.; Mencarelli, M.; Vicari, A. Deterioration of bioplastic carrier bags in the environment and assessment of a new recycling alternative. *Chemosphere* **2012**, *89*, 136–143. [CrossRef]
66. Jangong, O.; Gareso, P.; Mutmainna, I.; Tahir, D. Fabrication and characterization starch/chitosan reinforced polypropylene as biodegradable. *J. Phys. Conf. Ser.* **2019**, *1341*, 082022. [CrossRef]
67. Tosin, M.; Weber, M.; Siotto, M.; Lott, C.; Degli Innocenti, F. Laboratory test methods to determine the degradation of plastics in marine environmental conditions. *Front. Microbiol.* **2012**, *3*, 225. [CrossRef]
68. Wu, C.S. Preparation, characterization, and biodegradability of renewable resource-based composites from recycled polylactide bioplastic and sisal fibers. *J. Appl. Polym. Sci.* **2012**, *123*, 347–355. [CrossRef]
69. Adhikari, D.; Mukai, M.; Kubota, K.; Kai, T.; Kaneko, N.; Araki, K.S.; Kubo, M. Degradation of bioplastics in soil and their degradation effects on environmental microorganisms. *J. Agric. Chem. Environ.* **2016**, *5*, 23. [CrossRef]
70. Greene, J. *Marine Biodegradation of PLA, PHA, and Bio-Additive Polyethylene Based on ASTM D7081*; ACADEMIA: San Francisco, CA, USA, 2012.
71. Arcos-Hernandez, M.V.; Laycock, B.; Pratt, S.; Donose, B.C.; Nikolić, M.A.; Luckman, P.; Werker, A.; Lant, P.A. Biodegradation in a soil environment of activated sludge derived polyhydroxyalkanoate (PHBV). *Polym. Degrad. Stab.* **2012**, *97*, 2301–2312. [CrossRef]
72. Pérez-Arauz, A.; Aguilar-Rabiela, A.; Vargas-Torres, A.; Rodríguez-Hernández, A.-I.; Chavarría-Hernández, N.; Vergara-Porras, B.; López-Cuellar, M. Production and characterization of biodegradable films of a novel polyhydroxyalkanoate (PHA) synthesized from peanut oil. *Food Packag. Shelf Life* **2019**, *20*, 100297. [CrossRef]
73. Bagheri, A.R.; Laforsch, C.; Greiner, A.; Agarwal, S. Fate of so-called biodegradable polymers in seawater and freshwater. *Glob. Chall.* **2017**, *1*, 1700048. [CrossRef]
74. Vaverková, M.D.; Adamcová, D. Biodegrability of bioplastic materials in a controlled composting environment. *J. Ecol. Eng.* **2015**, *16*. [CrossRef]
75. Mostafa, N.; Farag, A.A.; Abo-dief, H.M.; Tayeb, A.M. Production of biodegradable plastic from agricultural wastes. *Arab. J. Chem.* **2018**, *11*, 546–553. [CrossRef]
76. Volova, T.; Gladyshev, M.; Trusova, M.Y.; Zhila, N. Degradation of polyhydroxyalkanoates in eutrophic reservoir. *Polym. Degrad. Stab.* **2007**, *92*, 580–586. [CrossRef]
77. Wei, Y.; Wei, Z.; Zhang, F.; Li, X.; Tan, W.; Xi, B. Role of humic acid chemical structure derived from different biomass feedstocks on Fe (III) bioreduction activity: Implication for sustainable use of bioresources. *Catalysts* **2019**, *9*, 450. [CrossRef]
78. Giangeri, G.; Morlino, M.S.; De Bernardini, N.; Ji, M.; Bosaro, M.; Pirillo, V.; Antoniali, P.; Molla, G.; Raga, R.; Treu, L. Preliminary investigation of microorganisms potentially involved in microplastics degradation using an integrated metagenomic and biochemical approach. *Sci. Total Environ.* **2022**, *843*, 157017. [CrossRef]
79. Kim, D.-W.; Ahn, J.-H.; Cha, C.-J. Biodegradation of plastics: Mining of plastic-degrading microorganisms and enzymes using metagenomics approaches. *J. Microbiol.* **2022**, *60*, 969–976. [CrossRef]
80. *6400-04*; Standard Specification for Compostable Plastics. ASTM International: West Conshohocken, PA, USA, 2004.
81. Kolstad, J.J.; Vink, E.T.; De Wilde, B.; Debeer, L. Assessment of anaerobic degradation of Ingeo™ polylactides under accelerated landfill conditions. *Polym. Degrad. Stab.* **2012**, *97*, 1131–1141. [CrossRef]
82. Albertsson, A.-C.; Hakkarainen, M. Designed to degrade. *Science* **2017**, *358*, 872–873. [CrossRef]
83. Vink, E.T.; Rábago, K.R.; Glassner, D.A.; Springs, B.; O'Connor, R.P.; Kolstad, J.; Gruber, P.R. The sustainability of NatureWorks™ polylactide polymers and Ingeo™ polylactide fibers: An update of the future. *Macromol. Biosci.* **2004**, *4*, 551–564. [CrossRef]
84. Tokiwa, Y.; Calabia, B.P. Biodegradability and biodegradation of polyesters. *J. Polym. Environ.* **2007**, *15*, 259–267. [CrossRef]
85. Castro-Aguirre, E.; Iniguez-Franco, F.; Samsudin, H.; Fang, X.; Auras, R. Poly (lactic acid)—Mass production, processing, industrial applications, and end of life. *Adv. Drug Deliv. Rev.* **2016**, *107*, 333–366. [CrossRef]
86. Hajighasemi, M.; Nocek, B.P.; Tchigvintsev, A.; Brown, G.; Flick, R.; Xu, X.; Cui, H.; Hai, T.; Joachimiak, A.; Golyshin, P.N. Biochemical and structural insights into enzymatic depolymerization of polylactic acid and other polyesters by microbial carboxylesterases. *Biomacromolecules* **2016**, *17*, 2027–2039. [CrossRef] [PubMed]
87. Kawai, F. Polylactic acid (PLA)-degrading microorganisms and PLA depolymerases. In *Green Polymer Chemistry: Biocatalysis and Biomaterials*; ACS Publications: Washington, DC, USA, 2010; pp. 405–414.
88. Bandini, F.; Misci, C.; Taskin, E.; Cocconcelli, P.S.; Puglisi, E. Biopolymers modulate microbial communities in municipal organic waste digestion. *FEMS Microbiol. Ecol.* **2020**, *96*, fiaa183. [CrossRef] [PubMed]
89. Mohanan, N.; Montazer, Z.; Sharma, P.K.; Levin, D.B. Microbial and enzymatic degradation of synthetic plastics. *Front. Microbiol.* **2020**, *11*, 580709. [CrossRef] [PubMed]

90. Ghorpade, V.M.; Gennadios, A.; Hanna, M.A. Laboratory composting of extruded poly (lactic acid) sheets. *Bioresour. Technol.* **2001**, *76*, 57–61. [CrossRef] [PubMed]
91. Chidambarampadmavathy, K.; Karthikeyan, O.P.; Heimann, K. Sustainable bio-plastic production through landfill methane recycling. *Renew. Sustain. Energy Rev.* **2017**, *71*, 555–562. [CrossRef]
92. Ho, K.-L.G.; Pometto, A.L.; Hinz, P.N. Effects of temperature and relative humidity on polylactic acid plastic degradation. *J. Environ. Polym. Degrad.* **1999**, *7*, 83–92. [CrossRef]
93. Nandakumar, A.; Chuah, J.-A.; Sudesh, K. Bioplastics: A boon or bane? *Renew. Sustain. Energy Rev.* **2021**, *147*, 111237. [CrossRef]
94. Gioia, C.; Giacobazzi, G.; Vannini, M.; Totaro, G.; Sisti, L.; Colonna, M.; Marchese, P.; Celli, A. End of life of biodegradable plastics: Composting versus Re/upcycling. *ChemSusChem* **2021**, *14*, 4167–4175. [CrossRef]
95. Mergaert, J.; Anderson, C.; Wouters, A.; Swings, J.; Kersters, K. Biodegradation of polyhydroxyalkanoates. *FEMS Microbiol. Rev.* **1992**, *9*, 317–321. [CrossRef]
96. Volova, T.; Boyandin, A.; Vasil'ev, A.; Karpov, V.; Kozhevnikov, I.; Prudnikova, S.; Rudnev, V.; Xuân, B.B.; Dũng, V.V.; Gitel'Zon, I. Biodegradation of polyhydroxyalkanoates (PHAs) in the South China Sea and identification of PHA-degrading bacteria. *Microbiology* **2011**, *80*, 252–260. [CrossRef]
97. Gunning, M.A.; Geever, L.M.; Killion, J.A.; Lyons, J.G.; Higginbotham, C.L. Mechanical and biodegradation performance of short natural fibre polyhydroxybutyrate composites. *Polym. Test.* **2013**, *32*, 1603–1611. [CrossRef]
98. Wang, S.; Lydon, K.A.; White, E.M.; Grubbs III, J.B.; Lipp, E.K.; Locklin, J.; Jambeck, J.R. Biodegradation of poly (3-hydroxybutyrate-co-3-hydroxyhexanoate) plastic under anaerobic sludge and aerobic seawater conditions: Gas evolution and microbial diversity. *Environ. Sci. Technol.* **2018**, *52*, 5700–5709. [CrossRef] [PubMed]
99. Nafchi, A.M.; Moradpour, M.; Saeidi, M.A.A.K. *Thermoplsatic Starches: Properties, Challenges and Prospects*; Islamic Azad University: Tehran, Iran; Food Biopolymer Research Group, University Sains Malaysia: George Town, Malaysia, 2013.
100. Tang, X.; Alavi, S. Recent advances in starch, polyvinyl alcohol based polymer blends, nanocomposites and their biodegradability. *Carbohydr. Polym.* **2011**, *85*, 7–16. [CrossRef]
101. Torres, F.; Troncoso, O.; Torres, C.; Díaz, D.; Amaya, E. Biodegradability and mechanical properties of starch films from Andean crops. *Int. J. Biol. Macromol.* **2011**, *48*, 603–606. [CrossRef]
102. Czaja-Jagielska, N.; Melski, K. Biodegradation of starch-based films in conditions of nonindustrial composting. *Pol. J. Environ. Stud.* **2013**, *22*, 1039.
103. Hillis, D.M.; Heller, H.C.; Hacker, S.D.; Hall, D.W.; Laskowski, M.J.; Sadava, D.E. *Life: The Science of Biology*; Macmillan Higher Education: New York, NY, USA, 2020.
104. Vikman, M.; Vartiainen, J.; Tsitko, I.; Korhonen, P. Biodegradability and compostability of nanofibrillar cellulose-based products. *J. Polym. Environ.* **2015**, *23*, 206–215. [CrossRef]
105. Amin, M.R.; Chowdhury, M.A.; Kowser, M.A. Characterization and performance analysis of composite bioplastics synthesized using titanium dioxide nanoparticles with corn starch. *Heliyon* **2019**, *5*, e02009. [CrossRef]
106. Goto, T.; Kishita, M.; Sun, Y.; Sako, T.; Okajima, I. Degradation of polylactic acid using sub-critical water for compost. *Polymers* **2020**, *12*, 2434. [CrossRef]
107. Javierre, C.; Sarasa, J.; Claveria, I.; Fernandez, A. Study of the biodisintegration on a painted bioplastic material waste. *Mater. Plast.* **2015**, *52*, 116–121.
108. Cafiero, L.M.; Canditelli, M.; Musmeci, F.; Sagnotti, G.; Tuffi, R. Assessment of disintegration of compostable bioplastic bags by management of electromechanical and static home composters. *Sustainability* **2020**, *13*, 263. [CrossRef]
109. Mihai, M.; Legros, N.; Alemdar, A. Formulation-properties versatility of wood fiber biocomposites based on polylactide and polylactide/thermoplastic starch blends. *Polym. Eng. Sci.* **2014**, *54*, 1325–1340. [CrossRef]
110. Balaguer, M.P.; Aliaga, C.; Fito, C.; Hortal, M. Compostability assessment of nano-reinforced poly (lactic acid) films. *Waste Manag.* **2016**, *48*, 143–155. [CrossRef] [PubMed]
111. Kale, G.; Auras, R.; Singh, S.P.; Narayan, R. Biodegradability of polylactide bottles in real and simulated composting conditions. *Polym. Test.* **2007**, *26*, 1049–1061. [CrossRef]
112. Tabasi, R.Y.; Ajji, A. Selective degradation of biodegradable blends in simulated laboratory composting. *Polym. Degrad. Stab.* **2015**, *120*, 435–442. [CrossRef]
113. Sun, Y.; Ren, X.; Rene, E.R.; Wang, Z.; Zhou, L.; Zhang, Z.; Wang, Q. The degradation performance of different microplastics and their effect on microbial community during composting process. *Bioresour. Technol.* **2021**, *332*, 125133. [CrossRef]

catalysts

MDPI

Article

7α and 7β Hydroxylation of Dehydroepiandrosterone by *Gibberella sp.* and *Absidia Coerulea* Biotransformation

Ming Song [1], Ruicheng Fu [2], Sulan Cai [1], Xuliang Jiang [3], Fuju Wang [4], Weizhuo Xu [2,*] and Wei Xu [2,*]

1 School of Functional Food and Wine, Shenyang Pharmaceutical University, Shenyang 110016, China
2 School of Life Sciences and Biopharmaceuticals, Shenyang Pharmaceutical University, Shenyang 110016, China
3 School of Pharmaceutical Engineering, Shenyang Pharmaceutical University, Shenyang 110016, China
4 Beijing Global Biotechnologies, Co., Ltd., Beijing 100193, China
* Correspondence: weizhuo.xu@syphu.edu.cn (W.X.); shxuwei8720@163.com (W.X.); Tel./Fax: +86-024-43520301 (W.X.); +86-024-43520307 (W.X.)

Abstract: The hydroxylation of dehydroepiandrosterone (DHEA) to 7α -hydroxy-5-androstene-17-one (7α-OH-DHEA) and 7β-hydroxy-5-androstene-17-one (7β-OH-DHEA) by *Gibberella sp.* CICC 2498 and *Absidia coerulea* CICC 41050 was investigated. The media ingredients were optimized. Single factors such as the DHEA concentration, culture time, medium volume, and inoculum rate were individually investigated to generate optimum biotransformation conditions. An orthogonal optimization process using a four-factor, three- level L_9 (3^3) experiment was designed and performed. Finally, the maximum production of 7β-OH-DHEA from DHEA biotransformation by *Absidia coerulea* is 69.61%. This strategy would provide a possible way to enhance the 7β-OH-DHEA yield in the pharmaceutical industry.

Keywords: DHEA; biotransformation; C7-hydroxylation; *Gibberella sp.*; *Absidia coerulea*

Citation: Song, M.; Fu, R.; Cai, S.; Jiang, X.; Wang, F.; Xu, W.; Xu, W. 7α and 7β Hydroxylation of Dehydroepiandrosterone by *Gibberella sp.* and *Absidia Coerulea* Biotransformation. *Catalysts* **2023**, *13*, 272. https://doi.org/10.3390/catal13020272

Academic Editors: Zhilong Wang and Tao Pan

Received: 13 December 2022
Revised: 19 January 2023
Accepted: 20 January 2023
Published: 25 January 2023

1. Introduction

Dehydroepiandrosterone (DHEA) is a major C19 steroid hormone produced by the adrenal cortex. Meanwhile, it is also produced in small quantities in the gonads and brain [1]. Due to its long half-life in plasma, most DHEA would become dehydroepiandrosterone sulfate ester (DHEAS), reserved, and converted into specific hormones when needed [2]. As a kind of important pharmaceutical steroid, hydroxylation at different positions would exhibit diversified biological activities. For example, hydroxylation at position 9α/16α is crucial for the bioactivities of glucocorticoids (dexamethasone, triamcinolone, etc.) [3,4]. Hydroxylation at position 11α is essential for anti-inflammatory activities (hydrocortisone, prednisolone) [5,6]. Hydroxylation at position 14α is vital for the production of the 21-acetoxy analog of proligestone, which is a prodrug of Promegestone [7]. Hydroxylation at position 15α is a key intermediate for the production of progesterone [8].

Previous biotransformation investigation towards DHEA-analog steroids had demonstrated a varieties of metabolites spectrum. Huang et.al reported that 15α-hydroxy-17 a-oxa-D-homo-androst-4-ene-3,17-dione and androst-4-en-3,17-dione, were produced by *Penicillium griseopurpureum* [9]. Kołek et al. used androstenediol as a substrate and produced di- and trihydroxylation products such as 3β,17β-Dihydroxyandrost-5-en-7-one, 3β,7α,17β-trihydroxyandrost-5-ene, 3β,7β,17β-Trihydroxyandrost-5-ene [10,11].

When 7-oxo-DHEA was used as substrate, 3β,16β-dihydroxy-androst-5-en-7,17-dione, 3β-hydroxy-17α-oxa-D-homo-androst-5-en-7,17-dione, and 3β-acetoxy-androst-5-en-7,17-dione could be produced by *Laetiporus sulphureus* AM498, *Fusicoccum amygdali* AM258 and *Spicaria divaricata* AM423 [12]. Incubation of DHEA with *Ulocladium chartarum* MRC 72584 produced seven DHEA derivatives, such as 3β-hydroxyandrost-5-en-7,17-dione, 3β,7β-dihydroxyandrost-5-en-17-one, 3β,7α-dihydroxyandrost-5-en-17-one, etc. [13]. A

7α,15α-dihydroxyl-DHEA product was reported by Li et al. using *Colletotrichum lini* [14]. Microbial transformation by using *Mortierella isabellina* AM212 produced 7-Oxo-DHEA, 7α-Hydroxy-DHEA, 7β-Hydroxy-DHEA [10], and *Backusella lamprospora* VKM F- 944 could transform DHEA into 7α-hydroxy-DHEA [15]. These single, double, and triple hydroxylation reactions greatly enriched the DHEA metabolite ingredients.

In this research, two out of twelve filamentous strains of fungi demonstrated their metabolic abilities for DHEA (Table 1, Figure S1). After the culture, extraction, and isolation, it could be identified that 7α-Hydroxy-DHEA was produced by *Gibberella sp.* CICC 2498 and 7β-Hydroxy-DHEA was produced by *Absidia coerulea* CICC 41050 (Figure 1). Previous literature had reported several optimizations for the production of 7α-hydroxy-DHEA. As far as we know, it is the first time that the 7β-Hydroxy-DHEA is obtained by *Absidia coerulea* CICC 41050 biotransformation. Thus, we focused on the optimization of the 7β-Hydroxy-DHEA in this study, and the optimized transformation rate of 7β-Hydroxy-DHEA is 69.61%.

Table 1. Ability of DHEA biotransformation by the tested microorganisms.

Microorganism	Ability *	Microorganism	Ability *
Absidia coerulea CICC 41050	+	*Cunninghamella elegans* 3.910	−
Aspergillus fumigatus	−	*Gibberella fujikuroi* CICC 40272	−
Aspergillus niger	−	*Gibberella sp.* CICC 2498	+
Caldariomyces fumago CGMCC 16373	−	*Paecilomyces lilacinus*	−
Cunninghamella blakesleeana 3.970	−	*Penicillium*	−
Cunninghamella echinulata 3.967	−	*Trichoderma virens* CICC 2535	−

* Ability of DHEA biotransformation: (+) able, (−) not able.

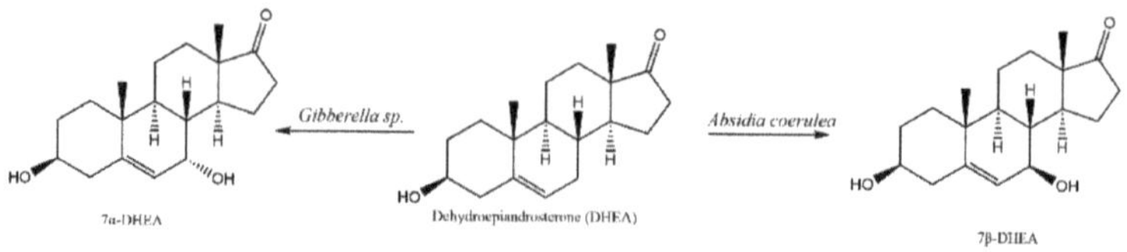

Figure 1. 7α- and 7β-hydroxylated DHEA obtained by *Gibberella sp.* CICC 2498 and *Absidia coerulea* CICC 41050.

2. Results and Discussion

2.1. Whole-Cell Biotransformation Results of DHEA

Thin layer chromatography (TLC) was used to identify whether *Gibberella sp.* CICC 2498 and *Absidia coerulea* CICC 41050 could transform dehydroepiandrosterone (DHEA). Figure 2 showed that the substrate control group (Group1, DHEA) has an ocher band, and the transformation groups (Group 4 and 6) have blue bands of different shades with good separation between bands, and almost no ocher substrate bands can be seen. The blue bands (products) are below the red band (DHEA), indicating that the product polarity is greater than DHEA. In addition to the major metabolite, some other products were also generated but could not be further identified due to their lower concentration.

2.2. HPLC Analysis of DHEA Transformed by Gibberella sp. CICC 2498 and Absidia Coerulea CICC 41050

Figure 3A showed the transformation results of DHEA by *Absidia coerulea* CICC 41050. It can be seen from the comparison between the transformation group 4, and the substrate control group 3 that the substrate (DHEA) peak in the transformation group was significantly reduced. This indicated that DHEA was transformed by *Absidia coerulea* CICC 41050. By comparing the results of transformation group 4, strain control group 1, and cosolvent control group 3, the increased peak in transformation group 4 was most likely the metabolite of DHEA transformed by *Absidia coerulea* CICC 41050, rather than the substance produced by microbial growth and metabolism. The types of metabolites of

DHEA transformed by *Absidia coerulea* CICC 41050 were few, and the content of metabolite I was significant, which was conducive to the later separation and purification. The retention time of metabolite I was 8.588 min (Figure 3A). In brief, incubation of *Absidia coerulea* CICC 41050 with DHEA (1 g/L) resulted in selective accumulation of the metabolite I.

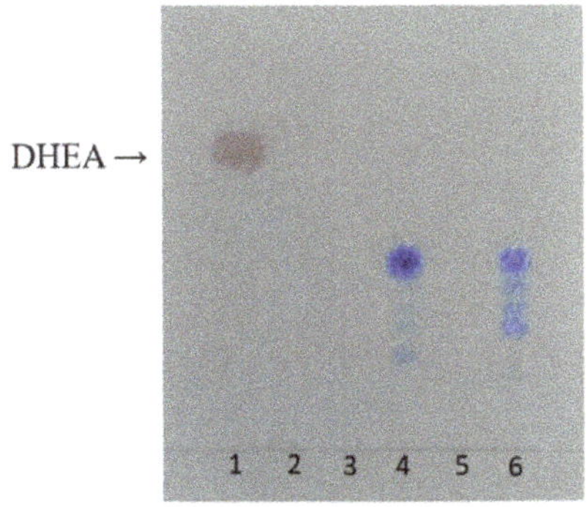

Figure 2. TLC analysis of microbial transformation of DHEA. Group 1, substrate (DHEA); Group 2, co-solvent (acetone); Group 3, *Absidia coerulea* CICC 41050; Group 4, *Absidia coerulea* CICC 41050 + DHEA; Group 5, *Gibberella sp.* CICC 2498; Group 6, *Gibberella sp.* CICC 2498 + DHEA. Colored with a 10% sulfuric acid-ethanol.

Figure 3B showed that the transformation of DHEA by *Gibberella sp.* CICC 2498. The separation of the metabolites of DHEA transformed by *Gibberella sp.* CICC 2498 was good, which was conducive to the later separation and purification. The main metabolite is named metabolite II, and the retention time of metabolite II is 11.211 min (Figure 3B).

2.3. Isolation, Purification and Structural Identification of Metabolites

The fermentation broth was further purified by semi-preparative HPLC. Figure 4 showed that the retention times of metabolites I and II were 15.198 min and 17.698 min, respectively. Metabolite I and II obtained by semi-preparative HPLC were confirmed by HPLC (retention times were 8.588 min and 11.211 min, respectively). Finally, metabolites I and II can be isolated from fermentation broth with purities of 94.0% and 96.0%, respectively.

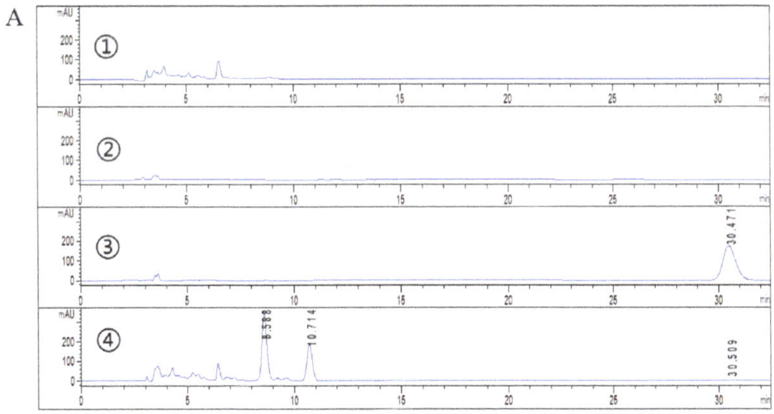

Figure 3. *Cont.*

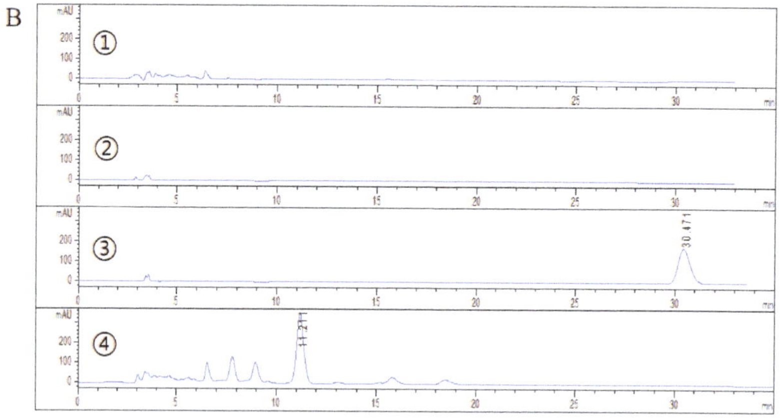

Figure 3. HPLC analysis of biotransformation of DHEA by *Absidia coerulea* CICC 41050 (**A**), and *Gibberella sp.* CICC 2498 (**B**). Group 1, strain; Group 2, cosolvent (acetone); Group 3, substrate (DHEA); Group 4, strain + substrate (DHEA).

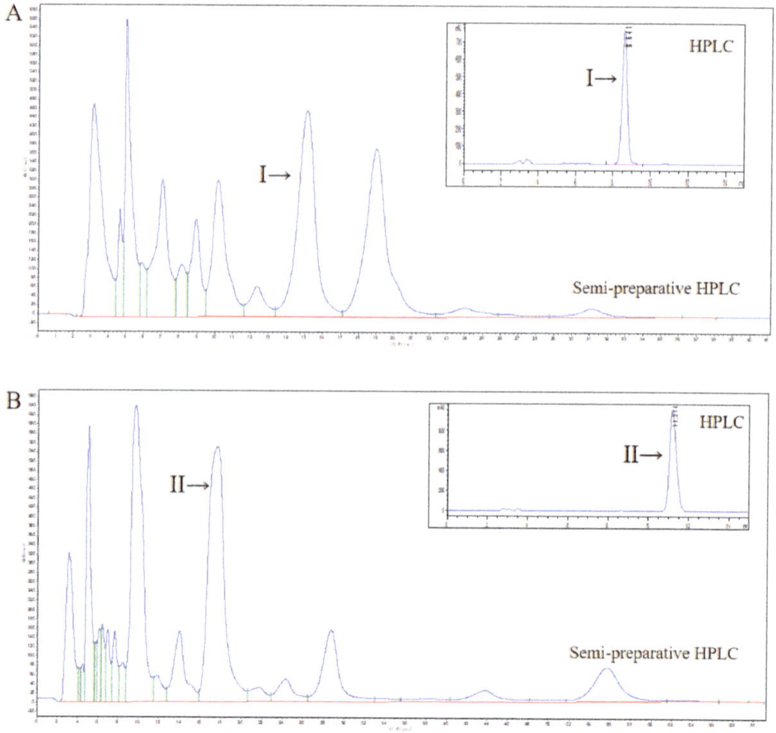

Figure 4. Semi-preparative HPLC separation diagram. (**A**), the semi-preparative separation of metabolite I from fermentation broth after incubation of *Absidia coerulea* CICC 41050 with DHEA (5 days, 1 g/L), metabolite I was identified by HPLC. (**B**), the semi-preparative separation of metabolite II from fermentation broth after incubation of *Gibberella sp.* CICC 2498 with DHEA (5 days, 1 g/L), metabolite II was identified by HPLC.

According to MS (ESI) m/z $[M+H]^+$ 305.1 and $[M-H]^+$ 303.2, the relative molecular weight of metabolite I is calculated to be 304. Compared with the relative molecular weight of 288 of the substrate DHEA ($C_{19}H_{28}O_2$), an oxygen atom is added, and the molecular formula is $C_{19}H_{28}O_3$ (Figures S2 and S3).

According to MS (ESI) m/z $[M+H]^+$ 305.1 and $[M-H]^+$ 303.2, the relative molecular weight of metabolite II is calculated to be 304. Compared with the relative molecular weight of 288 of the substrate DHEA ($C_{19}H_{28}O_2$), an oxygen atom is added, and the molecular formula is $C_{19}H_{28}O_3$ (Figures S4 and S5).

The position of the introduced hydroxyl group is analyzed according to the ^{13}C NMR and ^{1}H NMR spectral data (Figures S6–S9).

Metabolite I: 7β-Hydroxy-DHEA: ^{1}H-NMR (600 MHz, CDCl3) δH: 0.90 (3 H, s, 18-Me); 1.08 (3 H, s, 19-Me); 1.23–1.28 (1 H, m), 1.31 (1 H, d, J = 6.7 Hz, 6-H), 1.42–1.62 (7 H, m), 1.67–1.72 (1 H, m), 1.82–1.89 (4 H, m), 2.08–2.14 (1 H, m), 2.22–2.29 (2 H, m), 2.33–2.37 (1 H, m), 2.45–2.49 (1 H, m), 3.52–3.59 (1 H, m, 3α-H); 3.96 (1 H, dt, J_1 = 5.3 Hz, J_2 = 14.8 Hz, 7α-H); 5.31 (1 H, t, J = 1.8 Hz, 6-H). ^{13}C-NMR (151 MHz, CDCl3) δ_C: 13.5 (18-C), 19.1 (19-C), 20.3 (11-C), 24.1 (15-C), 31.2 (2-C), 31.4 (12-C), 35.9 (16-C), 36.6 (10-C), 36.8 (1-C), 40.4 (8-C), 41.6 (4-C), 47.7 (13-C), 48.2 (9-C), 51.1 (14-C), 71.2 (3-C), 72.8 (7-C), 125.4 (6-C), 143.7 (5-C), 221.1 (17-C).

Metabolite II: 7α-Hydroxy-DHEA: ^{1}H-NMR (600 MHz, CDCl3) δH: 0.89 (3 H, s, 18-Me), 1.02 (3 H, s, 19-Me), 1.10–1.15 (1 H, m), 1.26–1.31 (2 H, m), 1.50–1.61 (3 H, m), 1.66–1.73 (4 H, m), 1.78–1.91 (4 H, m), 2.09–2.19 (2 H, m), 2.28–2.33 (1 H, m), 2.35–2.39 (1 H, m), 2.45–2.50 (1 H, m), 3.55–3.61 (1 H, m, 3α-H), 3.98 (1 H, t, J = 4.7 Hz, 7β-H), 5.65 (1 H, d, J = 7.4 Hz, 6-H). ^{13}C-NMR (151 MHz, CDCl3) δ_C: 13.4 (18-C), 18.4 (19-C), 20.2 (11-C), 22.1 (15-C), 31.2 (2-C), 31.4 (12-C), 35.9 (16-C), 37.1 (10-C), 37.3 (1-C), 37.7 (8-C), 42.1 (4-C), 42.8 (13-C), 45.1 (9-C), 47.2 (14-C), 64.4 (7-C), 71.3 (3-C), 123.7 (6-C), 146.7 (5-C), 221.3 (17-C).The ^{1}H and ^{13}C NMR data of metabolites I and II are in agreement with those reported in the literature [10], which indicated that both 7 α/β hydroxylation DHEA were obtained (Figure 1) [10].

2.4. Optimization of 7β-OH-DHEA Production by Absidia Coerulea CICC 41050

2.4.1. Influence of Different Cosolvents

Figure 5A showed that, compared with the control group, using ethyl acetate, acetone, and ethanol as cosolvents can increase the transformation rate; when DMSO and chloroform were used as cosolvents, the transformation rate decreased. Therefore, acetone was selected as the best cosolvent. It can be seen from the results in Figure 5B that when the acetone concentration was 2%, the transformation rate was the highest. If the concentration of acetone was too low, the substrate could not be completely dissolved. However, with the increase in acetone concentration, acetone will have a toxic effect on fungi, inhibiting their growth and the activity of hydroxylase, thus affecting the conversion rate of substrate. Therefore, 2% acetone was selected as the cosolvent for the subsequent experiment.

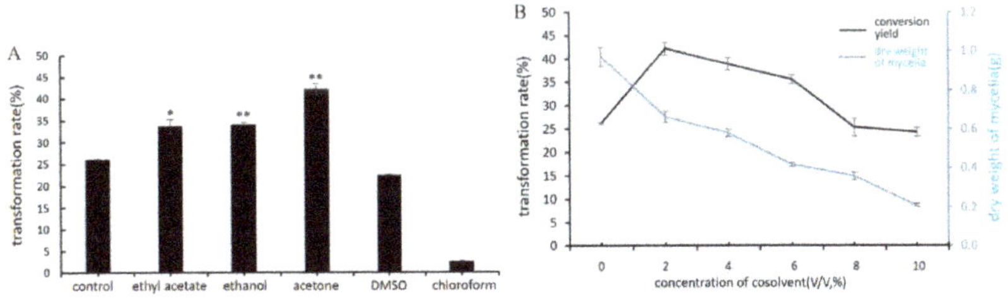

Figure 5. The effect of cosolvent type (**A**) and concentration (**B**) on transformation rate. (**A**), volume of cosolvent 1 mL (2%, V:V), concentration of DHEA 1 g/L, pH 6.5, 28 °C, 220 r/min. * $p < 0.05$, ** $p < 0.01$. (**B**), concentration of DHEA 1 g/L.

2.4.2. Effect of Key Nutrient Components and pH

It can be seen from the results in Figure 6A that the type of carbon source has a great impact on the transformation rate of 7β-OH-DHEA. When sucrose is the carbon source, the transformation rate is the highest, which is consistent with the type of carbon source in the transformation medium (Section 3.2). The production of 7β-OH-DHEA by *Absidia coerulea* CICC 41050 also depends on the concentration of sucrose. The concentration of sucrose (40 g/L) provides the highest yield of 7β-OH-DHEA, which is higher than that of 30 g/L in the original transformation medium (Section 3.2) (Figure 6B).

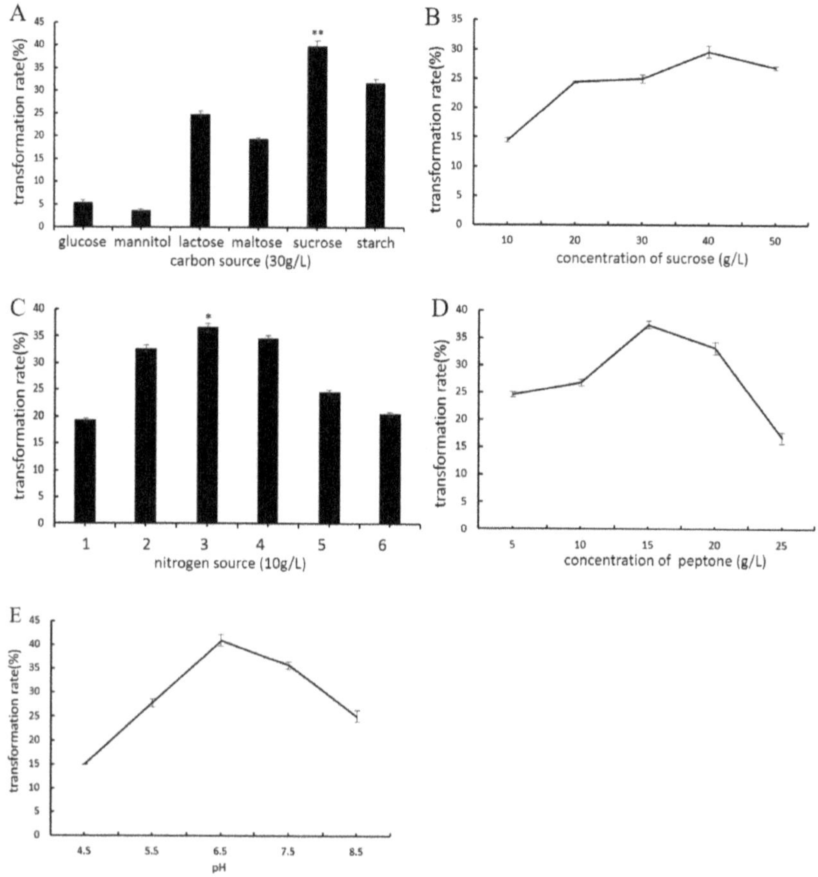

Figure 6. The effect of carbon source (**A**,**B**), nitrogen source (**C**,**D**) and initial pH (E) on transformation rate. (**A**,**B**), DHEA 1 g/L, acetone 2%, yeast extract 10 g/L, pH = 6.5, (**A**), 48 h (**B**), 24 h. ** $p < 0.01$. (**C**,**D**), 1: $NH_4\ NO_3$, 2: $(NH_4)_2\ SO_4$, 3: peptone, 4: yeast extract, 5: yeast extract powder, 6: beef extract, DHEA 1 g/L, acetone 2%, sucrose 30 g/L, pH = 6.5, (**C**), 48 h (**D**), 24 h. * $p < 0.05$. (**E**), DHEA 1 g/L, acetone 2%, yeast extract 10 g/L, sucrose 30 g/L.

The Impact of the alternative to yeast extract, nitrogen sources ($NH_4\ NO_3$, $(NH_4)_2\ SO_4$, peptone, yeast extract powder, and beef extract) on the improvement of 7β-hydroxylation catalyzed by *Absidia coerulea* CICC 41050 was investigated (Figure 6C). Replacement of yeast extract with peptone provided an 8–10% higher 7β-OH-DHEA yield (up to 37%). The results showed that peptone as a nitrogen source was superior to the original nitrogen source (yeast extract) of the transformation medium (Section 3.2). The effect of peptone at various concentrations on the transformation of DHEA by *Absidia coerulea* CICC 41050 was evaluated. The highest 7β-hydroxylase activity towards DHEA was reached at peptone content (15 g/L) (Figure 6D).

When studying the influence of pH of the transformation medium on DHEA conversion, it was shown that pH 6.5 provided the highest yield of 7β-OH-DHEA, while higher acidic or alkaline pH negatively affected the transformation rate.

According to the results of the single factor experiment, a three-factor and three-level orthogonal experiment were designed to explore the best medium composition (Table 2). The orthogonal experiment results are shown in Table 3.

Table 2. Orthogonal experiment factors and levels assignment for medium composition.

Level \ Factor	A/Sucrose (g/L)	B/Peptone (g/L)	C/Initial pH
1	30	10	5.5
2	40	15	6.5
3	50	20	7.5

Table 3. Orthogonal experimental design and results for medium composition.

No. \ Factor	A	B	C	Transformation Rate (%)
1	1	1	1	41.75
2	1	2	2	40.48
3	1	3	3	37.92
4	2	1	2	39.80
5	2	2	3	32.02
6	2	3	1	28.75
7	3	1	3	23.58
8	3	2	1	39.29
9	3	3	2	46.57
K1	40.05	35.04	36.60	
K2	33.52	37.26	42.28	
K3	36.48	37.75	31.17	
R	6.53	2.70	11.11	

Table 3 showed that the order of influence of the three factors on the transformation rate is C > A > B. Through range analysis, the optimal combination of the three factors is C_2 A_1 B_3: initial pH 6.5, sucrose 30 g/L, and peptone 20 g/L. However, the composition-optimized medium is not in Table 3, and verification experiments are required. Three parallel experiments were carried out. It was defined that an initial pH 6.5, sucrose 30 g/L, and peptone 20 g/L provided the maximum production (50.48%) of 7β-OH-DHEA by *Absidia coerulea* CICC 41050.

2.4.3. Effect of Biotransformation Conditions

Figure 7A shows that when the inoculum is less than 12%, the transformation rate increases with the increase of the inoculum, and when the inoculum is more than 12%, the transformation rate decreases. The production of 7β-OH-DHEA by *Absidia coerulea* CICC 41050 also depended on medium volume (Figure 7B). The transformation rate is highest when the medium volume is 60 mL in a 250 mL Erlenmeyer flask. When the volume of the medium is too large, the ventilation and dissolved oxygen in the medium are poor. The transformation rate reached its highest when the substrate was added for 48 h and became extremely low after 96 h (Figure 7C). It is speculated that the nutrients in the medium were consumed and the enzyme activity decreased. The effect of different concentrations of substrate (DHEA, 0.5–8.0 g/L) in the transformation medium was estimated. Figure 7D shows that 1 g/L DHEA can provide the highest transformation rate of 7β-OH-DHEA, while at more than 1 g/L, the transformation rate is declining; greater than 6 g/L, the transformation rate is very low, and the substrate is almost completely converted. The reason may be that the concentration of cosolvent increases with the increase in substrate concentration, and the toxicity of cosolvent inhibits the growth of fungi, thus negatively affecting DHEA conversion.

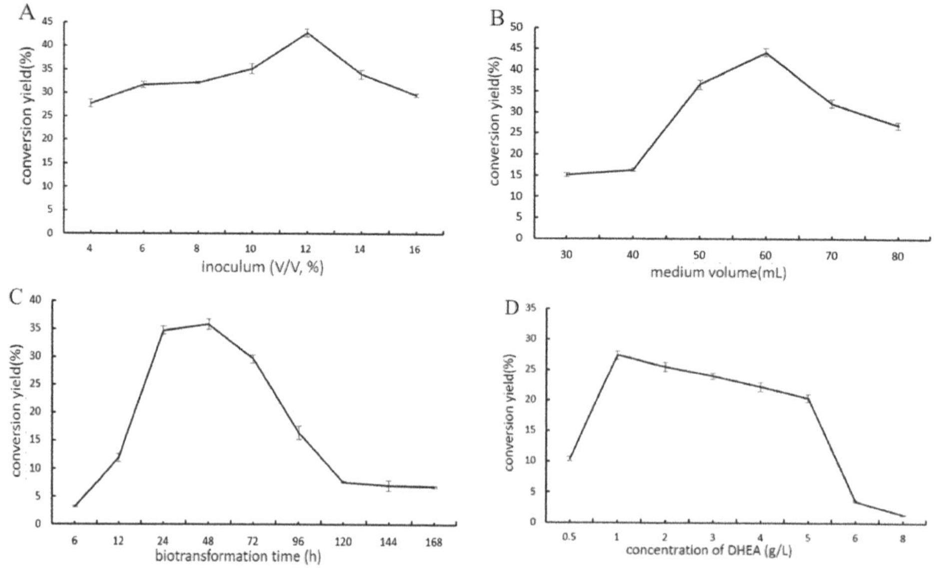

Figure 7. The effect of biotransformation conditions. (**A–D**), the composition-optimized medium: initial pH 6.5, sucrose 30 g/L, and peptone 20 g/L.

According to the results of the single-factor experiment, an orthogonal experiment with four factors and three levels was designed to explore the optimal biotransformation conditions (Table 4). The orthogonal experiment results are shown in Table 5.

Table 4. Orthogonal experiment factors and levels assignment for biotransformation conditions.

Factor / Level	A/DHEA (g/L)	B/Time (h)	C/Medium Volume (mL)	D/Inoculum Rate (V/V, %)
1	1	24	50	10
2	2	48	60	12
3	3	72	70	14

It can be seen from the results in Table 5 that the order of influence of the four factors in the orthogonal experiment of biotransformation conditions on the transformation rate is A > C > B > D. Through range analysis, the optimal combination of the four factors is A_1 C_2 B_2 D_1: DHEA 1 g/L, medium volume 60 mL, biotransformation time 48 h, and inoculum 10%. As the optimal biotransformation conditions are not listed in Table 5, validation tests are required.

Three parallel experiments were carried out. It was defined that sucrose 30 g/L, peptone 20 g/L, corn steep liquor 10 g/L, K_2 HPO_4 2 g/L, KH_2 PO_4 1.6 g/L, $MgSO_4$ 0.5 g/L, $FeSO_4$ 0.05 g/L, pH 6.5, DHEA 1 g/L, medium volume 60 mL, biotransformation time 48 h, and inoculum 10% provided maximum production (69.61%) of 7β-OH-DHEA by *Absidia coerulea* CICC 41050. Compared with the highest transformation rate of 62.81% in the orthogonal test and the primary transformation rate of 27.23%, transformation rate was increased by 6.80% and 42.38% respectively.

Table 5. Orthogonal experimental design and results for biotransformation conditions.

No. \ Factor	A	B	C	D	Transformation Rate (%)
1	1	1	1	1	53.62
2	1	2	2	2	61.28
3	1	3	3	3	62.81
4	2	1	2	3	38.94
5	2	2	3	1	47.61
6	2	3	1	2	20.07
7	3	1	3	2	30.03
8	3	2	1	3	28.10
9	3	3	2	1	42.73
K1	59.24	40.86	33.93	47.99	
K2	35.54	45.98	47.65	37.13	
K3	33.62	44.10	46.82	45.56	
R	19.98	4.01	16.62	3.31	

3. Materials and Methods

3.1. Chemicals

Dehydroepiandrosterone (DHEA) was obtained from Hubei Gongtong Pharmaceutical Co., Ltd. (Xiangyang city, Hubei, China). Methanol and acetonitrile were purchased from Concord Technology Co., Ltd. (Tianjin, Tianjin, China). Yeast extract was purchased from HopeBio Co., Ltd. (Qingdao, Shandong, China). All other chemical reagents were purchased from Yuwang Chemical Co., Ltd. (Shenyang, Liaoning, China).

3.2. Microorganism and Cultivation

Absidia coerulea 41050 and *Gibberella sp.* 2498 were purchased from the China Center of Industrial Culture Collection (CICC).

Potato dextrose agar (PDA) is composed of potatoes (200 g), glucose (20 g), agar (20 g), and 1000 mL distilled water. Seed culture media (g/L) are composed of potato starch (45 g), yeast extract (3 g), corn steep liquor (10 g), $CaCO_3$ (3 g), $MgSO_4$ (0.5 g), and FeSO4 (0.05 g). Transformation media (g/L): sucrose (30 g), yeast extract (10 g), corn steep liquor (10 g), K_2 HPO_4 (2 g), KH_2 PO_4 (1.6 g), $MgSO_4$ (0.5 g), $FeSO_4$ (0.05 g), pH 6.5.

The fungi were routinely maintained on PDA slants. To obtain first-generation mycelium, the spore suspension from one agar slant (1 week old) was inoculated aerobically in 50 mL of seed culture media on a rotary shaker (200 rpm) at 28 °C, for 48 h in Erlenmeyer flasks (250 mL). Then 5 mL of seed culture were inoculated into the transformation medium (50 mL in 250 mL Erlenmeyer flask) and cultured at 28 °C, 200 rpm for 5 days. Substrate controls were set without inoculating the fungi into the media and strain controls were set without adding the substrate into the media, with all other conditions remaining the same.

3.3. Sample Preparation

The cultivation broth was centrifuged (3000 r/min, 10 min) to obtain mycelia and transformation solution after 5 days. The mycelium and the transformation solution, with an equal volume of ethyl acetate, were extracted. After three extractions, the extraction solution was combined, evaporated under reduced pressure on the rotary evaporator, and then redissolved with 5 mL of methanol. Added an appropriate amount of anhydrous magnesium sulfate to dry and obtained the transformation sample for detection.

3.4. Thin Layer Chromatography (TLC)

The concentrated extract was analyzed by TLC. TLC on silica gel 60 F254 (25 aluminum sheets 20 × 20 cm; Merck, New York, NY, USA) with a solvent mixture of $CHCl_3$-CH_3 OH (10:1, v/v) was applied to separate the metabolites and stained by spraying the plates with H_2 SO_4/CH_3 CH_2 OH mixture (1:9, v/v). A UV light at 254 nm was used to visualize them.

3.5. HPLC Detection

A 0.1 mL of conversion product containing dehydroepiandrosterone was diluted five times with methanol and then filtrated by 0.45 μM organic membrane to obtain the sample solution. HPLC analysis was performed on a WondaSil C18 Superb column (5 μm, 4.6 mm × 250 mm, Shimadzu, Kyoto, Japan) with a methanol/water mixture (62:38, *v*/*v*), as mobile phase was at 30 °C with UV absorbance detection of 206 nm. Flow rate: 0.8 mL/min; injection volume: 10 μL.

3.6. Isolation and Identification of Major Metabolite

Isolation of the target metabolite was performed by semi-preparative HPLC. HPLC analysis was performed on a SinoChrom ODS-BP column (5 μm, 10 mm × 250 mm, Elite, China) with an acetonitrile/water mixture (30:70, *v*/*v*) as mobile phase at 36 °C with UV absorbance detection of 206 nm. Flow rate: 3.7 mL/min; injection volume: 100 μL.

Purified metabolites were identified by ESI-MS and NMR analysis under standard conditions. ^{1} H, ^{13} C NMR spectra were taken using a Brüker AVANCE III 400 instrument (Bruker Biospin AG, Fallanden, Switzerland). ^{1} H NMR spectra were recorded in $CDCl_3$ and DMSO-d_6 using tetramethylsilane (TMS) as an internal standard. Mass spectra were taken in ESI mode on an Agilent 1200 LC-MS (Agilent, Santa Clara, CA, USA).

3.7. Establishment of Standard Curve and Calculation of Transformation Rate

The 7β-OH-DHEA samples were dissolved in methanol and prepared into solutions with different concentration gradients (0.05, 0.10, 0.15, 0.20, 0.25 mg/mL). After filtering, carry out HPLC detection. Draw a standard curve with the concentration of DHEA as the abscissa and the peak area as the ordinate. The transformation rate is calculated as follows:

$$Transfomation\ rate = \frac{A \times M_b}{B \times M_a} \times 100\% \quad (1)$$

where *A* is the quantity of product (g), and *B* is the quantity of substrate (g). *Ma* and *Mb* are the relative molecular weight of the product (7β-OH-DHEA, 304.41) and the relative molecular weight of the substrate (DHEA, 288.43), respectively (Figure S10).

3.8. Optimization of DHEA Converted to 7β-OH-DHEA by Absidia Coerulea CICC 41050

3.8.1. Effect of the Type and Concentration of Cosolvent on Transformation Rate

Take 1 mL ethanol, acetone, ethyl acetate, dimethyl sulfoxide, and chloroform as cosolvent, respectively; dissolve 50 mg DHEA in the cosolvent, add the transformation medium, and use no cosolvent in the control group to calculate the transformation rate.

3.8.2. The Biotransformation Medium Was Studied by Single-Factor Experiment and Orthogonal Experiment

On the basis of the best cosolvent, the nitrogen source, carbon source, and initial pH were studied by a single-factor experiment. The 30 g/L sucrose in the transformation medium was replaced with glucose, mannitol, lactose, maltose, starch, etc. of the same concentration to explore the effect of carbon sources on the transformation yield. And then, the carbon source concentration in the transformation medium was set to 10 g/L, 20 g/L, 30 g/L, 40 g/L, 50 g/L to explore the effect of carbon source concentration on the transformation rate. The 10 g/L yeast extract in the transformation medium was replaced with sodium nitrate, ammonium sulfate, peptone, yeast extract powder, and beef extract of the same concentration to explore the effect of the nitrogen source on the transformation rate. The concentration of nitrogen source in the transformation medium was set to 5 g/L, 10 g/L, 15 g/L, 20 g/L, and 30 g/L to explore the effect of nitrogen source concentration on the transformation rate. Set the initial pH of the transformation medium to 4.5, 5.5, 6.5, 7.5, or 8.5 and explore the impact of different initial pH values on the transformation yield.

According to the results of the single-factor experiment, an orthogonal experiment with three factors and three levels L_9 (3^3) was designed.

3.8.3. Biotransformation Conditions Were Studied by Single-Factor Experiment and Orthogonal Experiment

On the basis of the best cosolvent and the best culture medium, the single-factor experiment of transformation conditions was conducted. Set inoculum at 4%, 6%, 8%, 10%, 12%, 14%, and 16% (v/v, 50 mL). Fill a 250 mL Erlenmeyer flask with 30 mL, 40 mL, 50 mL, 60 mL, 70 mL, and 80 mL of transformation medium, respectively. The biotransformation times are 6 h, 12 h, 24 h, 48 h, 72 h, 96 h, 120 h, 144 h, and 168 h. Add 25 mg, 50 mg, 100 mg, 150 mg, 200 mg, 250 mg, 300 mg, and 400 mg of DHEA to 50 mL of transformation medium, respectively. Inoculate volume, medium volume, biotransformation time, and concentration of substrate (DHEA) were investigated to explore the effect on the transformation rate.

The orthogonal experiment with four factors and three levels L_9 (3^4) was designed according to the inoculate volume, medium volume, biotransformation time, and concentration of substrate (DHEA) determined by the single-factor experiment.

3.9. *Statistical Analysis*

All the experiments were carried out in triplicate, and each presented value was the average of three independent experiments. Standard deviations (SD) were estimated using the following equation and shown as errors on the graphs:

$$SD = \sqrt{\frac{\sum|x - \mu|^2}{N}}$$

where $\sum$ means "sum of", x is a value in the data set, μ is the mean of the data set, and N is the number of data points in the population. SPSS 20.0 software was used to conduct a *t*-test on the data to determine the statistical difference; $p < 0.05$ was significant (*), $p < 0.01$ was extremely significant (**).

4. Conclusions

The aim of this study was to evaluate 7α and 7β hydroxylation of dehydroepiandrosterone (DHEA) by *Gibberella sp.* CICC 2498 and *Absidia coerulea* CICC 41050 biotransformation. The biotransformation products were analyzed by HPLC. The retention time of the main product of DHEA transformation by *Absidia coerulea* CICC 41050 was 8.588 min. The retention time of the main product of DHEA transformation by *Gibberella sp.* CICC 2498 was 11.211 min. A semi-preparative HPLC method was successfully established to separate the biotransformation products of DHEA. The purity of the two metabolites was 94% and 96%, respectively. The isolated products were identified by NMR and MS, and the product of DHEA transformation by *Absidia coerulea* CICC 41050 was 7 β- OH-DHEA, and the product of DHEA transformation by *Gibberella sp.* CICC 2498 was 7α-OH-DHEA. Determined by single-factor experiment and subsequent orthogonal experiment, the optimized media composition to produce 7 β-OH-DHEA by *Absidia coerulea* CICC 41050 was consisted of 30 g/L sucrose, 20 g/L peptone, 10 g/L corn steep liquor, 2 g/L K_2 HPO_4, 1.6 g/L KH_2 PO_4, 0.5 g/L $MgSO_4$, and 0.05 g/L $FeSO_4$ with pH 6.5. The transformation rate of 7β-OH-DHEA reached 50.48%. The optimal biotransformation conditions (DHEA 1 g/L, medium volume 60 mL, biotransformation time 48 h, and inoculum 10%) provided maximum production (69.61%) of 7β-OH-DHEA by *Absidia coerulea* CICC 41050. The transformation of the DHEA substrate by *Absidia coerulea* CICC 41050 was described for the first time. Meanwhile, the conversion period was shortened to 48 h.

Previous literature had reported that *Absidia griseolla* var. igachii could provide C6β, C7α/β and C14α hydroxylation on androst-4-ene-3,17-dione (4-AD) [4], and *Absidia coerulea* AM93 could generate C7α/β hydroxylation on androstenediol [11]. This work extended our knowledge of DHEA hydroxylation on C7 position to the *Absidia coerulea* CICC 41050. It seems that the *Absidia* species possesses the C7 hydroxylation potential. Further genomic and proteomic data mining are worthy fo exploration for the rest of the story.

Steroids are lipophilic compounds with a gonane skeleton and play an important role in higher organisms. Due to different hydroxylations of steroid molecules, they vary greatly in their mode of action [16]. Hydroxylation of dehydroepiandrosterone (DHEA) to positions 3,7, and15 is an essential step in the synthesis of many steroidal drugs [17]. However, low hydroxylation of DHEA production is a difficult issue that must be solved urgently in industry. At present, DHEA and other steroid substrates could be hydroxylated by cytochromes P450 [18]. However, cytochrome P450 is a membrane bound protein, that is not very easy to get in a purified form for extensive research. To overcome at least some of these drawbacks, whole-cell systems are the method of choice to accomplish hydroxylation of the DHEA. Thus, this strategy would provide a possible way to enhance the 7β-OH-DHEA yield in the pharmaceutical industry.

Supplementary Materials: The following supporting information can be downloaded at: https://www.mdpi.com/article/10.3390/catal13020272/s1, Figure S1. Morphology of *Absidia coerulea* and *Gibberella sp.* Figure S2. (ESI) m/z $[M+H]^+$ diagrams for the metabolite I. Figure S3. (ESI) m/z $[M+H]^-$ diagrams for the metabolite I. Figure S4. (ESI) m/z $[M+H]^+$ diagram for the metabolite II. Figure S5. (ESI) m/z $[M+H]^-$ diagram for the metabolite II. Figure S6. ^{1} H NMR diagram for the 7α-OH-DHEA. Figure S7. ^{13} C NMR diagram for the 7α-OH-DHEA. Figure S8. ^{1} H NMR diagram for the 7β-OH-DHEA. Figure S9. ^{13} C NMR diagram for the 7β-OH-DHEA. Figure S10. Standard curve of 7β-OH-DHEA by HPLC.

Author Contributions: M.S.: Data Curation, Investigation; R.F.: Investigation, Methodology; S.C. and X.J.: Data Curation; F.W.: Resource; W.X. (Weizhuo Xu) and W.X. (Wei Xu): Resources, Supervision, Writing-Review and Editing. All authors have read and agreed to the published version of the manuscript.

Funding: This research received no external funding.

Data Availability Statement: Data are available upon reasonable request.

Conflicts of Interest: The authors declare no conflict of interest.

References

1. EMBL-EBI. CHEBI:28689—Dehydroepiandrosterone. Available online: http://www.ebi.ac.uk/chebi/searchId.do?chebiId=CHEBI:28689 (accessed on 1 January 2020).
2. Corrigan, B. DHEA and Sport. *Clin. J. Sport Med.* **2002**, *12*, 236–241. [CrossRef] [PubMed]
3. Chuang, T.Y.; Cheng, A.J.; Chen, I.T.; Lan, T.Y.; Huang, I.H.; Shiau, C.W.; Hsu, C.L.; Liu, Y.W.; Chang, Z.F.; Tseng, P.H.; et al. Suppression of LPS-induced inflammatory responses by the hydroxyl groups of dexamethasone. *Oncotarget* **2017**, *8*, 49735–49748. [CrossRef] [PubMed]
4. Heidary, M.; Habibi, Z. Microbial transformation of androst-4-ene-3,17-dione by three fungal species *Absidia griseolla* var. igachii, *Circinella muscae* and *Trichoderma virens*. *J. Mol. Catal. B-Enzym.* **2016**, *126*, 32–36. [CrossRef]
5. Dragan, C.A.; Zearo, S.; Hannemann, F.; Bernhardt, R.; Bureik, M. Efficient conversion of 11-deoxycortisol to cortisol (hydrocortisone) by recombinant fission yeast *Schizosaccharomyces pombe*. *FEMS Yeast Res.* **2005**, *5*, 621–625. [CrossRef] [PubMed]
6. Dumas, B.; Cauet, G.; Lacour, T.; Degryse, E.; Laruelle, L.; Ledoux, C.; Spagnoli, R.; Achstetter, T. 11 beta-hydroxylase activity in recombinant yeast mitochondria. In vivo conversion of 11-deoxycortisol to hydrocortisone. *Eur. J. Biochem.* **1996**, *238*, 495–504. [CrossRef] [PubMed]
7. Andryushina, V.A.; Voishvillo, N.E.; Druzhinina, A.V.; Stytsenko, T.S.; Yaderets, V.V.; Petrosyan, M.A.; Zeinalov, O.A. 14 alpha-Hydroxylation of steroids by mycelium of the mold fungus *Curvularia lunata* (VKPM F-981) to produce precursors for synthesizing new steroidal drugs. *Pharm. Chem. J.* **2013**, *47*, 103–108. [CrossRef]
8. Faramarzi, M.A.; Tabatabaei Yazdi, M.; Amini, M.; Zarrini, G.; Shafiee, A. Microbial hydroxylation of progesterone with *Acremonium strictum*. *FEMS Microbiol. Lett.* **2003**, *222*, 183–186. [CrossRef] [PubMed]
9. Huang, L.-H.; Li, J.; Xu, G.; Zhang, X.H.; Wang, Y.G.; Yin, Y.L.; Liu, H.M. Biotransformation of dehydroepiandrosterone (DHEA) with *Penicillium griseopurpureum* Smith and *Penicillium glabrum* (Wehmer) Westling. *Steroids* **2010**, *75*, 1039–1046. [CrossRef] [PubMed]
10. Kolek, T.; Milecka, N.; Świzdor, A.; Panek, A.; Białońska, A. Hydroxylation of DHEA, androstenediol and epiandrosterone by *Mortierella isabellina* AM212. Evidence indicating that both constitutive and inducible hydroxylases catalyze 7 alpha- as well as 7beta-hydroxylations of 5-ene substrates. *Org. Biomol. Chem.* **2011**, *9*, 5414–5422. [CrossRef] [PubMed]
11. Milecka-Tronina, N.; Kołek, T.; Świzdor, A.; Panek, A. Hydroxylation of DHEA and its analogues by *Absidia coerulea* AM93. Can an inducible microbial hydroxylase catalyze 7alpha- and 7beta-hydroxylation of 5-ene and 5alpha-dihydro C19-steroids? *Bioorganic Med. Chem.* **2014**, *22*, 883–891. [CrossRef] [PubMed]

12. Lyczko, P.; Panek, A.; Ceremuga, I.; Świzdor, A. The catalytic activity of mycelial fungi towards 7-oxo-DHEA—An endogenous derivative of steroidal hormone dehydroepiandrosterone. *Microb. Biotechnol.* **2021**, *14*, 2187–2198. [CrossRef] [PubMed]
13. Yildirim, K.; Kuru, A.; Yılmazer Keskin, S.; Ergin, S. Microbial transformation of dehydroepiandrosterone (DHEA) by some fungi. *Biocatal. Biotransformation* **2021**, *39*, 465–474. [CrossRef]
14. Li, H.; Sun, J.; Xu, Z. Biotransformation of DHEA into 7α,15α-diOH-DHEA. In *Microbial Steroids: Methods and Protocols*; Barredo, J.-L., Herráiz, I., Eds.; Springer: New York, NY, USA, 2017; pp. 289–295.
15. Kollerov, V.; Shutov, A.; Kazantsev, A.; Donova, M. Hydroxylation of pregnenolone and dehydroepiandrosterone by zygomycete *Backusella lamprospora* VKM F-944: Selective production of 7alpha-OH-DHEA. *Appl. Microbiol. Biotechnol.* **2022**, *106*, 535–548. [CrossRef] [PubMed]
16. Schmitz, D.; Zapp, J.; Bernhardt, R. Steroid conversion with CYP106A2—Production of pharmaceutically interesting DHEA metabolites. *Microb. Cell Factories* **2014**, *13*, 81. [CrossRef] [PubMed]
17. Li, C.; Li, H.; Sun, J.; Zhang, X.; Shi, J.; Xu, Z. Production of 7alpha,15alpha-diOH-DHEA from dehydroepiandrosterone by Colletotrichum lini ST-1 through integrating glucose-feeding with multi-step substrate addition strategy. *Bioprocess Biosyst. Eng.* **2016**, *39*, 1259–1266. [CrossRef] [PubMed]
18. Sakaki, T. Practical application of cytochrome P450. *Biol. Pharm. Bull.* **2012**, *35*, 844–849. [CrossRef] [PubMed]

catalysts

MDPI

Article

Nysfungin Production Improvement by UV Mutagenesis in *Streptomyces noursei* D-3-14

Ming Song [1], Wubing He [2], Sulan Cai [1], Fuju Wang [3], Weizhuo Xu [2,*] and Wei Xu [2,*]

[1] School of Functional Food and Wine, Shenyang Pharmaceutical University, 103 Wenhua Road, Shenhe District, Shenyang 110016, China
[2] School of Life Sciences and Biopharmaceuticals, Shenyang Pharmaceutical University, 103 Wenhua Road, Shenhe District, Shenyang 110016, China
[3] Beijing Global Biotechnologies, Co. Ltd., No.99 Yuexiu Road, Haidian District, Beijing 100193, China
* Correspondence: weizhuo.xu@syphu.edu.cn (W.X.); shxuwei8720@163.com (W.X.); Tel.: +86-024-43520301 (W.X.); +86-024-43520307 (W.X.); Fax: +86-024-43520301 (W.X.); +86-024-43520307 (W.X.)

Abstract: *Streptomyces noursei* D-3-14 was taken as a starting strain and treated with UV (15 W, 30 cm) mutagenesis for 40 s for three consecutive rounds. High yielding strains were screened using chemical and biological potency determination, and the components of the fermentation products were detected using HPLC. Finally, the mutant strain *Streptomyces noursei* 72-22-1 with a chemical potency of 8912 (U/mL) and a biological potency of 5557 (U/mL) was obtained after the genetic stability evaluation. After optimization of the fermentation conditions, the chemical potency and biological potency of *Streptomyces noursei* 72-22-1 reached 14,082 U/mL and 10579 U/mL, respectively, which is 1.58 and 1.91 times those before optimization. HPLC analysis indicated that the mutant strain 72-22-1 displayed a higher content of polyfungin B. When equimolar nystatin A1, A3, and polyfungin B were tested for their fungicidal activities towards *Saccharomyces cerevisiae* ATCC 2061, polyfungin B exhibited a better efficacy than nystatin A1 and A3.

Keywords: *Streptomyces noursei*; nysfungin; nystatin A1; nystatin A3; polyfungin B; UV mutagenesis

Citation: Song, M.; He, W.; Cai, S.; Wang, F.; Xu, W.; Xu, W. Nysfungin Production Improvement by UV Mutagenesis in *Streptomyces noursei* D-3-14. *Catalysts* **2023**, *13*, 247. https://doi.org/10.3390/catal13020247

Academic Editors: Zhilong Wang and Tao Pan

Received: 30 December 2022
Revised: 17 January 2023
Accepted: 18 January 2023
Published: 21 January 2023

1. Introduction

Nystatin is a kind of polyene macrolide antibiotic [1–3] that was first isolated by Hazen and Brown from soil in Fauquier County, Virginia in the 1950s [4]. At that time, it was named Fungicidin and demonstrated fungistatic and fungicidal activities, without antibacterial action. In 1954, this Fungicidin was first commercialized by Bristol Myers Squibb and renamed nystatin, with its major ingredient being nystatin A1. Later, Chinese researchers also isolated and identified additional ingredients of actinomycetes derived from Fungicidin in Guangdong province soil [5,6]. In 1981, Thomas et al. compared pharmaceutical grade samples of Fungicidin from China, Hungary, Italy, US, and Russia, and identified that nystatin A1 represented the majority of the ingredients in the US, Italy, and Hungarian samples, with a 70% concentration [7], which coincides with the British Pharmacopoeia in 1980 [8]. Meanwhile, he also reported that only 12% nystatin A1 was founded in the Chinese samples, with other components nystatin A3 and polyfungin B ranging from 20 to 50% [9]. Since then, nystatin has usually been assigned as western nystatin, which has a high A1 content as the major ingredient. While nysfungin is normally used to name the Chinese-derived nystatin, besides the A1, which may also have A3 and polyfungin B.

Nystatin is effective against *Candida*, *Cryptococcus*, *Aspergillus*, *Histoplasma*, and *Blastomyces* [10,11]. Nystatin is one of the most commonly used topical antifungal drugs, with a high efficacy, low cost, and fewer side effects, due to not being absorbed from the gastrointestinal tract [12,13].

From their chemical structures, it could be seen that nystatin A3 has an additional digitoxose compared to nystatin A1 [9,14], while the polyfungin B has an absence of a C10 hydroxyl group compared with nystatin A3 [15]. All three components share the same macrolide polyene skeleton, as shown in Figure 1.

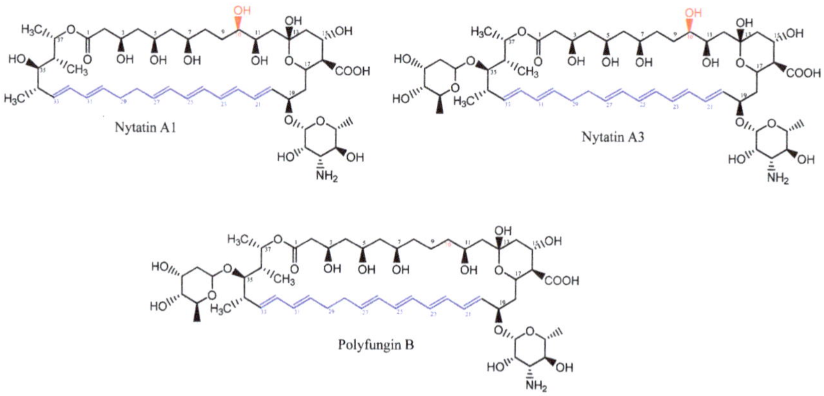

Figure 1. Chemical structures of nystatin A1, A3, and polyfungin B.

In the FDA approved nystatin drugs, the major ingredient is the single nystatin A1, with the CAS No. 34786-70-4. While in the Chinese pharmacopoeia [16], nystatin is still considered a multiple-ingredient drug, and was named Nysfungin, so as to differentiate it from the single major A1 nystatin.

Due to the actinomyces metabolic diversity, in this research, we used an environmentally isolated strain of *Streptomyces noursei* D-3-14 to perform ultraviolet (UV) mutagenesis, optimized the culture conditions, and finally obtained a higher production of Nysfungin. Mutation breeding is a powerful technique, in which microbial strains exposed to mutagen treatment are screened, to identify positive mutants with specific characteristics. This method has the advantages of simplicity, rapidity, and high efficiency [17]. UV is possibly one of the physical mutagens that causes genetic variation and enables the selection of traits as needed. UV has a strong genotoxic effect, producing DNA damage that leads to an altered DNA structure [18]. UV irradiation induces covalent crosslinks between neighboring pyrimidines. If left unrepaired, error-prone replication of this damaged DNA leads to an increased rate of mutagenesis and genome instability [19]. The aim of this study was to screen and characterize nystatin-producing *Streptomyces noursei* strains exposed to UV, to identify the strain that produced the highest activity of nystatin, thereby obtaining a high-activity and low-cost raw material for commercial nystatin production.

2. Results and Discussion

2.1. Determination of the Chemical and Biological Potency of the Starting Strain Streptomyces noursei D-3-14

According to the experimental process in Section 3.3, the chemical potency of the starting strain *Streptomyces noursei* D-3-14 was detected as 3464 U/mL, as shown in Table 1, and the biological potency was detected as 2703 U/mL, as shown in Table 2.

Table 1. Chemical potency detection of *Streptomyces noursei* D-3-14.

Expt. No	OD_{319}	Chemical Potency (U/mL)	Average Chemical Potency (U/mL)
1	0.295	3709	
2	0.249	3116	3464
3	0.284	3567	

Table 2. Biological potency detection of *Streptomyces noursei* D-3-14.

Expt. No	Inhibition Diameter (mm) 80 U/mL Reference	Inhibition Diameter (mm) 40 U/mL Reference	Inhibition Diameter (mm) 80 U/mL Sample	Inhibition Diameter (mm) 40 U/mL Sample	Biological Potency (U/mL)	Average Biological Potency (U/mL)
1	23.42	19.00	21.64	17.50	2658	
2	21.98	18.44	20.26	17.18	2538	2703
3	23.22	18.94	22.16	17.88	2914	

2.2. HPLC Analysis of the Starting Strain Streptomyces noursei D-3-14

According to the experimental process in Section 3.3, the HPLC analysis of D-3-14 was similar to the nysfungin reference, as seen in Figure 2.

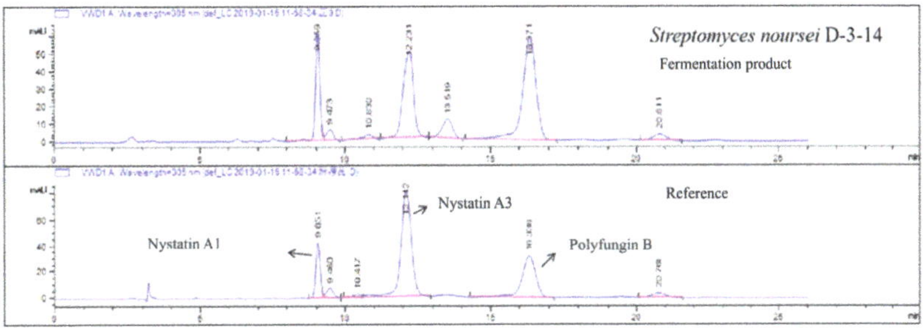

Figure 2. HPLC analysis of the fermentation products of *Streptomyces noursei* D-3-14 and the reference.

2.3. Determination of the UV Irradiation Duration of Mutagenesis

The lethal rate should increase with UV radiation duration. According to our experience, a 90% lethal rate is acceptable for further research. Figure 3 shows that the UV radiation duration was set to 40 s.

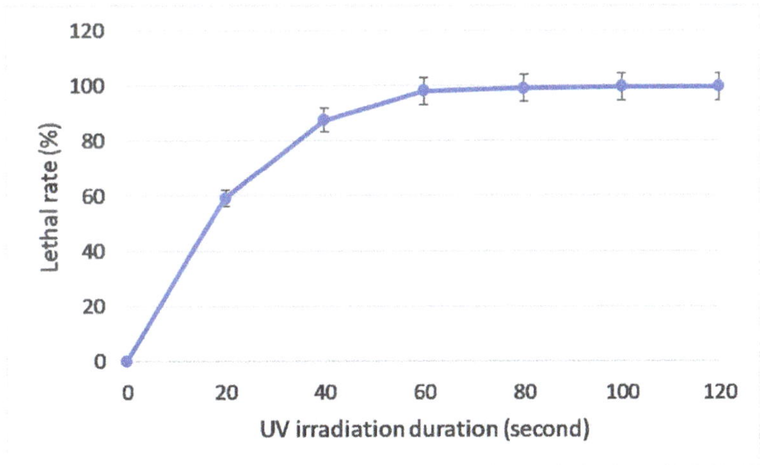

Figure 3. Lethality curve for UV irradiation duration.

2.4. Preliminary and Secondary Screening for Three Consecutive Rounds

According to the experimental process in Section 3.4, three consecutive rounds of preliminary and secondary screening were performed following the UV mutagenesis.

In the first round of preliminary screening, 74 mutants were obtained. Then 14 out of these 74 mutants were selected for the secondary screen. In this round, *Streptomyces noursei* 74-14-8 was identified as the resultant strain, with a chemical potency of 5783 U/mL and biological potency of 4626 U/mL, which exhibited about 1.67 and 1.71 times the chemical and biological potency of the starting strain, as seen in Table 3.

Table 3. Biological potency and Polyfungin B contents of various strains after the first round of UV mutagenesis.

Strain No.	Chemical Potency (U/mL)	Biological Potency (U/mL)	Polyfugin B Content (%)
74-14-8	5783	4626	33.63
74-14-61	5177	3467	27.64
74-14-67	5101	3825	21.61

In the second round of screening, 22 out of 72 preliminary screened mutants were selected. The resultant *Streptomyces noursei* 72-22-1 was identified as having a chemical and biological potency of 8912 U/mL and 5557 U/mL, exhibiting 1.54 and 1.20 times the chemical and biological potency of *Streptomyces noursei* 74-14-8 (Table 4).

Table 4. Biological potency and Polyfungin B contents of various strains after the second round of UV mutagenesis.

Strain No.	Chemical Potency (U/mL)	Biological Potency (U/mL)	Polyfugin B Content (%)
72-22-1	8912	5557	53.63
72-22-3	5328	3467	47.64
72-22-5	7214	5679	31.61
72-22-10	5912	3578	44.72
72-22-11	6879	4344	35.11
72-22-14	6020	4684	36.79
72-22-19	5829	3689	32.77
72-22-20	5741	4121	33.04
72-22-49	5311	3877	28.64

In the third round of screening, 24 out of 112 preliminary screened mutants were selected. The resultant *Streptomyces noursei* 112-24-63 was identified as having a chemical and biological potency of 11097 U/mL and 10751 U/mL, which was 1.25 and 1.93 times the chemical and biological potency of *Streptomyces noursei* 72-22-1 (Table 5).

Table 5. Biological potency and Polyfungin B contents of various strains after the third round of UV mutagenesis.

Strain No.	Chemical Potency (U/mL)	Biological Potency (U/mL)	Polyfugin B Content (%)
112-24-11	8692	6945	39.66
112-24-23	4463	5007	27.46
112-24-38	4353	5010	26.10
112-24-63	11,097	10,751	31.86
112-24-70	7837	9044	27.92
112-24-78	5545	8539	22.14
112-24-97	7340	2951	23.37

2.5. Genetic Stability Experiment

According to the experimental process in 3.5, genetic stability experiments were performed on *Streptomyces noursei* 72-22-1 and 112-24-63. The results can be seen in Figure 4.

It was observed that Strain 112-24-63 could not withstand the stability test, but 72-22-1 could endure five consecutive generations of culture. Later, *Streptomyces noursei* 72-22-1 underwent fermentation optimization for a better potency.

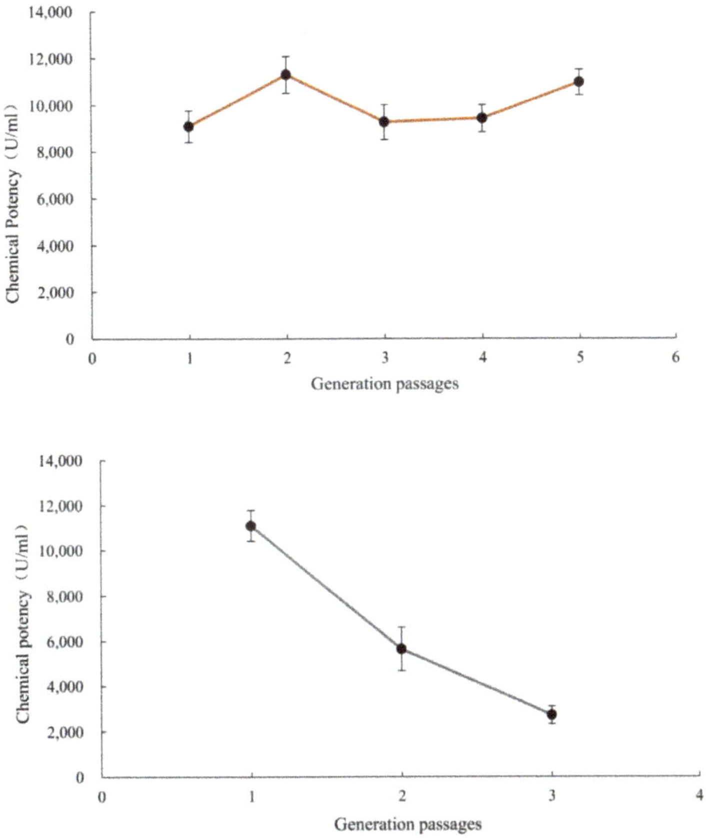

Figure 4. Genetic stability experiment results for *Streptomyces noursei* 72-22-1 and 112-24-63.

2.6. Single-Factor Evaluation for the Nysfungin Fermentation

Single factor glucose concentration, peanut meal concentration, starting pH, and inoculation volume ration were evaluated to generate the results in Figure 5. The results showed that when the content of glucose in the fermentation medium was 5.5%, the yield of nystatin was higher. A carbon source is one of the essential nutrients in a microbial culture medium, providing energy for the growth, reproduction, and metabolic activities of microorganisms. Our previous work found that during streptomycin fermentation, the glucose concentration in the fermentation broth must be controlled to be lower than a certain level in the later stage of fermentation. If the glucose concentration is higher than 10 mg/mL, the synthesis of mannosidase will be inhibited and the production of streptomycin will be significantly reduced. When the culture medium contains two or more carbon sources, microorganisms generally use glucose first and then the other carbon sources. These results indicate that secondary metabolites, such as antibiotics, are regulated by carbon metabolites, such as glucose.

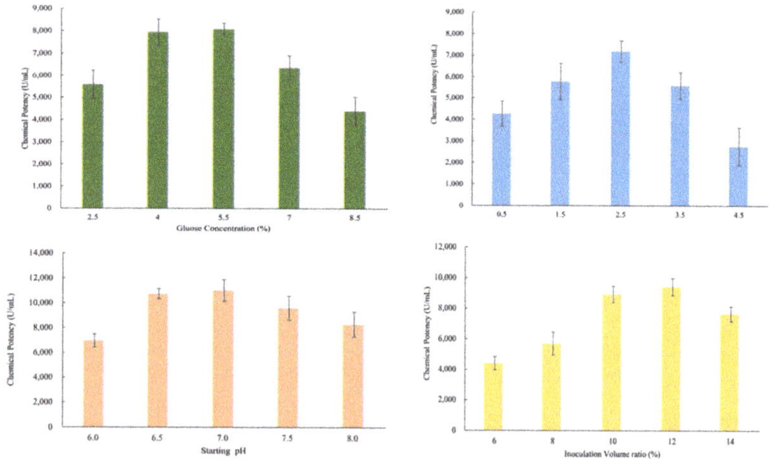

Figure 5. Evaluation results for single factors of glucose concentration, peanut meal concentration, starting pH, and inoculation volume ratio for fermentation.

The nitrogen source not only plays a nutritional role in the fermentation of microorganisms, but also contains inducers, precursors, and other substances required for the synthesis of secondary metabolites. When there are multiple nitrogen sources in the fermentation medium, microorganisms always use the simple nitrogen sources first, and then decompose complex nitrogen sources. Moreover, when the concentration of these simple nitrogen sources (such as ammonium ions, amino acids) is high, they synthesize few secondary metabolites. The results showed that when the content of peanut meal in the fermentation medium was 2.5%, the yield of nystatin was higher.

The experimental results showed that when the initial pH of the fermentation medium was 7.0, and the inoculation volume was 12%, the yield of nystatin fermentation was higher.

According to the above single-factor evaluation results, a four-factor and three-level L_9 (3^4) orthogonal experiment was designed, as seen in Table S1. Orthogonal experimental results and analysis (Table S2) showed that a glucose concentration of 5.5%, peanut melt concentration of 2.5%, initial pH of 7.5, and inoculation volume ratio of 14% were the optimal conditions. The final optimized chemical and biological potency results for *Streptomyces noursei* 72-22-1 were 14,082 U/mL and 10579 U/mL, which are 1.58 and 1.91 times those before optimization.

2.7. HPLC Analysis for Streptomyces noursei 72-22-1

According to the experimental process in 3.3, the fermentation product in the HPLC analysis of 72-22-1 was similar to the nysfungin reference, as seen in Figure 6.

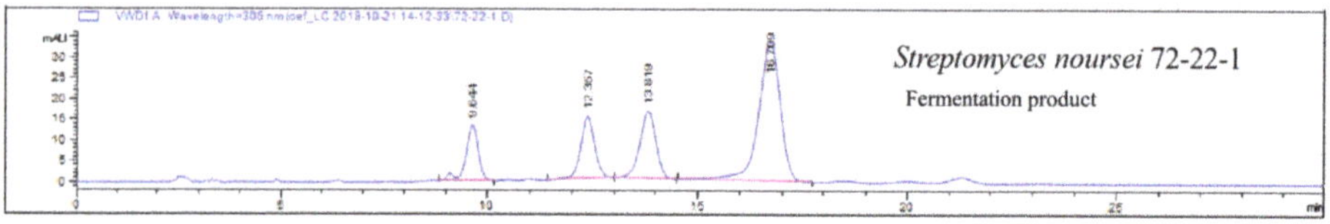

Figure 6. HPLC analysis of the fermentation products of *Streptomyces noursei* 72-22-1.

It can be seen that in the mutant *Streptomyces noursei* 72-22-1, nystatin A1, A3, and polyfungin B exhibited retention times of 9.085 min, 12.288 min, and 16.599 min. The ratio of polyfungin B was about 53%, which is much higher than the reference.

2.8. Fungicidal Activity of Nystatin A1, A3, and Polyfungin B towards Saccharomyces cerevisiae ATCC 2061

Single ingredients of nystatin A1, A3, and polyfungin B were extracted and isolated using semi-preparative HPLC, and their fungicidal activity was evaluated according to the experimental process in 3.3.

The results are summarized in Table 6. It can be seen that polyfungin B displayed a better fungicidal activity than nystatin A1 and A3, which is a novel result in nysfungin activity research.

Table 6. Results of nystatin A1, A3, and polyfungin B inhibiting *Saccharomyces cerevisiae* ATCC 2061.

Expt. No.	Inhibition Zone Diameter of Nystatin A1 (mm)	Inhibition Zone Diameter of Nystatin A3 (mm)	Inhibition Zone Diameter of Polyfugin B (mm)
1	20.68	16.44	21.62
2	19.26	16.84	21.18
3	18.56	15.84	19.12
4	19.76	16.60	20.18
5	19.14	16.52	20.08
Average Value	19.48 ± 0.71	16.45 ± 0.33	20.44 ± 0.88 **

** $p < 0.05$, highly significant.

3. Materials and Methods

3.1. Chemicals, Strains, Culture Media, and Growth Conditions

Due to the lack of a nysfungin standard, domestic commercial nysfungin tablets (Nysfungin, Zhenyuan Pharmaceuticals Co. Ltd., Shaoxing City, Zhejiang, China) were purchased and processed to be used as a reference. All other chemicals were analytically pure, unless expressed otherwise.

The starting *streptomyces noursei* D-3-14 was isolated and stored in our lab and had a chemical potency of 3464 U/mL and a biotic potency of 2703 U/mL. Strains were cultured at 28 °C for 7 days on Gause's synthetic solid medium slant (GA media). A loop of the freshly grown lawn was inoculated into 50.0 mL seed culture medium, and cultured at 150 rpm, 28 °C for 24 h. Then, 10% of this seed culture was inoculated into 50 mL fermentation culture medium and cultured at 200 rpm, 28 °C, for 86 h. The resultant culture could be used to extract nystatin.

Gause's synthetic solid medium consisted of (g/L) soluble starch, 20.0; KNO_3, 1.0; NaCl, 0.5; K_2HPO_4, 0.5; $MgSO_4{\cdot}7H_2O$, 0.5; $FeSO_4{\cdot}7H_2O$, 0.01 and agar, 15, pH 7.2. Seed culture medium consisted of (g/L) peanut meal, 20.0; soluble starch, 10; glucose, 10; $CaCO_3$, 6.0; soybean oil, 3.0; $(NH_4)_2SO_4$, 2.0; peptone 2.0; $MgSO_4{\cdot}7H_2O$, 0.5; K_2HPO_4, 0.2 and pH 7.2. Fermentation culture medium consisted of (g/L) glucose, 55.0; peanut meal, 25.0; $CaCO_3$, 10.0; $(NH_4)_2SO_4$, 3.0; peptone 3.0; dried silkworm chrysalis meal, 2.0; soybean oil, 1.0; KH_2PO_4, 0.02 and pH 7.2.

Saccharomyces cerevisiae ATCC 2061 was purchased from China General Microbiological Culture Collection Center (CGMCC), Beijing, China. It was cultured on a YM medium slant at 28 °C for 3 d. When was mature, 6.0 mL sterilized NaCl solution (0.9%) was added, and the lawn was scraped to obtain a suspension. The suspension was then transferred into Erlenmeyer flasks with glass beads and incubated 28 °C, 180 rpm for 30 min. The final suspension concentration was adjusted and calculated using a hemocytometer.

The YM culture medium consisted of (g/L) glucose, 10.0; peptone 5.0; yeast extract, 5.0; malt extract, 5.0; agar 15.0 and pH 6.2. The biological potency detection culture medium consisted of (g/L) maltose, 40.0; peptone 20.0; NaCl, 3.0; agar 15.0 and pH 7.0–7.2. The Gauss's culture medium was sterilized at 121 °C for 20 min, all other media were sterilized at 115 °C for 30 min.

3.2. UV Mutagenesis, Nysfungin Crude Extraction, and HPLC Detection

First, 3 mL of *Streptomyces noursei* single spore suspension was added to the 75 mm culture dishes and radiated with UV light (15 W, 30 cm) for 60 s. The dish covers were then opened and radiated for 20 s, 40 s, 60 s, 80 s, 100 s, and 120 s. Next, 12 mL seed culture medium was supplemented, and the dishes were cultured in black bags at 28 °C for 2 h. Both the UV radiated (Mutagenesis group) and untreated spore (Control group) suspensions were serially diluted and calculated using a hemocytometer. Then, 100 uL of spore suspension with 10^{-3}, 10^{-4}, and 10^{-5} concentrations were spread on the Gauss plates, each concentration was spread in three replicates, the plates were cultured at 28 °C for 5~7 days, the colony numbers were counted, and the lethal rate was calculated.

$$\text{Lethal rate } (\%) = \frac{(\text{cfu of control group} \times \text{dilution ratio}) - (\text{cfu of Mutagenesis group} \times \text{dilution ratio})}{(\text{cfu of control group} \times \text{dilution ratio})} \times 100\%$$

After 86 h, the fermentation broth was centrifugated at 3500 rpm for 10 min, to collect the wet mycelia. These wet mycelia were transferred into a brown beaker, 3 times the volume of 98% alcohol (*v/v*) was added and stirred for 40 min, then the solution was centrifugated at 3500 rpm for 10 min, the supernatant was collected and extracted with an equal volume of 98% alcohol (*v/v*) three times. All the supernatants were combined and vapored with reduced pressure to concentrate to 20 mL, and these solutions were kept at 4 °C overnight to obtain the nystatin suspension. These suspensions could be further processed at 4000 rpm for 6 min to eliminate the supernatant and to obtain the crude nystatin extract crystal and were stored at −20 °C before use.

Subsequently, 50.00 mg of the above crude nystatin crystal was dissolved in 5.00 mL methanol (chromatographic grade) and filtered using a 0.22 μm filter, to produce a 10 mg/mL solution. HPLC was performed using a WondaSil C18 Column (250 × 4.6 mm × 5 μm, Shimadzu, Kyoto, Japan), and the mobile phase consisted of methanol and acetonitrile, acetate buffer = 26:37:37, the detection wavelength was 305 nm, the column temperature was 30 °C, the loading speed was 1 ml/min, and the loading volume was 10 μL.

3.3. Determination of the Chemical and Biological Potencies

First, the nystatin reference was resolved in *N*,*N*-dimethylformamide to make a 1000 U/mL stock. Then, 0 mL, 0.2 mL, 0.4 mL, 0.6 mL, 0.8 mL, 1.0 mL, 1.2 mL, 1.4 mL, 1.6 mL, 1.8 mL, and 2.0 mL stock was dissolved and filled to 25.00 mL, and the absorption values were detected on OD319 to establish a standard curve.

Chemical potency was detected using the following process adapted from the European Pharmacopoeia 8.0 [20]: First, after 86 h, 5 mL well-dispersed culture broth was washed three times with ddH_2O and centrifugated to collect the mycelia. Then, the mycelia were well mixed with 10 mL methanol and left to stand for 2 h. Next, the supernatant was collected after 3500 rpm for 10 min, and extracted using 10 mL methanol twice. The combined methanol solution was filled to 50.00 mL. Then, 2.0 mL was used to detect OD319, and the chemical potency was evaluated by comparing with the above standard curve.

The biological potency was detected using the following process adapted from a USP monograph [21]. The first step was to evaluate the proper concentration of *Saccharomyces cerevisiae* ATCC 2061. Then, 1 mL of 10^{-1}, 10^{-2}, 10^{-3}, 10^{-4}, and 10^{-5} dilution suspension was added to 15 mL biological potency detection culture medium at 48 °C and mixed well. When the media become solid, one Oxford cup was placed on the center of the petri dish and 100 uL nystatin reference solution (80 U/mL) was supplemented and the cover was put on. Each concentration was performed in triple replicates and cultured at 28 °C for 14–18 h. The diameter of the inhibition zone was measured with a Vernier caliper. Empirically, a diameter of 18–22 mm is acceptable for biological potency detection.

Then, 250 uL of *Saccharomyces cerevisiae* ATCC 2061 was added to the 15 mL biological potency detection culture medium at 48 °C and mixed well. When the media became solid, four Oxford cups were symmetrically placed on the petri dish. Two cups on one diagonal line were supplemented with 100 uL nystatin reference solution (80 U/mL and 40 U/mL), while the other two cups on the other diagonal line were supplemented with 100 uL nystatin sample solution (80 U/mL and 40 U/mL). After being cultured at 28 °C for 14–18 h. The diameter of the inhibition zone was measured with a Vernier caliper. And the biological potency of the crude nystatin extracts was calculated using the following formula:

$$\theta = log^{-1}\left(\frac{T_2 - S_2 + T_1 - S_1}{S_2 + T_2 - S_1 - T_1} \times I\right)$$

in which,

S_1 = inhibition zone diameter of the low-dose control solution
S_2 = inhibition zone diameter of the high-dose control solution
T_1 = inhibition zone diameter of the low-dose sample solution
T_2 = inhibition zone diameter of the high-dose sample solution
I = lg(high-dose concentration/low-dose concentration)

3.4. Preliminary Screen and Secondary Screen for Mutants

Preliminary screening was performed using the following process: Single colonies on the UV radiation plates were inoculated onto a fresh Gauss's synthetic media plate and cultured at 28 °C for 5–7 days, to obtain viable mutants. A 1 × 1 cm^2 lawn was scraped and transferred into 50 mL fermentation broth and cultured at 28 °C and 180 rpm for 86 h. Then, the crude nystatin was extracted, to examine the chemical potency. Those strains with higher chemical potency than the starting strains were screened as the preliminary strains.

A secondary screen was performed on those strains with higher chemical potencies. Those strains were further detected by HPLC to determine their biological potencies. Only those strains exhibiting both higher chemical and biological potencies were screened and selected for the next steps.

3.5. Genetic Stability Evaluation

The selected strains with high chemical and biological potency were cultured for five consecutive generations, and the chemical potency was tested for each generation to evaluate the genetic stability.

3.6. Single Ingredient Evaluation for the Obtained Nysfungin

The obtained Nysfungin was detected using HPLC, each of the major peaks was further extracted and isolated using semi-preparative HPLC, and the resultant single ingredients were identified by MS and NMR and tested for their fungicidal activities using the cup-plate methods in 3.3.

3.7. Optimization of the Fermentation Conditions

The culture process was performed as described in 3.1. The glucose concentration was set at 2.5%, 4.0%, 5.5%, 7.0%, and 8.5%. The peanut meal concentration was set at 0.5%, 1.5%, 2.5%, 3.5%, and 4.5%. The starting pH value was set at 6.0, 6.5, 7.0, 7.5, and 8.0. The inoculation volume ratio was set at 6%, 8%, 10%, 12%, and 14%. Each inoculation volume was replicated three times.

The orthogonal experiment with four factors and three levels L_9 (3^4) was designed according to the glucose concentration, peanut meal, starting pH, and inoculate volume ratio, determined in a single factor experiment.

3.8. Statistical Analysis

All experiments were carried out in triplicate and each presented value is the average of three independent experiments. SPSS 20.0 software was used to conduct t-tests on the data, to determine the statistical difference, $p < 0.05$ was significant (*), $p < 0.01$ was extremely significant (**).

4. Conclusions and Discussion

As polyene macrolide antibiotics produced by the *Streptomyces noursei strain,* both nystatin and nysfungin have broad-spectrum antifungal effects, with the strongest inhibition of *Candida albicans*, and have been widely used in clinical practice [3]. In this study, *Streptomyces noursei* D-3-14, a nystatin producing strain, was treated with UV mutagenesis for three rounds and screened for high yield mutants. Finally, *Streptomyces noursei* 72-22-1, a genetically stable strain with enhanced nystatin production, was obtained. Its chemical potency was 8912 U/mL and its biological potency was 5557 U/mL, 2.57 times and 2.06 times of those of the original strain, respectively. After optimizing the fermentation conditions, the chemical potency and biological potency of *Streptomyces noursei* 72-22-1 were 14,082 U/mL and 10579 U/mL, respectively, 4.07 and 3.92 times those of the starting strain.

Admittedly, although antibiotics have been applied for many years, there are only a few clinical options for polyene macrolide ingredients. In the FDA Orange Book dated December 2022, there are only four polyene macrolide fungicidal antibiotics listed. The only discontinued ingredient is Candicidin (Vanobid), which was approved on 1 January 1982 and manufactured by Sanofi Aventis US. The Eyevance Pharmeceuticals-manufactured Natamycin (Natacyn) was approved on the same day as Candicidin and is the only listed and prescribed Natamycin. Ten out of the 15 entries of Amphotericin B have been discontinued and only five manufacturers are still producing it. Nystatin has 123 entries, with 62 discontinued and 61 approved manufacturers [22].

All these data suggest that the diversity of polyene macrolide antifungal antibiotics remains limited. When equimolar single ingredients of nystatin A1, A3, and polyfungin B were tested, polyfugin B exhibited a better fungicidal activity than nystatin A1 and A3. These promising results warrant further extensive metabolic and biological studies of polyfugin B, nystatin A3, and other structurally similar ingredients in the near future.

Supplementary Materials: The following supporting information can be downloaded at https://www.mdpi.com/article/10.3390/catal13020247/s1, Table S1. Orthogonal experimental factors and level assignments. Table S2. Orthogonal experimental results and analysis.

Author Contributions: M.S. (data curation, investigation); W.H. (investigation, methodology); S.C. (data curation); F.W. (resources); W.X. (Weizhuo Xu) and W.X. (Wei Xu) (resources, supervision, writing—review and editing). All authors have read and agreed to the published version of the manuscript.

Funding: This research received no external funding.

Data Availability Statement: Data are available upon reasonable request.

Conflicts of Interest: The authors declare no conflict of interest.

References

1. Lyu, X.; Zhao, C.; Yan, Z.M.; Hua, H. Efficacy of Nystatin for the Treatment of Oral Candidiasis: A Systematic Review and Meta-Analysis. *Drug Des. Devel. Ther.* **2016**, *10*, 1161–1171. [CrossRef] [PubMed]
2. Baldino, M.E.L.; Medina-Silva, R.; Sumienski, J.; Figueiredo, M.A.; Salum, F.G.; Cherubini, K. Nystatin effect on chlorhexidine efficacy against Streptococcus mutans as planktnic cells and mixed biofilm with Candida albicans. *Clin. Oral Investig.* **2022**, *26*, 633–642. [CrossRef] [PubMed]
3. Brescansin, E.G.; Portilho, M.; Pessine, F.B.T. Physical and chemical analysis of commercial nystatin. *Acta Sci. Health Sci.* **2013**, *35*, 215–221. [CrossRef]
4. Hazen, E.L.; Brown, R. Fungicidin, an Antibiotic Produced by a Soil Actinomycete. *Proc. Soc. Exp. Biol. Med.* **1951**, *76*, 93–97. [CrossRef] [PubMed]
5. Ling, D.; Chen, S.; Wang, S.; Sun, Z.; Ma, J. Structure identification of two major ingredients in the domestic Fungicidin. *Acta Pharm. Sin.* **1986**, *21*, 454–457.
6. Ma, J.; Liu, Y.; Wen, D.; Zhu, F.; Wang, S.; Ling, D. Isolation, preparation, identification and rename of the Chinese nystatin multi major ingredients. *Antibiotics* **1987**, *12*, 83–90.
7. Thomas, A.H.; Pharm, B.; Newland, P.; Quinlan, G.J. Identification and determination of the qualitative composition of nystatin using thin-layer chromatography and high-performance liquid chromatography. *J. Chromatogr. A* **1981**, *216*, 367–373. [CrossRef]
8. Paterson, G.R. The british pharmacopoeia 1980. *Drug Ther. Bull.* **1980**, *18*, 100.

9. Thomas, A.H.; Newland, P.; Sharma, N.R. The heterogeneous composition of pharmaceutical-grade nystatin. *Analyst* **1982**, *107*, 849–854. [CrossRef] [PubMed]
10. Caffrey, P.; Hogan, M.; Song, Y.H. New Glycosylated Polyene Macrolides: Refining the Ore from Genome Mining. *Antibiotics* **2022**, *11*, 334. [CrossRef] [PubMed]
11. Helal, S.H.; Abdel-Aziz, H.M.M.; El-Zayat, M.M.; Hasaneen, M.N.A. Preparation, characterization and properties of three different nanomaterials either alone or loaded with nystatin or fluconazole antifungals. *Sci. Rep.* **2022**, *12*, 22110. [CrossRef] [PubMed]
12. Peng, D.-S.; Lo, C.-H.; Tseng, Y.-L.; Kuo, S.L.; Chiang, C.-P.; Chiang, M.-L. Efficacy of oral nystatin treatment for patients with oral mucosal dysesthesia but without objective oral mucosal manifestations and necessity of Candida culture test before oral nystatin treatment. *J. Dent. Sci.* **2022**, *17*, 1802–1813. [CrossRef] [PubMed]
13. Quindós, G.; Gil-Alonso, S.; Marcos-Arias, C.; Sevillano, E.; Mateo, E.; Jauregizar, N.; Eraso, E. Therapeutic tools for oral candidiasis: Current and new antifungal drugs. *Med. Oral Patol. Oral Cir. Buccal* **2019**, *24*, e172–e180. [CrossRef] [PubMed]
14. Chong, C.N.; Rickards, R.W. Macrolide antibiotic studies. XVI. The structure of nystatin. *Tetrahedron Lett.* **1970**, *11*, 5145–5148. [CrossRef] [PubMed]
15. Zielinski, J.; Jereczek, E.; Sowinski, P.; Falkowski, L.; Rudowski, A.; Borowski, E. The structure of a novel sugar component of polyene macrolide antibiotics: 2,6-dideoxy-L-ribohexopyranose. *J. Antibiot.* **1979**, *32*, 565–568. [CrossRef] [PubMed]
16. Department of Health. *Pharmacopoeia of the People's Republic of China*; China Medical Science Press: Beijing, China, 1990; pp. 36–38.
17. Ruiz, N.Q.; Campo, Y.C.; Stashenko, E.E.; Fuentes, J.L. Antigenotoxic Effect Against Ultraviolet Radiation-induced DNA Damage of the Essential Oils from *Lippia* Species. *Photochem. Photobiol.* **2017**, *93*, 1063–1072. [CrossRef] [PubMed]
18. Wang, S.; Zhang, L.; Yang, G.; Han, J.; Thomsen, L.; Pan, K. Breeding 3 elite strains of Nannochloropsis oceanica by nitrosoguanidine mutagenesis and robust screening. *Algal Res.* **2016**, *19*, 104–108. [CrossRef]
19. Hauer, M.H.; Gasser, S.M. Chromatin and nucleosome dynamics in DNA damage and repair. *Genes Dev.* **2017**, *31*, 2204–2221. [CrossRef] [PubMed]
20. Council of Europe. *European Pharmacopoeia*, 10th ed.; Council of Europe: Strasbourg, France, 2019; pp. 3401–3402.
21. The United States Pharmacopeial Convention. *United State Pharmacopeia*, 40th ed.; The United States Pharmacopeial Convention: Rockville, MD, USA, 2017; pp. 5400–5401.
22. *Orange Book: Approved Drug Products with Therapeutic Equivalence Evaluations*; FDA: Silver Spring, MD, USA, 2022.

catalysts

Article

Identification of New Amylolytic Enzymes from Marine Symbiotic Bacteria of *Bacillus* Species

Mohammad Reza Erfanimoghadam and Ahmad Homaei *

Department of Marine Biology, Faculty of Marine Science and Technology, University of Hormozgan, Bandar Abbas P.O. Box 3995, Iran
* Correspondence: a.homaei@hormozgan.ac.ir or a.homaei@gmail.com; Tel.: +98-76-33711000-11; Fax: +98-76-33670716

Abstract: α-amylases are one of the most common and important industrial enzymes widely used in various industries. The present study was conducted with the aim of isolating and identifying symbiotic α-amylase enzyme-producing bacteria in the intestine of *Silago Sihama* and *Rasterliger Canagorta* fish living in Qeshm Island, Hormozgan. The intestinal symbiotic bacteria of these species were isolated using nutrient agar culture medium; then, α-amylase producing bacteria were screened using a special culture medium containing starch and the Lugol's solution test. The α-amylase enzyme activity of enzyme-producing bacteria was measured using the starch substrate. Finally, bacteria with the highest enzyme activity were selected and identified by the 16S rRNA gene sequence analysis. The results showed that out of 22 isolated bacteria, 10 were able to grow in a special culture medium, and 5 strains of these 10 bacteria had the ability to produce relatively stronger halos. The four bacterial strains belonging to the genus *Bacillus* that had the highest α-amylase enzyme activity were identified and registered in the NCBI gene database as *B. subtilis* strains HR13, HR14, HR15, and HR16. Among these four strains, two strains of *B. subtilis*, HR13 and HR16, displayed high enzyme activity and maximum activity at 60 °C at pH values of 5 and 7, respectively. α-Amylase enzymes isolated from marine symbiotic bacteria of *Bacillus* species can be considered potential candidates for application in various industries.

Keywords: α-amylase; starch hydrolysis; marine enzymes; Persian Gulf

Citation: Erfanimoghadam, M.R.; Homaei, A. Identification of New Amylolytic Enzymes from Marine Symbiotic Bacteria of *Bacillus* Species. *Catalysts* **2023**, *13*, 183. https://doi.org/10.3390/catal13010183

Academic Editors: Zhilong Wang and Tao Pan

Received: 4 December 2022
Revised: 6 January 2023
Accepted: 11 January 2023
Published: 13 January 2023

1. Introduction

Oceans cover 71% of the earth's surface and are home to a diverse range of species, such as algae, bacteria, fungi, sponges, and fish. The sea is a challenging place to inhabit because it has both deep and shallow areas with different temperatures and pressures, salinity changes, different pHs, light, hydrostatic pressure, and the distribution of different nutrients, all of which lead to a wide variety of marine organisms with unique characteristics [1–3].

Marine microorganisms are a promising source for discovering novel enzymes because of their distinctive natural environments, physiological traits, distinctive metabolic processes, and use of varied nutrients [4,5]. Marine bacteria are a diverse group of marine microorganisms that have developed physiological adaptations in response to various environmental factors and evolutionary processes. They also produce a variety of hydrolyzing enzymes, such as amylases, lipases, and proteases, which may have applications in contemporary biotechnology [6]. The main benefits of employing microorganisms for enzyme synthesis over plants and animals are rapid growth, huge production capacity, and simple enzyme extraction from bacteria [5,6].

The buoyancy of fish in water causes bacteria to cover the outer surface of their bodies, and as a result, fish are in continuous contact with the microorganisms that cover their bodies. Some of these microorganisms do not live on the surface of the fish but are instead found as part of the microbes that live within the fish body, such as the oral cavity [7]. The combination of all environmental factors for fish symbiotic bacteria results in the development of enzymes with exceptional features of the bacteria in this milieu [8,9].

Amylases are one of the most important sources of industrial enzymes with many applications in various industries, such as pharmaceuticals, food industries, auxiliary food preparation, detergents, papermaking, and textiles. They provide the hydrolysis of starch into small sugar units of dextrin or smaller glucose polymers and are classified into three groups, namely α-amylases, β-amylases, and γ-amylases. α-amylases cleave the bond (α (1→4)) between adjacent glucose units in the linear chain of amylose in starch carbohydrates [10]. Since α-amylase has multiple cleavage sites, α-amylase is faster than β-amylase. In addition, α-amylase enzymes have the necessary stability against high temperatures, especially α-amylase synthesized from *Bacillus subtilis* and *Archaea* bacteria species, which are resistant to heat [11]. Various physical and chemical factors, such as temperature, pH, incubation period, carbon sources as inducers, surfactants, nitrogen sources, phosphate, and different metal ions, affect the production of α-amylase [12].

Although the production of α-amylase enzyme by Gram-negative bacterial strains such as *B.* sp., *Aeromonas* sp., and *Stenotrophomonas* sp. has been reported, the production of α-amylase isolated from fish intestinal bacteria is still ambiguous and has not been widely studied [13]. So far, no research has been performed on the isolation of α-amylase enzyme-producing bacteria coexisting with the digestive systems of *Sillago sihamas* and *Rastrelliger Kanagurta* fish from the southern coasts of Iran. In this study, the symbiotic bacteria of the intestine of *Sillago sihamas* and *Rastrelliger Kanagurta* fish were isolated, and then α-amylase enzyme-producing bacteria were screened, and those with the highest potential to produce α-amylase were identified by biochemical, morphological, and molecular methods, and finally, some biochemical characteristics of the enzyme were investigated.

2. Results and Discussion

2.1. Primary and Secondary Screening of α-Amylase-Producing Bacteria

In this study, 22 colonies were isolated from the collected samples of *Rastrelliger kanagurta* and *Sillago sihama*, of which 12 colonies were isolated from the intestine of *Sillago sihama* and 10 colonies were isolated from the intestine of *Rastrelliger kanagurta*. Among the twenty-two isolated bacterial strains, ten bacterial strains were able to grow in a special culture medium containing starch, five bacterial strains were able to produce strong halos by Lugol's test, and other bacterial strains had a weak or moderate halo creation ability by Lugol's test (Figure 1). The results revealed that these bacterial strains produce the α-amylase enzyme. In this experiment, starch was employed as a carbon source, and only those bacteria that had α-amylase enzyme were able to hydrolyze starch in this culture medium and grow in this specific medium (Tables 1 and 2). For secondary screening, among the bacteria isolated from the intestines of *Sillago sihama* and *Rastrelliger kanagurta*, five bacterial strains with the highest degrees of halo formation (strong and medium) were chosen in the primary screening stage to measure enzyme activity. Meanwhile, four strains of *B. subtilis* strain HFBP08, *B. subtilis* strain ZIM3, *B. subtilis* strain SXK, and B. subtilis strain soil G2B had the highest amount of enzyme activity with the activity levels of 0.074, 0.048, 0.061, and 0.072, respectively. Finally, these four potential α-amylase enzyme-producing strains were selected for further analyses and molecular identification. The results indicated that *Rastrelliger kanagurta* and *Sillago sihama* fish species have rich sources of α-amylase-producing bacteria. The most potent bacterium among those producing enzymes was *B. subtilis*. In general, based on our findings, this bacterial strain can be a suitable source for the production of heat-tolerant amylase enzymes with high functional stability for use in industry. For example, α-amylase generated by these *Bacillus* species is used in the confectionery industry because of its temperature resistance, as well as starch liquefaction, processing, and application. They have a wide range of commercial applications and are especially useful in the food and pastry industries. Earlier research, similar to the current study, used a specialized growth medium containing starch for the first screening of bacteria that produce α-amylase. [14,15].

a **b**

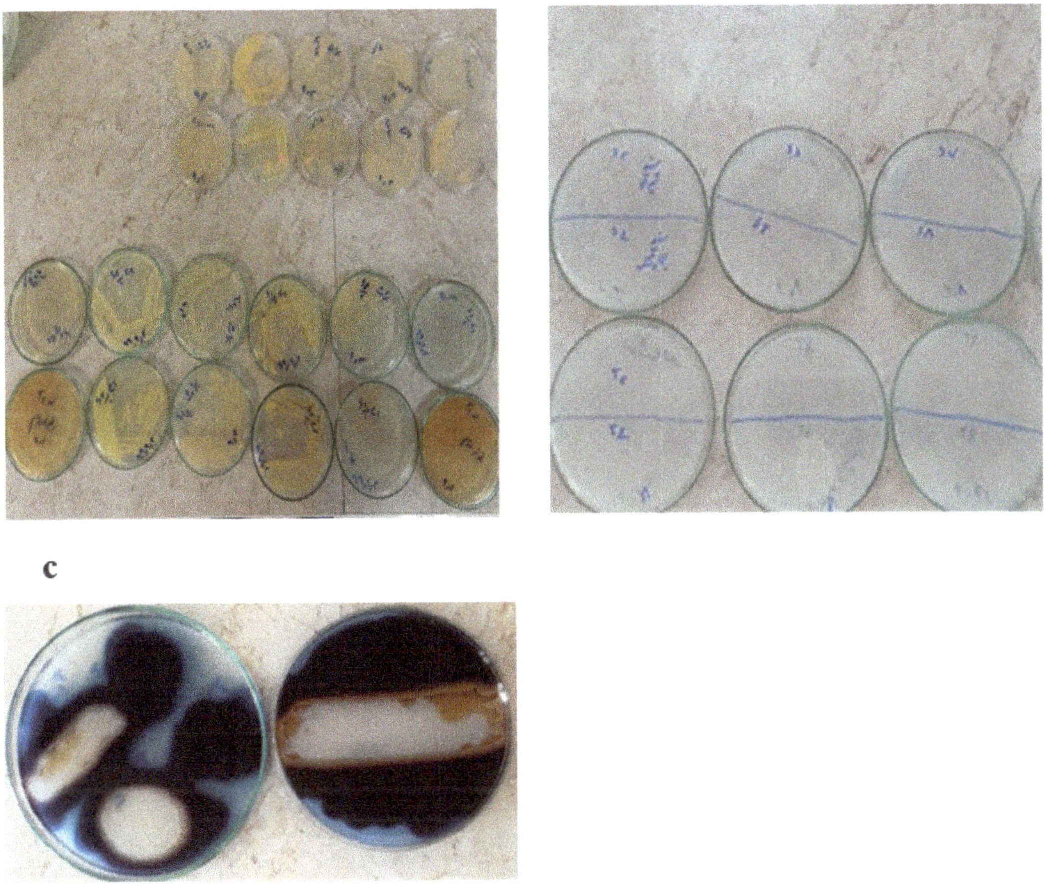

c

Figure 1. Pure culture of isolated bacteria in nutrient agar medium (**a**), growth of α-amylase-producing bacteria in a specific solid culture medium containing starch (**b**), and a clear halo around the colony of α-amylase-producing bacteria using Lugol's solution (**c**).

Table 1. Evaluation of bacterial strains capable of producing α-amylase from the intestines of *Sillago sihama*.

Number	Isolated Bacterial Code	Investigating the Growth of Bacteria in the Culture Medium Containing Starch	The Diameter of the Halo Formed after Lugol's Addition
1	T1	Positive	Strong
2	T2	Negative	No halo
3	T3	Negative	No halo
4	T4	Negative	No halo
5	T5	S5 Positive	Strong
6	T6	Negative	Weak
7	T7	Positive	Medium
8	T8	Negative	No halo

Table 2. Assessment of bacterial strains capable of generating α-amylase from the intestines of *Rastrelliger kanagurta*.

Number	Isolated Bacterial Code	Investigating the Growth of Bacteria in the Culture Medium Containing Starch	The Diameter of the Halo Formed after Lugol's Addition
1	S1	Positive	Weak
2	S2	Negative	No halo
3	S3	Negative	No halo
4	S4	Negative	No halo
5	S5	S5 Positive	Strong
6	S6	Negative	No halo
7	S7	Positive	Medium
8	S8	Negative	No halo
9	S9	Negative	No halo
10	S10	S Negative	No halo
11	S11	Positive	Medium
12	S12	Positive	Strong

2.2. Molecular Identification of Potential α-Amylase-Producing Bacterial Strains

Identification techniques based on molecular studies are essential and accurate instruments for the proper characterization of microbial species. In this regard, four bacterial strains (HR13, HR16, HR15, and HR14) that had the highest α-amylase activity were identified by 16S rRNA gene analysis (Figure 2). All four strains with the capacity to produce α-amylase enzyme aligned most closely to the *Bacillus* genus, Bacillaceae family, and *Bacillus subtilis* species, according to the NCBI database's analysis of the nucleotide sequence of the 16S rRNA gene of isolated bacteria. Based on these results, the bacterial strains (HR13, HR16, HR15, and HR14) have the highest similarity (99%) with *B. subtilis* strains HFBP08, ZIM3, SXK, and soilG2B. According to the results of 16S rRNA gene sequences, *B. subtilis* strain HR13, *B. subtilis* strain HR14, *B. subtilis* strain HR15 B., and *B. subtilis* strain HR16 were registered in the NCBI database with accession numbers MZ571841, MZ571838, MZ571839, and MZ571840, respectively. The phylogenetic tree of *B. subtilis* strains HR13, HR14, HR15 B., and HR16 isolated from the fish intestine in this study, as well as sequences available in NCBI for *B. Methanobacterium formicicum*, *B. infantis* strain C4, *B. amyloliquefaciens* strain TS.18 S.BK, *B. velezensis* strain CB02999, *B. tequilensis* strain HYM43, and *B. mojavensis* strain WSE-KSU305, were drawn and their evolutionary relationships were investigated (Figure 3). The evolutionary relationships of the strains obtained in this research with other *Bacillus* species, including *B. infantis*, *B. amyloliquefaciens*, *B. velezensis*, *B. tequilensis*, and *B. mojavensis*, are shown in a phylogenetic tree. Therefore, the present study shows that the strains belonging to *B. subtilis* are potential bacteria for the production of the α-amylase enzyme.

In 2014, Castro et al. isolated endophytic microorganisms symbiotic with leaves and branches of mangrove trees in Brazil from two mangrove species, namely *Rhizophora mangle* and *Avicennianitida*. *Bacillus* was the most isolated genus from mangrove trees in this region [16]. Similar to the present study, bacteria belonging to the *Bacillus* genus, including *B. subtilis*, *B. stearothermophilus*, and *B. amyloliquefaciens*, have been identified as potential α-amylase enzyme-producing bacteria [17–19]. *B. amyloliquefaciens* produces the most α-amylase enzyme in the world, and the α-amylase generated by this bacterium was the first enzyme employed in the industry for starch sweetening and liquefaction. [20–22]. Therefore, these findings demonstrate that *bacilli* have the unique capacity to synthesize α-amylase enzyme.

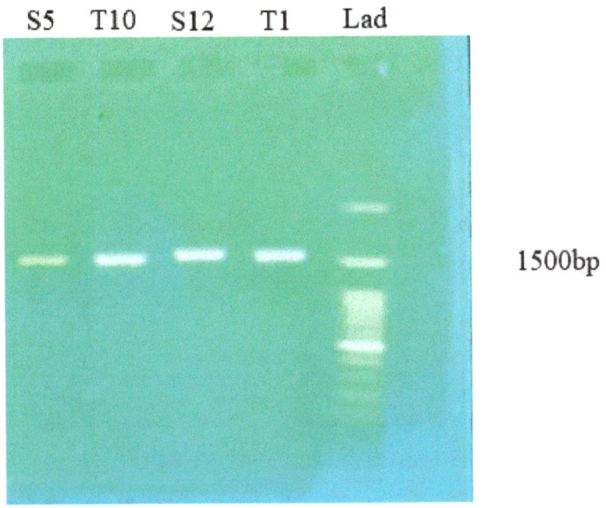

Figure 2. Identification of 16S rRNA gene bands of enzyme-producing bacteria with a molecular weight of 1500 bp; Lad is a molecular marker.

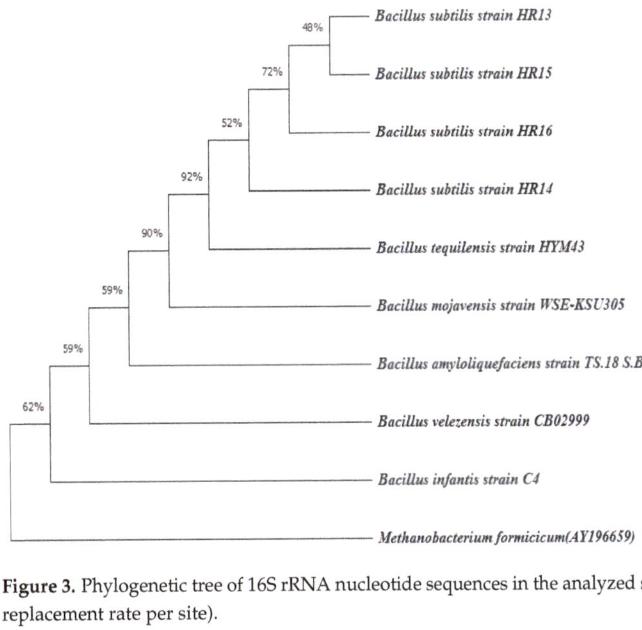

Figure 3. Phylogenetic tree of 16S rRNA nucleotide sequences in the analyzed strains (0.05 nucleotide replacement rate per site).

2.3. The Effect of Temperature and pH on the Activity and Stability of the α-Amylase Enzyme

The effect of temperature on α-amylase enzyme activity in *B. subtilis* strains HR13 and HR16 showed that both strains display maximum activity at 60 °C (Figure 4). At temperatures higher than 80 °C, the enzyme activity of both strains suddenly drops, so that at a temperature of 90 °C, the enzyme activity of the HR16 strain reaches almost zero.

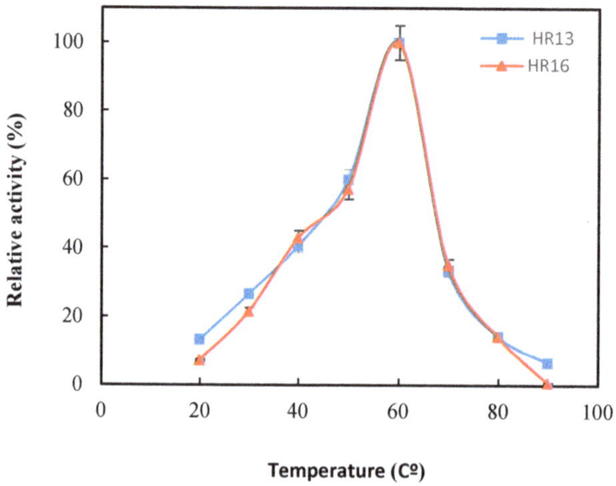

Figure 4. The effect of temperature on the enzyme activity levels of HR13 (■) and HR16 (▲) strains; the activity at the optimal temperature was taken as 100%.

The effect of different pH values on the enzyme activity of two strains, *B. subtilis* HR13 and *B. subtilis* HR16, shows their maximum enzyme activity at pH 5 and 7, respectively (Figure 5). The α-amylase enzyme activity of HR16 had a sharp drop after the optimum pH so that at pH values of 8 and 9, 35% and 28% of the initial activity of the enzyme was retained, respectively. In the case of the HR13 strain, the activity of the enzyme did not significantly decrease after reaching the optimal pH, and the enzyme maintained more than 75% of its initial activity in the range of pH 6–8.

Irreversible thermal inactivation of enzymes isolated from *B. subtilis* strains HR13 and HR16 at temperatures of 80 and 90 °C demonstrated that at 80 °C with increasing incubation time (Figure 6), the amount of enzyme activity decreased, and after temperature incubation for up to 10 min, the enzyme activity in the HR13 B. *subtilis* strain reaches less than half. However, the enzyme of *B. subtilis* strain HR16 maintained > 50% of its initial activity after a temperature incubation for 20 min at 80 °C. Notably, after 60 min of temperature incubation, the activity of both enzyme strains reached almost zero.

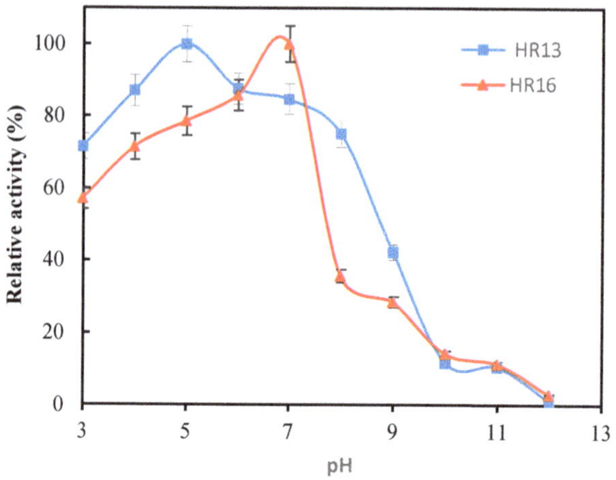

Figure 5. The impact of pH on the enzyme activity levels of HR13 (■) and HR16 (▲); the activity at the optimal pH was taken as 100%.

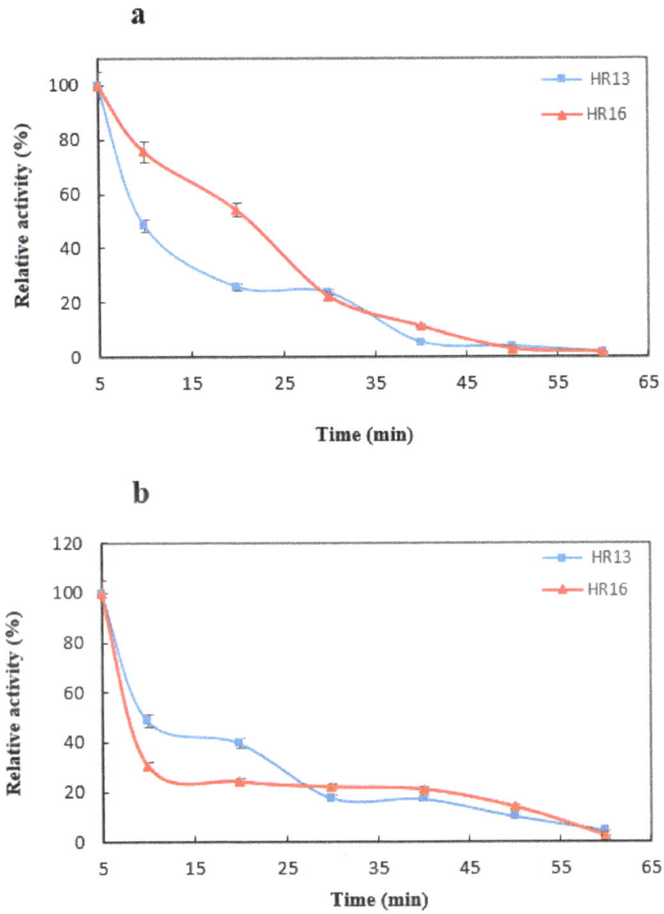

Figure 6. Thermal stability of the α-amylase enzyme activity of HR13 (■) and HR16 (▲) after incubating the enzymes in a water bath at 80 °C (**a**) and 90 °C (**b**) for different time periods.

At 90 °C, irreversible thermal inactivation of the enzyme decreases enzyme activity substantially faster than at 80 °C. After 30 min of temperature incubation, the enzyme activity of *B. subtilis* strain HR13 reaches less than 20% of its initial activity. About 22% of the enzyme's initial activity was still present in the *B. subtilis* strain HR16 enzyme after 30 min of incubation at 90 °C. The irreversible inactivation of enzymes isolated from HR13 and HR16 was investigated and compared at an alkaline pH of 8 (Figure 7). Both forms of the enzyme exhibit a notable decline in activity at pH 8 with extended incubation times. The enzymes of strains HR13 and HR16 preserved 56% and 63% of their initial activity after 30 min of incubation at pH 8, respectively. At pH 12, the activity of the free enzyme decreased as the incubation time increased. After 60 min of incubation at pH 8, the free enzyme in both enzyme forms was reversible to about 18% of its initial activity.

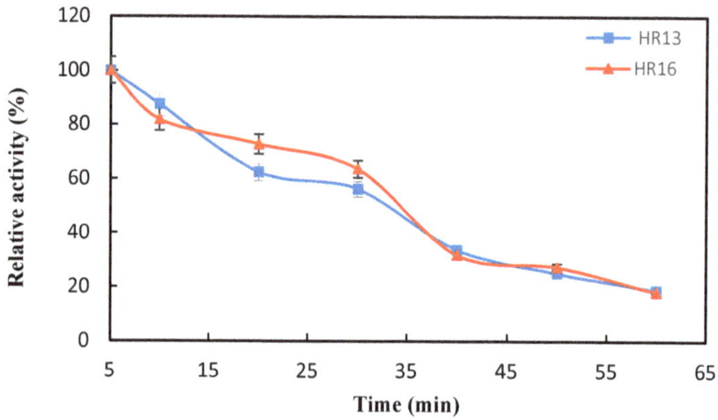

Figure 7. Irreversible inactivation of the α-amylase enzyme in HR13 (■) and HR16 (▲) after different time periods of incubation at pH 8.

3. Materials and Methods

3.1. Materials Used for the Collection of Short Fish Samples

All reagents were purchased from Merck Co. (Darmstadt, Germany).

3.2. Collection of Sillago sihama and Rastrelliger kanagurta Fish

Freshly caught fish samples of *Sillago sihama* and *Rastrelliger kanagurta* were purchased from the Qeshm fishmongers' market and transferred to the laboratory in a flask containing ice at a temperature of 4–10 °C.

3.3. Isolation of Intestinal Bacteria from Sillago sihama and Rastrelliger kanagurta Fish

First, the abdominal surface of the fish was cleaned with 70% alcohol. After opening the stomach of the fish with a surgical blade, the intestines were removed under sterile conditions. After homogenizing and diluting the samples with physiological serum, they were cultured in nutrient agar culture medium and kept in a greenhouse at 30 °C for 48 h in order to isolate bacteria. After the incubation period, the plates were examined morphologically (color and appearance) under the laminar hood, and the colonies were purified.

3.4. Primary Screening of α-Amylase-Producing Bacteria

Bacteria isolated from the intestines of *Sillago sihama* and *Rastrelliger kanagurta* were cultured on a special culture medium containing starch (1%) and kept in a greenhouse at 30 °C for 48 h. After the greenhouse period, the growth of bacteria on the starch culture medium was analyzed, and the samples exhibiting the ability to grow in this medium were selected for further analyses. At this stage, Lugol's solution was poured onto the special culture medium containing starch; then, based on the diameter of the clear halo around the bacterial colony, which indicates starch hydrolysis and enzyme production by bacteria, the α-amylase enzyme-producing colonies were isolated and selected.

3.5. Secondary Screening of α-Amylase-Producing Bacteria and Measurement of Enzyme Activity

The α-amylase enzyme-producing bacteria, which had produced a stronger halo in the primary screening stage, were selected for secondary screening and enzyme activity measurement. At this stage, bacteria were cultured in nutrient broth and incubated in a shaker incubator at 30 °C and 100 rpm. After 24 h, the culture medium containing the grown bacteria was centrifuged at 12,000 rpm for 20 min, and the supernatant was used as a solution containing the α-amylase enzyme in the next stages of the experiment [23]. The Bernfeld method was used to measure alpha-amylase enzyme activity [24]. For this purpose, 400 μL of phosphate buffer solution, 100 μL of α-amylase enzyme extract, and

500 μL of 1% starch were incubated at 60 °C for 20 min. Then, 1 mL of DNS solution was added to the test tube and incubated for 5 min in a boiling water bath. After cooling, 1 mL of distilled water was added to it. After stirring the contents of the absorption tube, the sample was read at a wavelength of 540 nm.

3.6. Identification of Potential α-Amylase Enzyme-Producing Bacteria

In this study, the identification of potential bacteria producing the α-amylase enzyme was carried out by examining phenotypic and biochemical characteristics as well as using 16S rRNA gene analysis. In order to extract the bacterial DNA contents, the boiling method was used [25,26]. Polymerase chain reaction (PCR) was applied to amplify a 1500-bp fragment of the 16S rRNA gene using the extracted DNA of potential α-amylase-producing bacteria (HR13, HR16, HR15, and HR14) by means of forward (5′-AGAGTTTGATCCTGGCTCAG-3′) and reverse (5′-AAGGAGGTGATCCAGCC-3′) primers at a final volume of 50 μL, containing 2 μL of extracted bacterial strain DNA, 0.5 μL of forward primer, 0.5 μL of reverse primer, 25 μL of amplicon master mix solution, and 22 μL of distilled water. The thermal cycling was carried out as follows: an initial denaturation at 95 °C for 1 min, followed by 30 cycles of denaturation at 95 °C for 20 s, annealing at 63 °C for 30 s, extension at 72 °C for 1 min, and the final extension at 72 °C for 5 min.

After checking the quality and quantity of PCR products on 1% agarose gel electrophoresis, they were sent to FAZA Pajooh Co. for double-sided sequencing. The sequences were compared with the 16S rRNA gene sequences of bacteria registered in the NCBI database using the BLAST tool available on the NCBI website. Then, the closest strain was selected based on the S rRNA16 gene sequence and biochemical tests [27,28].

In order to analyze the phylogenetic relationships, the target sequences were assessed and compared in accordance with the NCBI gene bank. In the end, using MEGA4 software and the Neighbor-Joining (NJ) algorithm with bootstrapping, 1000 repetitions of the phylogenetic tree were drawn.

3.7. Effect of pH and Temperature on α-Amylase Activity

The α-amylase enzyme activity was evaluated at a temperature range of 20–90 °C in a 20 mM phosphate buffer (pH 7.4). In order to examine the enzyme activity at any temperature, both the substrate (1% starch) and the enzyme solution (buffer and crude enzyme extract) must reach equilibrium at that temperature before measuring enzyme activity. The relative enzyme activity was measured at different pH values of 2–12 at room temperature. For this purpose, a mixed buffer (containing 25 mM tris-base, glycine, sodium phosphate, and sodium acetate) was prepared and adjusted using NaOH and HCl solutions at different pH values from 2 to 12. In this experiment, a 1% starch solution was also prepared as a substrate, and the enzyme activity was evaluated in the above buffer at different pH values.

3.8. Effects of pH and Temperature on Enzyme Stability

To measure the temperature stability of the enzyme, first, the crude enzyme extracts of both bacteria were placed at 80 and 90 °C for different time intervals (5, 10, 20, 30, 40, 50, and 60 min). Then, 500 μL of the substrate was added to each of the samples, and the resulting mixture was incubated in the assay conditions. Finally, α-amylase activity was stopped using dinitrosalicylic acid reagent and the absorbance of the reaction mixture was read at a wavelength of 540 nm.

In order to measure the pH stability of the enzyme, first, the crude enzyme extracts of both bacteria were exposed to pH values of 8 and 9 for different times (5, 10, 20, 30, 40, 50, and 60 min). Then, 500 μL of the substrate was added to each of the samples, and the resulting mixture was incubated in the assay conditions. Finally, α-amylase activity was measured using the dinitrosalicylic acid reagent and then reading the absorbance of the reaction mixture at a wavelength of 540 nm.

4. Conclusions

The current research aims to isolate and identify regional bacteria generating symbiotic α-amylase enzymes in the intestines of short and talal fishes from Qeshm Island waters. In this study, the symbiotic bacteria *Bacillus* sp. HR13 and *Bacillus* sp. HR16 found in the intestines of *Sillago sihamas*, and *Rastrelliger Kanagurta* had the highest α-amylase enzyme activity at 60 °C. This implies they could be great choices for purification, mass production, and commercialization for use in the food and detergent industries. In general, the findings of this study show that the Persian Gulf is a rich source of bacterial strains that produce the widely used amylase enzyme, and the fish species *Sillago sihamas* and *Rastrelliger Kanagurta*, two of the most abundant species in the Persian Gulf, are rich in bacteria that adapt to the unique conditions of this region.

Author Contributions: Conceptualization and research design, A.H.; experimental, M.R.E.; data analysis, M.R.E. and A.H.; manuscript—writing, M.R.E.; manuscript—editing, A.H. All authors have read and agreed to the published version of the manuscript.

Funding: This research received no external funding.

Data Availability Statement: All data were included in the manuscript.

Acknowledgments: The authors express their gratitude to the research council of the University of Hormozgan for financial support during the course of this project.

Conflicts of Interest: The authors declare no conflict of interest.

References

1. Qeshmi, F.I.; Homaei, A.; Fernandes, P.; Hemmati, R.; Dijkstra, B.W.; Khajeh, K. Xylanases from marine microorganisms: A brief overview on scope, sources, features and potential applications. *Biochim. Biophys. Acta (BBA) Proteins Proteom.* **2020**, *1868*, 140312. [CrossRef] [PubMed]
2. Homaei, A. Immobilization of Penaeus merguiensis alkaline phosphatase on gold nanorods for heavy metal detection. *Ecotoxicol. Environ. Saf.* **2017**, *136*, 1–7. [CrossRef] [PubMed]
3. Homaei, A. Purification and biochemical properties of highly efficient alkaline phosphatase from *Fenneropenaeus merguiensis* brain. *J. Mol. Catal. B Enzym.* **2015**, *118*, 16–22. [CrossRef]
4. Homaei, A.; Qeshmi, F.I. Purification and characterization of a robust thermostable protease isolated from *Bacillus subtilis* strain HR02 as an extremozyme. *J. Appl. Microbiol.* **2022**, *133*, 2779–2789. [CrossRef]
5. Izadpanah Qeshmi, F.; Homaei, A.; Khajeh, K.; Kamrani, E.; Fernandes, P. Production of a Novel Marine *Pseudomonas aeruginosa* Recombinant L-Asparaginase: Insight on the Structure and Biochemical Characterization. *Mar. Biotechnol.* **2022**, *24*, 559–613. [CrossRef]
6. Pasalari, A.; Homaei, A. Isolation and Molecular Identification of Xylanase-Producing Bacteria from *Ulva flexuosa* of the Persian Gulf. *Processes* **2022**, *10*, 1834. [CrossRef]
7. Cahill, M.M. Bacterial flora of fishes: A review. *Microb. Ecol.* **1990**, *19*, 21–41. [CrossRef]
8. Balaji, N.; Rajasekaran, K.M.; Kanipandian, N.; Vignesh, V.; Thirumurugan, R. Isolation and screening of proteolytic bacteria from freshwater fish *Cyprinus carpio*. *Int. Multidiscip. Res. J.* **2012**, *2*, 56–59.
9. Beygmoradi, A.; Homaei, A.; Hemmati, R.; Santos-Moriano, P.; Hormigo, D.; Fernández-Lucas, J. Marine chitinolytic enzymes, a biotechnological treasure hidden in the ocean? *Appl. Microbiol. Biotechnol.* **2018**, *102*, 9937–9948. [CrossRef]
10. Homaei, A.; Ghanbarzadeh, M.; Monsef, F. Biochemical features and kinetic properties of α-amylases from marine organisms. *Int. J. Biol. Macromol.* **2016**, *83*, 306–314. [CrossRef]
11. Reddy, N.; Nimmagadda, A.; Rao, K.S. An overview of the microbial α-amylase family. *Afr. J. Biotechnol.* **2003**, *2*, 645–648.
12. Sivaramakrishnan, S.; Gangadharan, D.; Nampoothiri, K.M.; Soccol, C.R.; Pandey, A. α-Amylases from microbial sources–An overview on recent developments. *Food Technol. Biotechnol.* **2006**, *44*, 173–184.
13. Makhdoumi, A. Bacterial diversity in south coast of the Caspian Sea: Culture-dependent and culture-independent survey. *Casp. J. Environ. Sci.* **2018**, *16*, 259–269.
14. Jabir, T.; Jesmi, Y.; Vipindas, P.V.; Hatha, A.M. Diversity of nitrogen fixing bacterial communities in the coastal sediments of southeastern Arabian Sea (SEAS). *Deep. Sea Res. Part II Top. Stud. Oceanogr.* **2018**, *156*, 51–59. [CrossRef]
15. Jackson, C.R.; Churchill, P.F.; Roden, E.E. Successional changes in bacterial assemblage structure during epilithic biofilm development. *Ecology* **2001**, *82*, 555–566. [CrossRef]
16. Castro, R.A.; Quecine, M.C.; Lacava, P.T.; Batista, B.D.; Luvizotto, D.M.; Marcon, J.; Ferreira, A.; Melo, I.S.; Azevedo, J.L. Isolation and enzyme bioprospection of endophytic bacteria associated with plants of Brazilian mangrove ecosystem. *SpringerPlus* **2014**, *3*, 382. [CrossRef]

17. Ktari, N.; Ben Khaled, H.; Nasri, R.; Jellouli, K.; Ghorbel, S.; Nasri, M. Trypsin from zebra blenny (*Salaria basilisca*) viscera: Purification, characterisation and potential application as a detergent additive. *Food Chem.* **2012**, *130*, 467–474. [CrossRef]
18. Debashish, G.; Malay, S.; Barindra, S.; Joydeep, M. Marine enzymes. *Mar. Biotechnol. I* **2005**, *96*, 189–218.
19. Stoll, B.J.; Hansen, N.I.; Sánchez, P.J.; Faix, R.G.; Poindexter, B.B.; Van Meurs, K.P. Early onset neonatal sepsis: The burden of group *B Streptococcal* and *E. coli* disease continues. *Pediatrics* **2011**, *127*, 817–826. [CrossRef]
20. Ye, F.; Yang, R.; Hua, X.; Shen, Q.; Zhao, W.; Zhang, W. Modification of stevioside using transglucosylation activity of *Bacillus amyloliquefaciens* α-amylase to reduce its bitter aftertaste. *LWT—Food Sci. Technol.* **2013**, *51*, 524–530. [CrossRef]
21. Pérez-Carrillo, E.; Serna-Saldívar, S.O.; Alvarez, M.M.; Cortes-Callejas, M.L. Effect of sorghum decortication and use of protease before liquefaction with thermoresistant α-amylase on efficiency of bioethanol production. *Cereal. Chem.* **2008**, *85*, 792–798. [CrossRef]
22. Demirkan, E.S.; Mikami, B.; Adachi, M.; Higasa, T.; Utsumi, S. α-Amylase from *B. amyloliquefaciens*: Purification, characterization, raw starch degradation and expression in *E. coli*. *Process Biochem.* **2005**, *40*, 2629–2636. [CrossRef]
23. Dong, G.; Vieille, C.; Savchenko, A.; Zeikus, J.G. Cloning, sequencing, and expression of the gene encoding extracellular alpha-amylase from *Pyrococcus furiosus* and biochemical characterization of the recombinant enzyme. *Appl. Environ. Microbiol.* **1997**, *63*, 3569–3576. [CrossRef] [PubMed]
24. Bahri, S.; Homaei, A.; Mosaddegh, E. Zinc sulfide-chitosan hybrid nanoparticles as a robust surface for immobilization of *Sillago sihama* α-amylase. *Colloids Surf. B Biointerfaces* **2022**, *218*, 112754. [CrossRef] [PubMed]
25. Fajingbesi, A.O.; Anzaku, A.A.; Akande, M.; Emmanuel, I.A.; Akwashiki, O. Production of protease enzyme from fish guts using *Pseudomonas fluorescens*, *Enterobacter cloacae* and *Bacillus megaterium*. *J. Clin. Path. Lab. Med.* **2018**, *2*, 1–7.
26. Farha, A.K.; Tr, T.; Purushothaman, A.; Salam, J.A.; Hatha, A.M. Phylogenetic diversity and biotechnological potentials of marine bacteria from continental slope of eastern Arabian Sea. *J. Genet. Eng. Biotechnol.* **2018**, *16*, 253–258. [CrossRef] [PubMed]
27. Ramezani-Pour, N.; Badoei-Dalfard, A.; Namaki-Shoushtari, A.; Karami, Z. Nitrile-metabolizing potential of *Bacillus cereus* strain FA12; Nitrilase production, purification, and characterization. *Biocatal. Biotransformation* **2015**, *33*, 156–166. [CrossRef]
28. Izadpanah Qeshmi, F.; Javadpour, S.; Malekzadeh, K.; Tamadoni Jahromi, S.; Rahimzadeh, M. Persian gulf is a bioresource of potent L-asparaginase producing bacteria: Isolation & molecular differentiating. *Int. J. Environ. Res.* **2014**, *8*, 813–818.

catalysts

Article

Correlation Relationship between Phase Inversion of Pickering Emulsions and Biocatalytic Activity of Microbial Transformation of Phytosterols

Wenyu Zhao [1], Haisheng Xie [1], Xuehong Zhang [2] and Zhilong Wang [1,*]

1 State Key Laboratory of Microbial Metabolism and Engineering Research Center of Cell & Therapeutic Antibody, Ministry of Education, School of Pharmacy, Shanghai Jiao Tong University, 800 Dongchuan, Shanghai 200240, China
2 State Key Laboratory of Microbial Metabolism, School of Life Science and Biotechnology, Shanghai Jiao Tong University, 800 Dongchuan, Shanghai 200240, China
* Correspondence: zlwang@sjtu.edu.cn

Abstract: Microbial transformation of hydrophobic phytosterols into the pharmaceutical steroid precursors AD (androst-4-ene-3, 17-dione) and ADD (androst-4-diene-3, 17-dione) in a water–plant oil two-phase system by *Mycolicibacterium neoaurum* is a paradigm of interfacial biocatalysis in Pickering emulsions stabilized by bacterial cells. In the present work, phase inversion of Pickering emulsions—i.e., Pickering emulsions turning from water-in-oil (W/O) emulsions into oil-in-water (O/W) ones—was observed during microbial transformation in the presence of high concentrations of crystal phytosterols. It was found that there is a correlation relationship between the phase behaviors of Pickering emulsions and the biocatalytic activity of utilizing *M. neoaurum* as a whole-cell catalyst. Efficient microbial transformation under the high crystal phytosterol loadings was achieved due to the formation of O/W emulsions where interfacial biocatalysis took place. Under the optimal conditions (volume ratio of soybean oil to water: 15:35 mL, phytosterols concentration in the soybean oil: 80 g/L, glucose as co-substrate in the aqueous culture medium: 10 g/L), the concentrations of AD and ADD reached 4.8 g/L based on the whole broth (16 g/L based on the oil phase) after microbial transformation for 9 days.

Keywords: microbial transformation; phytosterols; Pickering emulsion; phase inversion; plant oil

Citation: Zhao, W.; Xie, H.; Zhang, X.; Wang, Z. Correlation Relationship between Phase Inversion of Pickering Emulsions and Biocatalytic Activity of Microbial Transformation of Phytosterols. *Catalysts* **2023**, *13*, 72. https://doi.org/10.3390/catal13010072

Academic Editor: David D. Boehr

Received: 17 November 2022
Revised: 23 December 2022
Accepted: 28 December 2022
Published: 30 December 2022

1. Introduction

Using *Mycolicibacterium neoaurum* as whole-cell biocatalyst, microbial transformation of phytosterols into the corresponding steroid precursors—such as AD (androst-4-ene-3, 17-dione) and ADD (androst-4-diene-3, 17-dione)—is an important route in the steroid pharmaceutical industry [1–4]. The industrial process is a paradigm for microbial transformation of crystal substrates due to the limited solubility of phytosterols in aqueous solutions, such as the solubility of β-sitosterol—a major component of phytosterols (approximately 2 mg/L in an aqueous solution [5], approximately 68 mM (28 g/L) in BEHP (bis-(2-ethylhexyl) phthalate) [6])—and the solubility of soybean sterols (approximately 13 g/L in soybean oil and 8 g/L in coconut oil [5]). There are two major strategies for enhancing mass transport during microbial transformation of phytosterols. One method is the solubilization of phytosterols in an aqueous solution by addition of biocompatible solubilizers, such as hydroxypropyl-β-cyclodextrin [7,8]. Another is microbial transformation in a water–organic solvent two-phase system [4,5]. In this case, it is believed that the organic solvent phase acts as a reservoir of phytosterols and microbial transformation takes place in the aqueous solution phase, where diffusion of phytosterols from the organic solvent phase to the aqueous solution is more effective than that from phytosterol crystals to the aqueous solution [9]. Many other two-phase systems have also been exploited for microbial transformation of phytosterols, such as water–ionic liquid two-phase systems [10,11], polymer-based aqueous two-phase systems [12], and nonionic-surfactant-based cloud point systems [13,14].

Rather than focusing on the diffusion of phytosterols in the water–organic solvent two-phase system, the location of bacterial cells in the medium may be more important. Recently, a novel mechanism for *M. neoaurum* to utilize phytosterols solubilized in the BEHP (bis-(2-ethylhexyl) phthalate) phase of a water–BEHP two-phase system has been proposed. During microbial culture using cholesterol as the sole carbon source, the water–BEHP two-phase system turns into Pickering emulsions, where the hydrophobic bacterial cells act not only as solid emulsifiers by attachment on the oil–water interfaces, but also as biocatalysts for microbial transformation of cholesterol into AD and ADD [15]. This process is known as interfacial biocatalysis, because the bacteria attached to the oil–water interfaces assimilate both hydrophobic nutrients solubilized in the oil phase and hydrophilic ones solubilized in the water phase. By loading pre-cultured bacterial cells in the water–BEHP two-phase system (i.e., bacterial-cell-stabilized Pickering emulsions), microbial transformation of phytosterols also produces AD and ADD successfully [4]. However, both microbial growth and microbial transformation to produce AD and ADD in the water–BEHP two-phase system fail to utilize phytosterols as the sole carbon source. Conversely, the addition of glucose as a co-substrate for microbial growth and microbial transformation of phytosterols successfully produces AD and ADD in the bacterial-cell-stabilized Pickering emulsions [16]. This demonstrates that a certain amount of bacterial cells acting as solid emulsifiers is necessary for the formation of bacterial-cell-stabilized Pickering emulsions and, subsequently, the efficient microbial transformation of phytosterols to produce AD and ADD via interfacial biocatalysis. At the same time, crystal substrate inhibition—i.e., decreasing phytosterol degradation with the increase in crystal phytosterol loadings—is also observed during microbial transformation of phytosterols in the bacterial-cell-stabilized Pickering emulsions [16]. Unfortunately, the crystal substrate inhibition severely limits the efficient microbial transformation under the condition of high phytosterol loading.

Plant oils have been applied widely as natural solvents for extractive microbial transformation or fermentation due to their biocompatibility and low cost [17,18]. Water–plant oil two-phase systems have also been studied extensively for the microbial transformation of phytosterols in both academic research [5,19–22] and industrial application. In the present work, the influence of operational parameters (such as phytosterol concentrations, volume ratio of oil to water, and glucose concentration) on the basic structures of Pickering emulsions (such as the fraction of emulsion and the microscopic morphology) was investigated by using the microbial transformation of phytosterols in a water–soybean oil two-phase system as an example. Keeping interfacial biocatalysis in Pickering emulsions [15,16] in mind, the relationship between the emulsion structures and the bioactivity of microbial transformation under various conditions—especially with different phytosterol loadings—was correlated.

2. Results

2.1. Plant Oil Acting Both as Co-Substrate and as Solvent

Microbial culture was carried out in the aqueous culture medium in the absence of glucose while in the presence of various volumes of soybean oil as the sole carbon source (Figure 1A). It was observed that the biomass increased with the increase in soybean oil volume, indicating that the soybean oil served as the sole carbon source for microbial growth. In other words, microbial growth occurred in the water–soybean oil two-phase system. Furthermore, microbial transformation was also carried out in the water–soybean oil two-phase system by the addition of phytosterols (no phytosterol crystal was observed with 25 g/L of phytosterols in soybean oil) (Figure 1B). Phytosterols were converted into the corresponding AD and ADD. Meanwhile, the water–soybean oil two-phase system turned into W/O Pickering emulsions (data not shown). The microbial growth and microbial transformation occurred at the same time in the absence of glucose as a co-substrate. The phenomenon was the same as the microbial transformation of phytosterols in a water–BEHP two-phase system by the addition of glucose as a co-substrate for microbial growth [16].

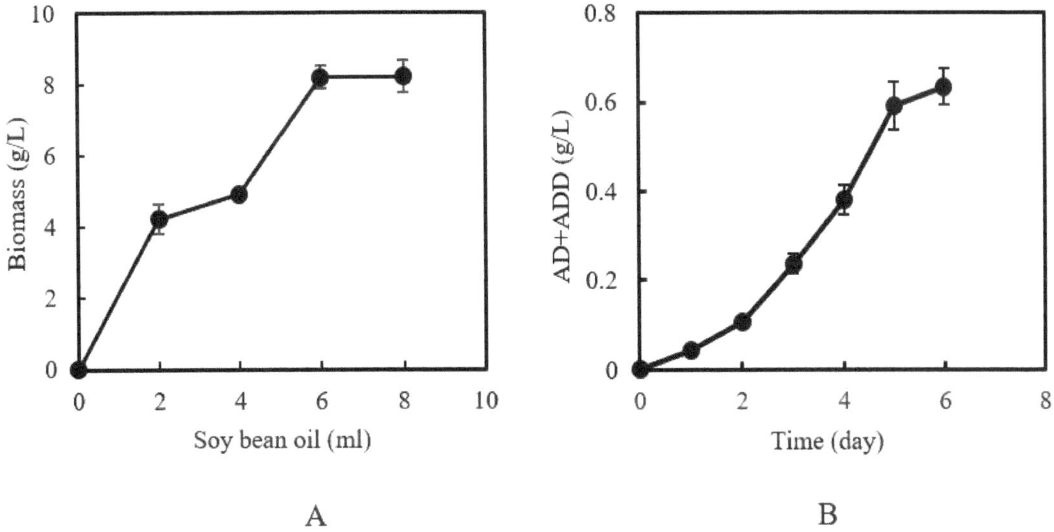

Figure 1. Soybean oil acting as both a substrate for microbial growth and a solvent for solubilized phytosterols: (**A**) Microbial culture with soybean oil as the sole carbon source. The microbial culture was carried out in the aqueous culture medium (40 mL) in the absence of glucose and with various volumes of soybean oil for 6 days. (**B**) Time course of microbial transformation of phytosterols in a water–soybean oil two-phase system. The basic condition was the aqueous culture medium in the absence of glucose (40 mL) and soybean oil (10 mL) with 25 g/L phytosterols.

2.2. *Effect of Volume Ratio of Oil to Water*

The plant oil, acting as both a co-substrate for microbial growth and an organic solvent for Pickering emulsions, influenced the structures of the Pickering emulsions and further influenced the accumulation of AD and ADD during microbial transformation in a water–soybean oil two-phase system (Figure 2). After microbial transformation, the water–soybean oil two-phase system turned into W/O Pickering emulsions. With the increase in the soybean oil volume from 5 to 15 mL (No. 1 to 3), the volume of the emulsion phase increased (i.e., the volume of the excess water phase decreased). The Pickering emulsions became very viscous with the increase in the soybean oil volume. With further increases in the oil volume (No. 4 and 5), the whole water–soybean oil two-phase system was emulsified completely, and no excess water phase was observed (Figure 2A). However, the flowability of the Pickering emulsions improved markedly at a high volume ratio of oil to water (data no shown, only direct observation of the shaken flasks during microbial transformation). Correspondingly, the microscopic morphology of the Pickering emulsions indicated that the sizes of the droplets decreased with the increase in the volume ratio of oil to water, i.e., relatively large droplets at a low volume ratio of oil to water (No. 1–3) but small droplets at a high volume ratio of oil to water (No. 5) (Figure 2B). The changes in the emulsion structures dramatically influenced the microbial transformation of the phytosterols (Figure 2C). The concentrations of AD and ADD increased dramatically with the increase in the oil volume (the concentration of phytosterols in the oil was kept at 25 g/L) and then reached their maximum of 2.1 g/L at the oil volume of 15 mL (No. 3), while further increases in the oil volume led to decreases in the AD and ADD concentrations.

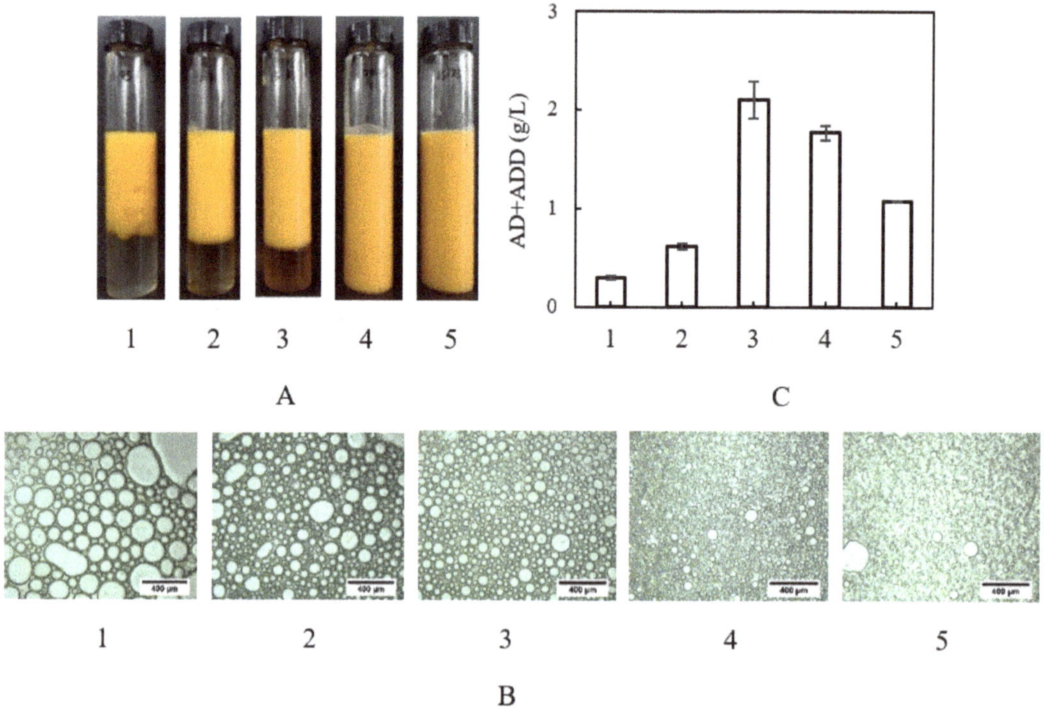

Figure 2. Effects of the volume ratio of soybean oil to water: (**A**) Emulsion fraction. (**B**) Microscopic emulsions (bar = 400 μm). (**C**) AD and ADD concentrations. In the basic experimental condition, the total volume of the soybean oil and the aqueous culture medium was 50 mL. There was no glucose in the aqueous culture medium. The concentration of phytosterols in the soybean oil was 25 g/L. The microbial transformation time was 7 days. No. 1, 2, 3, 4, and 5 corresponded to volume ratios of soybean oil to the aqueous culture medium of 5:45, 10:40, 15:35, 20:30, and 25:25 mL, respectively.

2.3. *Influence of Glucose as a Co-Substrate*

Glucose acting as a co-substrate enhanced the microbial growth, which influenced the Pickering emulsion structures and the subsequent accumulation of AD and ADD (Figure 3). Excess water phase was observed in the Pickering emulsions in the absence of glucose (No. 1). With the increase in the glucose concentration, the excess water phase disappeared (No. 2–4). Further increases in the glucose concentration caused the excess water phase to appear again (No. 5), whereupon the flowability of the emulsions became very good (Figure 3A). Microscopic observation showed that homogeneous small water droplets were present at the glucose concentration of 10 g/L (No. 2), while further increases in the glucose concentration led to the formation of heterogeneous irregular water droplets (No. 3 and 4). In particular, multiple oil-in-water-in-oil (O/W/O) emulsions were formed at the glucose concentration of 60 g/L (No. 5), where the bacterial cells were dispersed in the oil phase (Figure 3B). Correspondingly, the accumulation of AD and ADD achieved the maximum of 2.5 g/L at a glucose concentration of 10 g/L. Both the absence of glucose and high glucose concentrations were unfavorable for the accumulation of AD and ADD (Figure 3C).

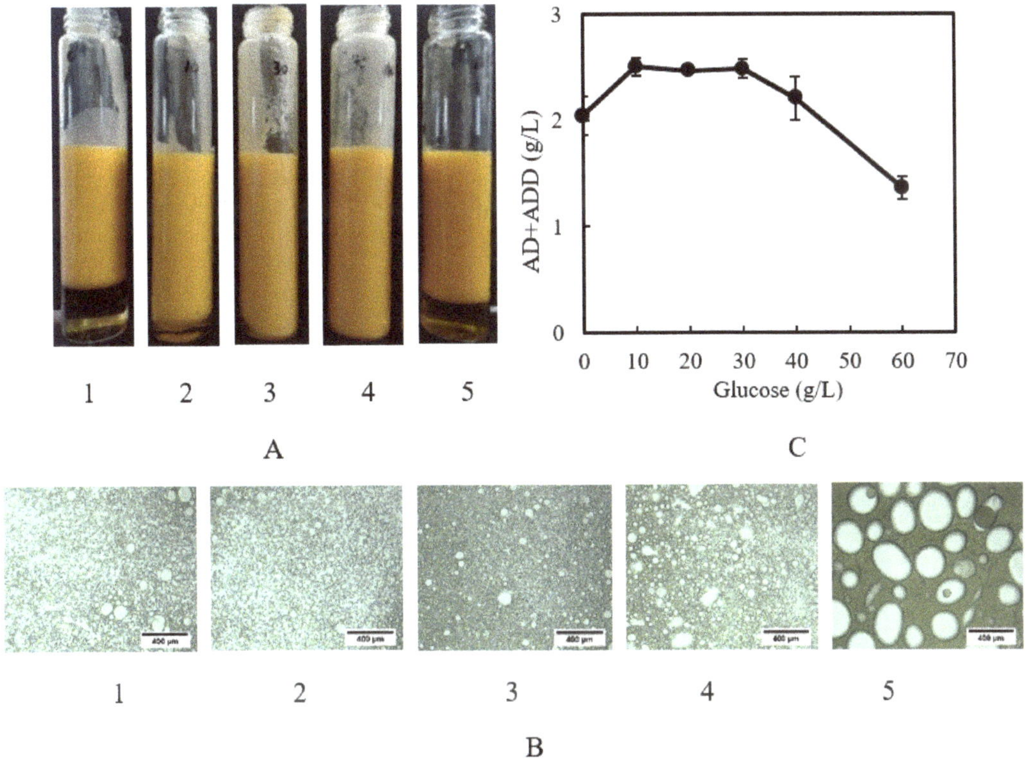

Figure 3. Influence of glucose acting as a co-substrate: (**A**) Emulsion fraction. (**B**) Microscopic emulsions (bar = 400 μm). (**C**) AD and ADD concentration. In the basic experimental condition, the volume ratio of soybean oil to the aqueous culture medium was 15:35 mL. The concentration of phytosterols in the soybean oil was 25 g/L. The microbial transformation time was 7 days. No. 1, 2, 3, 4, and 5 corresponded to glucose concentrations in the aqueous culture medium of 0, 10, 30, 40, and 60 g/L, respectively.

2.4. *Microbial Transformation of Crystal Phytosterols*

Under the optimal volume ratio of oil to water (15:35 mL) and glucose concentration (10 g/L) in the aqueous culture medium, the effect of phytosterol loading on the Pickering emulsion structures, and then on the accumulation of AD and ADD, was examined (Figure 4). No phytosterol crystals were observed at phytosterol concentrations below 25 g/L in the soybean oil. No excess water phase was observed when the phytosterol concentration was 10, 25, or 40 g/L (No. 1–3), while the flowability of the Pickering emulsions improved markedly at the phytosterol concentration of 40 g/L (No. 3). Further increases in the phytosterol concentration caused an excess water phase to appear, where some microaggregates of phytosterol crystals were observed with the adsorbed bacterial cells. The formation of bacteria–phytosterol crystal microaggregates was confirmed by confocal laser scanning microscopy (CLSM) in our previous work [16]. At the same time, the flowability of the Pickering emulsions further increased (No. 4 and 5) (Figure 4A). The microscopic morphology indicated that some homogeneous small water droplets were observed at the low phytosterol concentrations where no phytosterol crystals were observed (No. 1 and 2). Some heterogeneous irregular water droplets were observed and some bacterial cells were partitioned in the continuous oil phase at the phytosterol concentration of 40 g/L (No. 3). When increasing the phytosterol concentration to 50 g/L (No. 4), multiple accompanying O/W/O emulsions with good flowability low stability and were found, characteristic of the phase inversion of Pickering emulsions. Subsequently, the Pickering emulsions turned from W/O into O/W type (known as phase inversion of Pickering emulsions [23]) at very high phytosterol concentrations (e.g., 80 g/L, No. 5). In the O/W Pickering emulsions, no bacterial cells were observed in the continuous water phase (Figure 4B). Corresponding to the various structures of the Pickering emulsions, the accumulation of AD and ADD also changed dramatically (Figure 4C). The product concentration increased with the increase in the phytosterol concentration to 25 g/L and then decreased to the minimum value at the phytosterol concentration of 40 g/L. Very interestingly, the product concentration increased with the further increase in the phytosterol concentration above 40 g/L. The highest product concentration (4.2 g/L) was achieved at the phytosterol concentration of 80 g/L, which provided a method for microbial transformation with high phytosterol loadings.

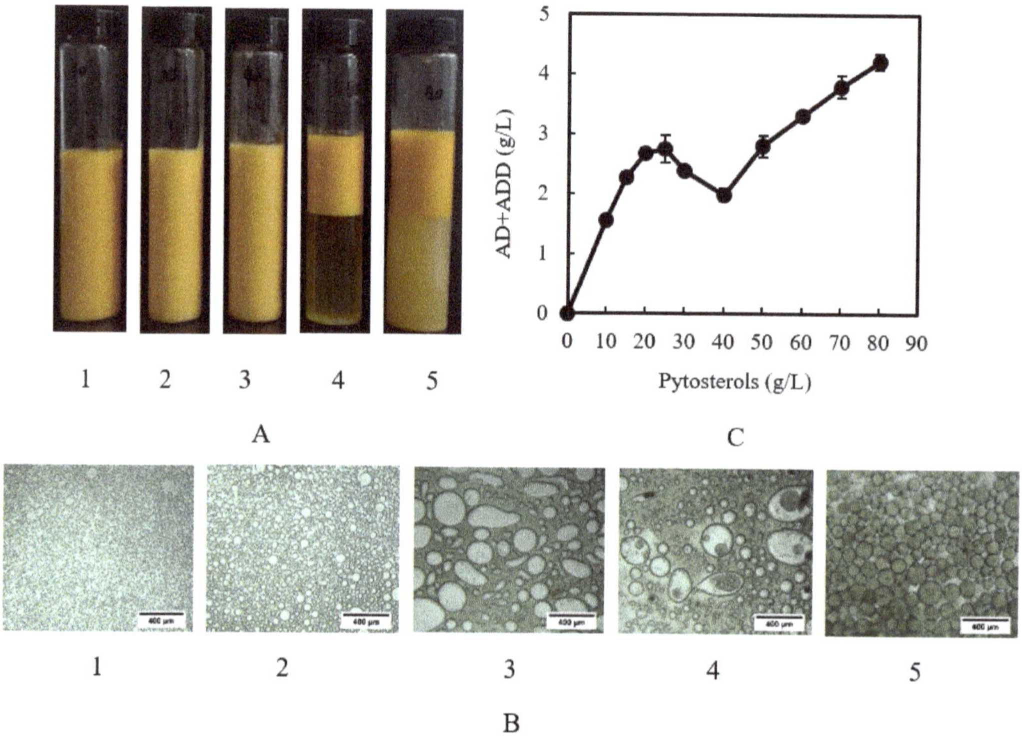

Figure 4. Microbial transformation with various phytosterol loadings: (**A**) Emulsion fraction. (**B**) Microscopic emulsions (bar = 400 μm). (**C**) AD and ADD concentrations. In the basic experimental condition, the volume ratio of soybean oil to the aqueous culture medium was 15:35 mL. The glucose concentration in the aqueous culture medium was 10 g/L. The microbial transformation time was 7 days. No. 1, 2, 3, 4, and 5 corresponded to phytosterol concentrations in the soybean oil of 10, 25, 40, 50, and 80 g/L, respectively.

2.5. *Time Course of Microbial Transformation*

The time course of microbial transformation with high phytosterol loadings was further examined (Figure 5). At the beginning of the microbial transformation, a W/O Pickering emulsion was observed (No. 1), and it had already been confirmed that the Pickering emulsions were stabilized by phytosterol crystals as solid emulsifiers [16]. Very recently, phytosterol crystals have also been utilized as emulsifiers for the fabrication of eatable W/O Pickering emulsions in the food field [24]. With the progress of microbial transformation, the volume of the excess water phase increased and the flowability of the Pickering emulsions improved. At the same time, some bacteria–phytosterol crystal microaggregates were also observed in the excess water phase (No. 2–5) (Figure 5A). The microscopic morphology demonstrated that the shape of the water droplets in the continuous oil phase became irregular, and some bacteria–phytosterol crystal microaggregates were partitioned in the continuous oil phase within 1–3 days (No. 1 and 2). On the 4th day, multiple O/W/O emulsions were observed, where bacterial cells and phytosterol crystals were significantly partitioned in the oil phases (No. 3). Phase inversion occurred on the 5th day, where the multiple emulsions turned into O/W Pickering emulsions (No. 4). The O/W Pickering emulsions were maintained until the end of the microbial transformation (the 9th day, No. 5) (Figure 5B). Corresponding to the formation of O/W Pickering emulsions, rapid accumulation of AD and ADD was found, and 4.8 g/L of AD and ADD was achieved on the 9th day (Figure 5C).

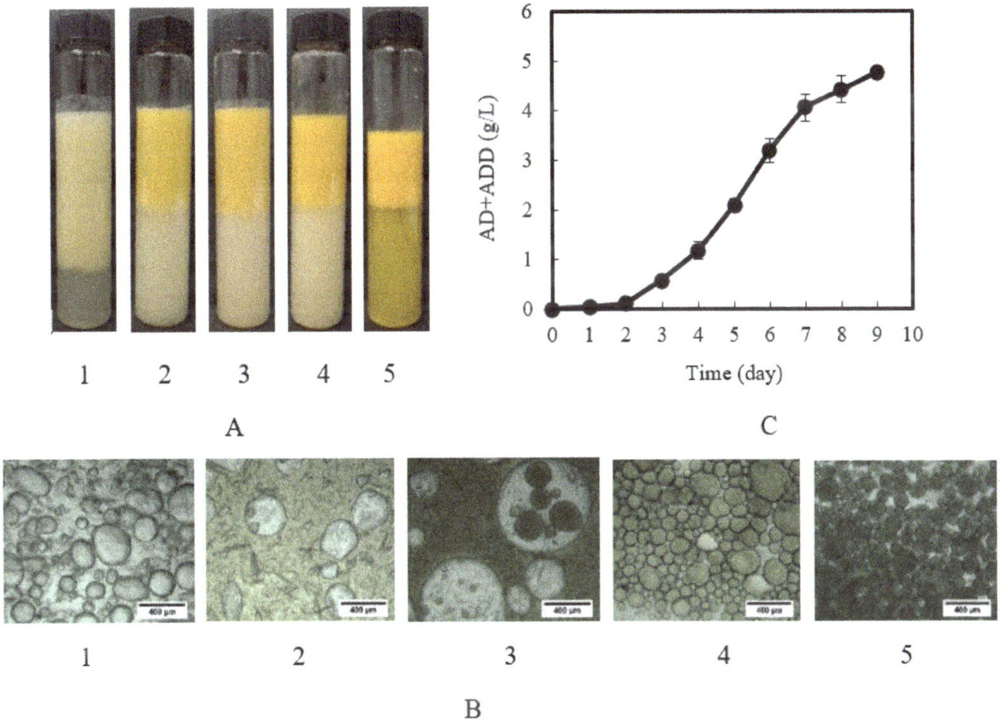

Figure 5. Time course of microbial transformation with high loading of crystal phytosterols: (**A**) Emulsion fraction. (**B**) Microscopic emulsions (bar = 400 μm). (**C**) AD and ADD concentrations. In the basic experimental condition, the volume ratio of soybean oil to the aqueous culture medium was 15:35 mL. The glucose concentration in the aqueous culture medium was 10 g/L. The concentration of phytosterols in the soybean oil was 80 g/L. No. 1, 2, 3, 4, and 5 corresponded to microbial transformation times of 1, 3, 4, 5, and 9 days, respectively.

3. Discussion

Pickering emulsions turning from W/O to O/W type, or vice versa, is called phase inversion [25,26]. Around the phase inversion point, Pickering emulsions usually present as multiple unstable emulsions with good flowability. Methods of adjusting phase inversion range from the co-stabilization of Pickering emulsions by adding other soft or hard particles [25,27], to the modification of solid particles' wettability by physical adsorption [16,28,29], chemical graft modification [30,31], or even genetic manipulation of bacterial cells [32]. In addition to wettability, the concentration of solid emulsifiers is also a major factor influencing the phase inversion of Pickering emulsions [23]. Under the condition of high particle concentrations, self-aggregation causes the wettability of solid particles to change and phase inversion to occur [23,33,34]. It has also been reported that phase inversion of Pickering emulsions occurs with the increase in the inner phase fraction [35], and the corresponding demulsification effect is also applied for the downstream process of microbial fermentation [26]. In our previous work, we confirmed that hydrophobic *M. neoaurum* cells act as solid particles to stabilize W/O Pickering emulsions [15]. It has also been reported that phytosterol crystals act as solid particles to stabilize W/O Pickering emulsions [24]. Furthermore, adsorption of *M. neoaurum* cells onto phytosterol crystals to form crystal phytosterol–bacteria microaggregates, which has been confirmed by CLSM analysis, exhibits wettability unlike that of crystal phytosterol particles or bacteria [16]. The altered wettability of crystal phytosterol–bacteria microaggregates leads to the instability of Pickering emulsions. In the present work, the irregular shape of water droplets, multiple emulsions, and phase inversion of Pickering emulsions were observed with the increase in the phytosterol concentration, due to the altered wettability of the crystal phytosterol–bacteria microaggregates (Figure 4B). Vice versa, a similar phenomenon was also observed with the increase in the bacterial cell concentration during microbial growth (Figure 5B). Very high cell concentrations, as was achieved by the addition of high concentrations of glucose as a co-substrate, may lead to the formation of irregularly shaped water droplets or even multiple emulsions (Figure 3B). This may share the same mechanism of self-aggregation resulting from the high concentration of solid particles [23,33,34]. However, only an irregular shape of water droplets was observed at a very high volume ratio of oil to water (Figure 2B). Although high cell concentrations (caused by the high fraction of soybean oil; Figure 1A) favored the phase inversion from W/O to O/W, this trend may be counteracted by the high volume fraction of the inner phase (the water phase, not the soybean oil phase) in the W/O Pickering emulsions (favorable to the phase inversion from W/O to O/W [26]).

Hydrophobic solid particles act as emulsifiers of Pickering emulsions by attaching to the oil–water interfaces. When the solid particles are functionalized with catalytic activity, chemical reactions occur at the oil–water interfaces, and the solid particles act as both catalysts for the chemical reactions and solid emulsifiers for the Pickering emulsions, known as interfacial catalysis [36–38]. *M. neoaurum* cells catalyze the microbial transformation of phytosterols into AD and ADD (Figure 1B). At the same time, hydrophobic *M. neoaurum* cells can stabilize Pickering emulsions in a water–BEHP two-phase system [15,16] as well as in a water–soybean oil two-phase system (Figure 2A). In this case, interfacial biocatalysis provides a novel strategy for microbial transformation of hydrophobic chemicals, such as cholesterol [15] and phytosterols [16]. In the present work, we further confirmed the correlation relationship between the morphological structures of Pickering emulsions and the efficiency of microbial transformation. Around the phase inversion point of Pickering emulsions (No. 4 of Figure 4 with phytosterol concentration 40 g/L), multiple unstable emulsions with good flowability were observed, and the microbial transformation exhibited the lowest AD and ADD concentrations. With phytosterol concentrations below 40 g/L, the AD and ADD concentrations increased and then decreased with the increase in the phytosterol concentration, reaching a maximum volume of 2.8 g/L for AD and ADD at a phytosterol concentration of 25 g/L. This phenomenon is known as crystal substrate inhibition. The mechanism behind this phenomenon can be attributed to the unstable Pickering emulsions due to the formation of crystal phytosterol–bacteria microaggregates [16]. At phytosterol concentrations above 40 g/L, relatively stable O/W Pickering emulsions were formed, and the microbial transformation maintained a relative high activity. In addition, the crystal phytosterol–bacteria microaggregates in the excess water phase (Figure 4A) may also provide an additional route for microbial transformation of phytosterols via direct interaction between phytosterol crystals and bacteria [39]. In this way, efficient microbial transformation with a high phytosterol loading becomes possible. At the same time, bacterial cell concentration is also an important factor affecting the morphology of Pickering emulsions (Figures 2 and 3). Thus, process parameters—such as glucose concentration [40] or the volume ratio of oil to water [5,19–21,41]—also strongly affect the microbial transformation of phytosterols in Pickering emulsions. Under the condition of a relatively high phytosterol loading, the time course of microbial transformation indicated that AD (ADD) accumulation was achieved at a relatively high rate (Figure 5). Microbial transformation of phytosterols involves the cascade processes of substrate phytosterol transport in the extracellular medium, transport of phytosterol across the cell membrane, and enzymatic catalysis in the intracellular environment. It should be pointed out that engineering of *Mycobacterium* sp. for enhancing transport of phytosterol across the cell membrane as well as the enzymatic degradation of phytosterols is only efficient [42,43] when phytosterol transport in the extracellular medium is not a rate-limited step.

4. Materials and Methods

4.1. Materials

Mycolicibacterium neoaurum (China Center of Industrial Culture Collection, CICC 21097)—a strain for accumulation of the steroid synthon AD by microbial transformation of sterols—was used in this study. Soybean oil (the major fatty acid constituents of which are C16:0, C18:1, and C18:2) was purchased from a local supermarket (Shanghai, China). AD and ADD from Sigma-Aldrich (St. Louis, MO, USA) were utilized as standard substances for HPLC analysis. Phytosterols (major components including β-sitosterol 40%, campesterol 20%, stigmasterol 17%, and brassicasterol 5%) from pilot plants were utilized as substrates for microbial transformation to produce the steroid synthons AD and ADD. Other chemicals were of analytical grade.

4.2. Microbial Transformation

M. neoaurum was maintained on a nutrient agar plate (10 g of yeast extract, 10 g of peptone, 10 g of glucose, and 15 g of agar per liter of water) at 4 °C. An isolated colony from the agar plate was inoculated into 50 mL of inoculum medium, i.e., 15 g of glucose, 5 g of $NaNO_3$, 5.4 g of K_2HPO_4, 2.6 g of KH_2PO_4, and 0.5 g of $MgSO_4{\cdot}7H_2O$ per liter of tap water, and an initial pH of 7, cultured in a 250 mL Erlenmeyer flask at 200 rpm and 30 °C for 3 days.

The aqueous culture medium consisted of 10 g of $NaNO_3$, 5.4 g of K_2HPO_4, 2.6 g of KH_2PO_4, 0.5 g of $MgSO_4{\cdot}7H_2O$, 0.1 g of $CaCl_2$, and 0.05 g of $FeCl_3{\cdot}6H_2O$ per liter of tap water, in the presence or absence of 10 g/L glucose as a co-substrate. Next, 15 mL of soybean oil with 25 g/L phytosterols (no phytosterol crystals were observed), 35 mL of aqueous culture medium, and 2 mL of inoculum were combined in a 250 mL Erlenmeyer flask, which was incubated at 30 °C and 200 rpm for microbial transformation for 7 days (unless otherwise specified). During the microbial transformation, the water–soybean oil two-phase system turned into Pickering emulsions due to the growing bacterial cells acting as solid-particle Pickering emulsifiers.

4.3. Analysis of AD and ADD Concentrations

After microbial transformation, the whole transformation medium was also used for the analysis of AD and ADD concentrations. First, 30 mL of ethyl acetate was added to each tube and shaken at 200 rpm for 1 h. The AD and ADD, residual phytosterols, and soybean oil were extracted into the ethyl acetate layer. The total mixture was centrifuged at 10,000 rpm for 10 min. Aliquots of the ethyl acetate layer were withdrawn for AD and ADD analysis.

Aliquots (400 μL) of the ethyl acetate layer (containing AD and ADD, phytosterols, and soybean oil) were placed in a 70 °C oven to completely evaporate the ethyl acetate. The residues (oil and solids) were extracted 4 times with 1 mL of aqueous methanol solution (80%, V/V). All of the extracts of each sample were combined for HPLC analysis, which was performed using a Shimadzu LC-20AT system (Shimadzu, Kyoto, Japan) equipped with an InertSustain C18 column (250 × 4.6 mm, 5 μm). The column was eluted with methanol and water (80:20, *V*/*V*) at a flow rate of 1 mL/min. The major products of microbial transformation were AD and ADD. The mass ratio of AD to ADD was approximately 5:1. AD and ADD were detected at 254 nm with retention times of 4.4 and 5.2 min, respectively. The corresponding concentration of the products AD and ADD was achieved based on the whole volume of microbial transformation.

4.4. Estimation of Microbial Growth

A certain volume of soybean oil (2, 4, 6, or 8 mL) was added to 40 mL of the aqueous culture medium (in the absence of glucose) as the sole carbon source. Then, 2 mL of inoculum was added, which was incubated at 30 °C and shaken at 200 rpm for 6 days for microbial culture. After the microbial culture, 30 mL of ethyl acetate was added to the broth, which was transferred into a 100 mL tube and centrifuged at 10,000 rpm for 10 min. The sediments in the aqueous solution phase were collected and dried at 70 °C for over 24 h until to a constant weight, before being utilized for the estimation of biomass.

4.5. Investigation of Pickering Emulsions

After microbial transformation, the Pickering emulsions and some excess water were transferred into a glass tube (80 mL) and stood for 2 h to allow for complete phase separation, which could be separated into an emulsion phase and an excess water phase. The phase separation system was captured using a digital camera. The emulsion fraction was deduced from the volume of the emulsion phase.

The emulsion type was inferred from the drop test, i.e., dilution of a drop of emulsion in either pure oil or water. The water-in-oil (W/O) emulsion was characterized by droplets that dispersed rapidly in oil while remaining in an agglomeration in the water, or vice versa for the oil-in-water (O/W) emulsion [44]. A small amount of the emulsion phase was picked up by a pipette and evenly spread on a glass slide, and microscopic images of the emulsion droplets were captured using an optical microscope (Olympus BX53M, Shinjuku, Tokyo, Japan) equipped with a high-speed CMOS camera (Basler acA2440-35uc, Ahrensburg, Germany).

5. Conclusions

Phase inversion of Pickering emulsions occurs during microbial transformation in the presence of a high crystal phytosterol loading. Efficient microbial transformation at high phytosterol loadings was achieved, where stable O/W Pickering emulsions were formed. The results confirmed the correlation relationship between the emulsion structures and the biocatalytic activity of the bacterial cells. In addition, process parameters such as the glucose concentration and the volume ratio of oil to water should also be optimized during microbial transformation, due to the influence of bacterial cell concentrations on the structures of Pickering emulsions.

Author Contributions: W.Z. performed the experiments and data analysis as well as writing the manuscript. H.X. assisted in the experiments and data analysis as well as the writing of the manuscript. X.Z. and Z.W. conceived and reviewed the manuscript. All authors have read and agreed to the published version of the manuscript.

Funding: The financial support from the National Natural Science Foundation of China (No: 22178218) is acknowledged.

Data Availability Statement: Data sharing is not applicable to this article as no datasets were generated or analyzed during this study.

Conflicts of Interest: The authors declare no competing financial interest.

References

1. Feng, J.; Wu, Q.; Zhu, D.; Ma, Y. Biotransformation enables innovations toward green synthesis of steroidal pharmaceuticals. *ChemSusChem* **2022**, *15*, e202102399. [CrossRef] [PubMed]
2. Fernandes, P.; Cruz, A.; Angelova, B.; Pinheiro, H.M.; Cabral, J.M.S. Microbial conversion of steroid compounds: Recent developments. *Enzyme Microb. Technol.* **2003**, *32*, 688–705. [CrossRef]
3. Liu, N.; Feng, J.; Zhang, R.; Chen, X.; Li, X.; Yao, P.; Wu, Q.; Ma, Y.; Zhu, D. Efficient microbial synthesis of key steroidal intermediates from bio-renewable phytosterols by genetically modified *Mycobacterium fortuitum* strains. *Green Chem.* **2019**, *21*, 4076–4083. [CrossRef]
4. Lorena, F.-C.; Beatriz, G.; José, L. New insights on steroid biotechnology. *Front Microbiol.* **2018**, *9*, 958. [CrossRef]
5. Phase, N.; Patil, S. Natural oils are better than organic solvents for the conversion of soybean sterols to 17-ketosteroids by *Mycobacterium fortuitum*. *World J. Microbiol. Biotechnol.* **1994**, *10*, 228–229. [CrossRef]
6. Cruz, A.; Fernandes, P.; Cabral, J.M.S.; Pinheiro, H.M. Whole-cell bioconversion of β-sitosterol in aqueous-organic two-phase systems. *J. Mol. Catal. B* **2001**, *11*, 579–585. [CrossRef]
7. Donova, M.V.; Dovbnya, D.V.; Koshcheyenko, K.A. Modified CDs-mediated enhancement of microbial sterol-side-chain degradation. In Proceedings of the Eighth International Symposium on Cyclodextrin, Budapest, Hungary, 31 March–2 April 1996; pp. 527–530.
8. Su, L.; Xu, S.; Shen, Y.; Xia, M.; Ren, X.; Wang, L.; Shang, Z.; Wang, M. The sterol carrier hydroxypropyl-β-cyclodextrin enhances the metabolism of phytosterols by *Mycobacterium neoaurum*. *Appl. Environ. Microbiol.* **2020**, *86*, e00441-20. [CrossRef]
9. Heipieper, H.J.; Neumann, G.; Cornelissen, S.; Meinhardt, F. Solvent-tolerant bacteria for biotransformation in two-phase fermentation systems. *Appl. Microbiol. Biotechnol.* **2007**, *74*, 961–973. [CrossRef]
10. Xiao, X.; He, J.-K.; Guan, Y.-X.; Yao, S.-J. Effect of cholinium amino acids ionic liquids as cosolvents on the bioconversion of phytosterols by *Mycobacterium sp.* resting cells. *ACS Sustain. Chem. Eng.* **2020**, *8*, 17124–17132. [CrossRef]
11. Yuan, J.-J.; Guan, Y.-X.; Wang, Y.-T.; Wang, H.-Q.; Yao, S.-J. Side-chain cleavage of phytosterols by *Mycobacterium* sp. MB 3683 in a biphasic ionic liquid/aqueous system. *J. Chem. Technol. Biotechnol.* **2016**, *91*, 2631–2637. [CrossRef]
12. Carvalho, F.; Marques, M.P.C.; de Carvalho, C.C.C.R.; Cabral, J.M.S.; Fernandes, P. Sitosterol bioconversion with resting cells in liquid polymer based systems. *Biores. Technol.* **2009**, *100*, 4050–4053. [CrossRef] [PubMed]
13. Wang, Z.; Zhao, F.; Chen, D.; Li, D. Biotransformation of phytosterol to adustra-diene-dione by resting cells of *Mycobacterium* in cloud point system. *Process Biochem.* **2006**, *41*, 557–561. [CrossRef]
14. Wang, Z.; Zhao, F.; Hao, X.; Chen, D.; Li, D. Microbial transformation of hydrophobic compound in cloud point system. *J. Mol. Catal. B* **2004**, *27*, 147–153. [CrossRef]
15. Xie, H.; Zhao, W.; Ali, D.C.; Zhang, X.; Wang, Z. Interfacial biocatalysis in bacteria-stabilized Pickering emulsions for microbial transformation of hydrophobic chemicals. *Catal. Sci. Technol.* **2021**, *11*, 2816–2826. [CrossRef]
16. Zhao, W.; Xie, H.; Zhang, X.; Wang, Z. Crystal substrate inhibition during microbial transformation of phytosterols in Pickering emulsions. *Appl. Microbiol. Biotechnol.* **2022**, *106*, 2403–2414. [CrossRef]
17. Lemos, D.A.; Sonego, J.L.S.; Boschiero, M.V.; Araujo, E.C.C.; Cruz, A.J.G.; Badino, A.C. Selection and application of nontoxic solvents in extractive ethanol fermentation. *Biochem. Eng. J.* **2017**, *127*, 128–135. [CrossRef]
18. Li, N.; Han, Z.; O'Donnell, T.J.; Kurasaki, R.; Kajihara, L.; Williams, P.G.; Tang, Y.; Su, W.W. Production and excretion of astaxanthin by engineered *Yarrowia lipolytica* using plant oil as both the carbon source and the biocompatible extractant. *Appl. Microbiol. Biotechnol.* **2020**, *104*, 6977–6989. [CrossRef]
19. Li, H.; Fu, Z.; Li, H.; Zhang, X.; Shi, J.; Xu, Z. Enhanced biotransformation of dehydroepiandrosterone to 3β,7α,15α-trihydroxy-5-androsten-17-one with *Gibberella intermedia* CA3-1 by natural oils addition. *J. Ind. Microbiol. Biotechnol.* **2014**, *41*, 1497–1504. [CrossRef]
20. Stefanov, S.; Yankov, D.; Beschkov, V. Biotransformation of phytosterols to androstenedione in two phase water-oil systems. *Chem. Biochem. Eng. Q* **2006**, *20*, 421–427.
21. Xu, Y.-G.; Guan, Y.-X.; Wang, H.-Q.; Yao, S.-J. Microbial side-chain cleavage of phytosterols by *Mycobacteria* in vegetable oil/aqueous two-phase system. *Appl. Biochem. Biotechnol.* **2014**, *174*, 522–533. [CrossRef]
22. Zhou, X.; Zhang, Y.; Shen, Y.; Zhang, X.; Xu, S.; Shang, Z.; Xia, M.; Wang, M. Efficient production of androstenedione by repeated batch fermentation in waste cooking oil media through regulating NAD+/NADH ratio and strengthening cell vitality of *Mycobacterium neoaurum*. *Bioresour. Technol.* **2019**, *279*, 209–217. [CrossRef] [PubMed]
23. Binks, B.P.; Fletcher, P.; Holt, B.; Beaussoubre, P.; Wong, K. Phase inversion of particle-stabilised perfume oil-water emulsions: Experiment and theory. *Phys. Chem. Chem. Phys.* **2010**, *12*, 11954–11966. [CrossRef] [PubMed]
24. Lan, M.; Zheng, J.; Huang, C.; Wang, Y.; Hu, W.; Lu, S.; Liu, F.; Ou, S. Water-in-oil Pickering emulsions stabilized by microcrystalline phytosterols in oil: Fabrication mechanism and application as a salt release system. *J. Agric. Food Chem.* **2022**, *70*, 5408–5416. [CrossRef]
25. Binks, B.P.; Lumsdon, S.O. Transitional phase inversion of solid-stabilized emulsions using particle mixtures. *Langmuir* **2000**, *16*, 3748–3756. [CrossRef]
26. Glonke, S.; Sadowski, G.; Brandenbusch, C. Applied catastrophic phase inversion: A continuous non-centrifugal phase separation step in biphasic whole-cell biocatalysis. *J. Ind. Microbiol. Biotechnol.* **2016**, *43*, 1527–1535. [CrossRef]

27. Xie, H.; Zhao, W.; Zhang, X.; Wang, Z. Demulsification of bacteria-stabilized Pickering emulsion using modified silica nanoparticle. *ACS Appl. Mater. Interfaces* **2022**, *14*, 24102–24112. [CrossRef]
28. Jiang, H.; Liu, L.; Li, Y.; Yin, S.; Ngai, T. Inverse Pickering emulsion stabilized by binary particles with contrasting characteristics and functionality for interfacial biocatalysis. *ACS Appl. Mater. Interfaces* **2020**, *12*, 4989–4997. [CrossRef]
29. Nallamilli, T.; Binks, B.P.; Mani, E.; Basavaraj, M.G. Stabilization of Pickering emulsions with oppositely charged latex particles: Influence of various parameters and particle arrangement around droplets. *Langmuir* **2015**, *31*, 11200–11208. [CrossRef]
30. Chen, Z.; Zhao, C.; Ju, E.; Ji, H.; Ren, J.; Binks, B.P.; Qu, X. Design of surface-active artificial enzyme particles to stabilize Pickering emulsions for high-performance biphasic biocatalysis. *Adv. Mater.* **2016**, *28*, 1682–1688. [CrossRef]
31. Meng, T.; Bai, R.; Wang, W.; Yang, X.; Guo, T.; Wang, Y. Enzyme-loaded mesoporous silica particles with tuning wettability as a Pickering catalyst for enhancing biocatalysis. *Catalysts* **2019**, *9*, 78. [CrossRef]
32. Jia, X.; Zhang, S.; Li, J.; Xia, J.; Yao, R.; Zhao, X.; Wu, B.; Bai, F.; Xiao, Y. Engineered bacterial biofloc formation enhancing phenol removal and cell tolerance. *Appl. Microbiol. Biotechnol.* **2020**, *104*, 1187–1199. [CrossRef] [PubMed]
33. Binks, B.P.; Philip, J.; Rodrigues, J.A. Inversion of silica-stabilized emulsions induced by particle concentration. *Langmuir* **2005**, *21*, 3296–3302. [CrossRef] [PubMed]
34. Whitby, C.P.; Parthipan, R. Influence of particle concentration on multiple droplet formation in Pickering emulsions. *J. Colloid Interf. Sci.* **2019**, *554*, 315–323. [CrossRef] [PubMed]
35. Lei, C.; Xie, Y.; Wu, Y.; Li, Y.; Li, B.; Pei, Y.; Liu, S. Properties and stability of water-in-water emulsions stabilized by microfibrillated bacterial cellulose. *Food Hydrocoll.* **2022**, *130*, 107698. [CrossRef]
36. Pera-Titus, M.; Leclercq, L.; Clacens, J.M.; de Campo, F.; Nardello-Rataj, V. Pickering interfacial catalysis for biphasic systems: From emulsion design to green reactions. *Angew. Chem. Int. Ed.* **2015**, *54*, 2006–2021. [CrossRef] [PubMed]
37. Wang, Z.; van Oers, M.C.M.; Rutjes, F.P.J.T.; van Hest, J.C.M. Polymersome colloidosomes for enzyme catalysis in a biphasic system. *Angew. Chem. Int. Ed.* **2012**, *51*, 10746–10750. [CrossRef]
38. Rodriguez, A.M.B.; Binks, B.P. Catalysis in Pickering emulsions. *Soft Matter* **2020**, *16*, 10221–10243. [CrossRef]
39. Achberger, A.M.; Doyle, S.M.; Mills, M.I.; Holmes, C.P.; Quigg, A.; Sylvan, J.B. Bacteria-oil microaggregates are an important mechanism for hydrocarbon degradation in the marine water column. *mSystems* **2021**, *6*, e01105-21. [CrossRef]
40. Shao, M.; Zhang, X.; Rao, Z.; Xu, M.; Yang, T.; Li, H.; Xu, Z.; Yang, S. A mutant form of 3-ketosteroid-Δ1 -dehydrogenase gives altered androst-1, 4-diene-3, 17-dione/androst-4-ene-3, 17-dione molar ratios in steroid biotransformations by *Mycobacterium neoaurum* ST-095. *J. Ind. Microbiol. Biotechnol.* **2016**, *43*, 691–701. [CrossRef]
41. Angelova, B.; Fernandes, P.; Cruz, A.; Pinheiro, H.M.; Mutafov, S.; Cabral, J.M.S. Hydroxylation of androstenedione by resting *Rhodococcus* sp cells in organic media. *Enzyme Microb. Technol.* **2005**, *37*, 718–722. [CrossRef]
42. He, K.; Sun, H.; Song, H. Engineering phytosterol transport system in *Mycobacterium sp.* strain MS136 enhances production of 9a-hydroxy-4-androstene-3, 17-dione. *Biotechnol. Lett.* **2018**, *40*, 673–678. [CrossRef] [PubMed]
43. Sun, H.; Yang, J.; He, K.; Wang, Y.-P.; Song, H. Enhancing production of 9a-hydroxy-androst-4-ene-3, 17-dione (9-OH-AD) from phytosterols by metabolic pathway engineering of *Mycobacteria*. *Chem. Eng. Sci.* **2021**, *230*, 116195. [CrossRef]
44. Song, N.; Wang, A.; Li, J.; Zhu, Z.; Shi, H.; Ma, X.; Sun, D. Study on influencing factors of Pickering emulsions stabilized by hydroxyapatite nanoparticles with nonionic surfactants. *Soft Matter* **2018**, *14*, 3889–3901. [CrossRef] [PubMed]

catalysts

MDPI

Article

Enhancement of α-Ketoglutaric Acid Production by *Yarrowia lipolytica* Grown on Mixed Renewable Carbon Sources through Adjustment of Culture Conditions

Ludwika Tomaszewska-Hetman *, Anita Rywińska, Zbigniew Lazar and Waldemar Rymowicz

Department of Biotechnology and Food Microbiology, Wrocław University of Environmental and Life Sciences, Chełmońskiego Str. 37, 51-630 Wroclaw, Poland

* Correspondence: ludwika.tomaszewska-hetman@upwr.edu.pl

Abstract: α-Ketoglutaric acid (KGA) is a valuable compound with a wide range of applications, e.g., in the cosmetics, pharmaceutical, chemical and food industries. The present study aimed to enhance the efficiency of KGA production by *Yarrowia lipolytica* CBS146773 from renewable carbon sources. In the investigation, various factors that may potentially affect KGA biosynthesis were examined in bioreactor cultures performed on a simple medium containing glycerol (20 g/L) and fed with four portions of a substrate mixture (15 + 15 g/L of glycerol and rapeseed oil). It was found that the process may be stimulated by regulation of the medium pH and aeration, application of selected neutralizing agents, supplementation with thiamine and addition of sorbitan monolaurate, whereas presence of biotin and iron ions had no positive effect on KGA biosynthesis. Adjustment of the parameters improved the process efficiency and allowed 82.4 g/L of KGA to be obtained, corresponding to productivity of 0.57 g/L h and yield of 0.59 g/g. In addition, the production of KGA was characterized by a low level (≤6.3 g/L) of by-products, i.e., citric and pyruvic acids. The results confirmed the high potential of renewable carbon sources (glycerol + rapeseed oil) for effective KGA biosynthesis by *Yarrowia lipolytica*.

Keywords: α-ketoglutaric acid biosynthesis; *Yarrowia lipolytica*; glycerol; rapeseed oil; culture conditions optimization

Citation: Tomaszewska-Hetman, L.; Rywińska, A.; Lazar, Z.; Rymowicz, W. Enhancement of α-Ketoglutaric Acid Production by *Yarrowia lipolytica* Grown on Mixed Renewable Carbon Sources through Adjustment of Culture Conditions. *Catalysts* **2023**, *13*, 14. https://doi.org/10.3390/catal13010014

Academic Editors: Zhilong Wang and Tao Pan

Received: 28 November 2022
Revised: 19 December 2022
Accepted: 20 December 2022
Published: 22 December 2022

1. Introduction

α-Ketoglutaric acid (KGA) is a valuable compound with many demonstrated applications in the cosmetic, pharmaceutical and food industries or as an intermediate in chemical syntheses [1]. The chemical industry uses KGA as a substrate for synthesis of biopolymers [2] and heterocyclic compounds [3]. In food production it is applied as a functional additive in beverages and a nutraceutical in functional food [4–6]. In medicine it has been used in the treatment of various diseases and for the synthesis of pharmaceuticals [7–11]. The health benefits of KGA consumption have been demonstrated not only in humans but also in animals; therefore, its preparations are of interest to feed producers [12–15].

Although KGA is present in the central metabolism of every living cell, it is an intermediate molecule that is synthesized only in an amount that meets the needs of the cells. Therefore, KGA cannot be obtained from easily accessible sources such as food [16]. Industrial production of KGA is achieved mainly by a multi-step chemical method that uses diethyl oxalate and diethyl succinate as the substrates [3,17]. The process yields 75% efficiency, but it causes environmental hazards and has low economic attractiveness of production because of application of raw materials derived from depleting petrochemical resources, use of harmful reagents and generation of toxic wastes and high by-production of impurities. Moreover, the product obtained by this method may be excluded from use in certain applications (e.g., food production) [1,18,19].

The possibility of converting various carbon sources to KGA using microbial fermentation has attracted scientists for several decades [1,4,19]. A wide variety of microorganisms have been found to be capable of synthesizing KGA. Nevertheless, the limiting factors for the use of most of them on a commercial scale are low titer, low yield, formation of by-products, intolerance to low pH and inhibitor presence and high nutritional requirements. Microorganisms that appear to be particularly effective in the biosynthesis of KGA are yeast, among which *Yarrowia lipolytica* has received special attention. In comparison to other yeast species, *Y. lipolytica* enables more efficient production of KGA from a very wide range of substrates and the use of simple culture media that meet low nutritional and vitamin supplementation requirements [1,19,20].

Previously we presented the genetically engineered strain of *Y. lipolytica* as a prospective producer of KGA from mixed renewable carbon sources (glycerol + rapeseed oil), along with simple technology of product recovery from post-culture broth [21]. Moreover, the biomass of this strain was proved to be of nutritional and health-beneficial characteristics especially desired in food and feed applications. The aim of the present work was to increase the efficiency of KGA production from mixed glycerol/rapeseed-oil-based media by the transformant strain *Y. lipolytica* CBS146773 by optimizing the conditions of biosynthesis.

2. Results and Discussion

In our earlier investigation, the transformant strain *Yarrowia lipolytica* CBS146773, formerly named 1.31.GUT1/6.CIT1/3.E34672 [21], was obtained by genetic engineering. This strain was characterized by overexpression of genes encoding glycerol kinase (*GUT1*), citrate synthase (*CIT1*) and mitochondrial acid transporter (*YALI0E34672g*) and was found to be a good producer of KGA in the process with synergistic co-feeding of glycerol and rapeseed oil [21]. The use of glycerol and rapeseed oil as substrates for *Y. lipolytica* was dictated by ecological and economic reasons. Both substrates are renewable carbon sources suitable for KGA biosynthesis; however, their applications differ in the efficiency of KGA production. The yield of KGA biosynthesis obtained using rapeseed oil might exceed 100%, whereas glycerol application results in by-formation of pyruvic acid (PA), which is linked to the decreased production of KGA and negatively affects the selectivity of the process. However, in practice, the possibility of even partial replacement of the water-insoluble substrate with hydrophilic glycerol can greatly facilitate unit operations during industrial-scale processes [21,22]. Since the use of a mixed-substrates feeding strategy has already been demonstrated to be suitable for the process with *Y. lipolytica* CBS146773, in this work we focused on the selection of the conditions of the KGA biosynthesis that are aimed at fully exploiting the genetic potential of the selected strain.

2.1. Maintenance of pH

The effect of pH-control strategy on the growth and biosynthesis of KGA by the yeast strain *Y. lipolytica* CBS146773 was studied in two aspects: examination of pH value and comparison of different neutralizing agents used for maintaining pH at the appropriate level. The literature includes information about the positive effect of acidic pH on KGA biosynthesis by *Y. lipolytica* yeast [23–27]. In the reports, the pH level optimal for KGA production varied from 2.79 to 4.5 depending on the strain and substrate used in a specific process. In this study the pH impact was tested in the range of 3.0–4.5, using KOH as neutralizing agent. As presented in Figure 1A, the yeast strain presented good growth in the examined pH range. However, the increase in pH value resulted in an increased level of biomass from 18.6 g/L at pH 3.0 to 22.4 g/L when pH 4.5 was applied. It was found that the formation of KGA was dependent on pH. The highest KGA production of 56.8 g/L corresponding to the yield (Yp/s) of 0.41 g/g was obtained in the culture conducted at pH 3.5. Simultaneously, in the same conditions, PA content in the post-culture broth was the lowest and reached 13.9 g/L, whereas in other cultures the concentration of the acid was significantly higher (23.0–38.0 g/L). It was also observed that increasing the pH level from 3.0 to 4.5 resulted in increased citric acid (CA) by-product formation from 1.0 to 13.2 g/L.

Taking into account overall production of the acid pool, the highest selectivity of the KGA biosynthesis process (0.76 = 76%) was noted when the pH was maintained at 3.5. Some authors reported that not only the pH value applied but also its control strategy, in which the pH level set at the growth phase was lowered and was controlled at the low level to the end of the cultivation process, enhancing KGA biosynthesis [24,27]. In the process conducted on glycerol with such a two-step pH control strategy, *Y. lipolytica* WSH-Z06 produced 32.5% more KGA (53.4 g/L) than in the process in which a one-step strategy was applied [27].

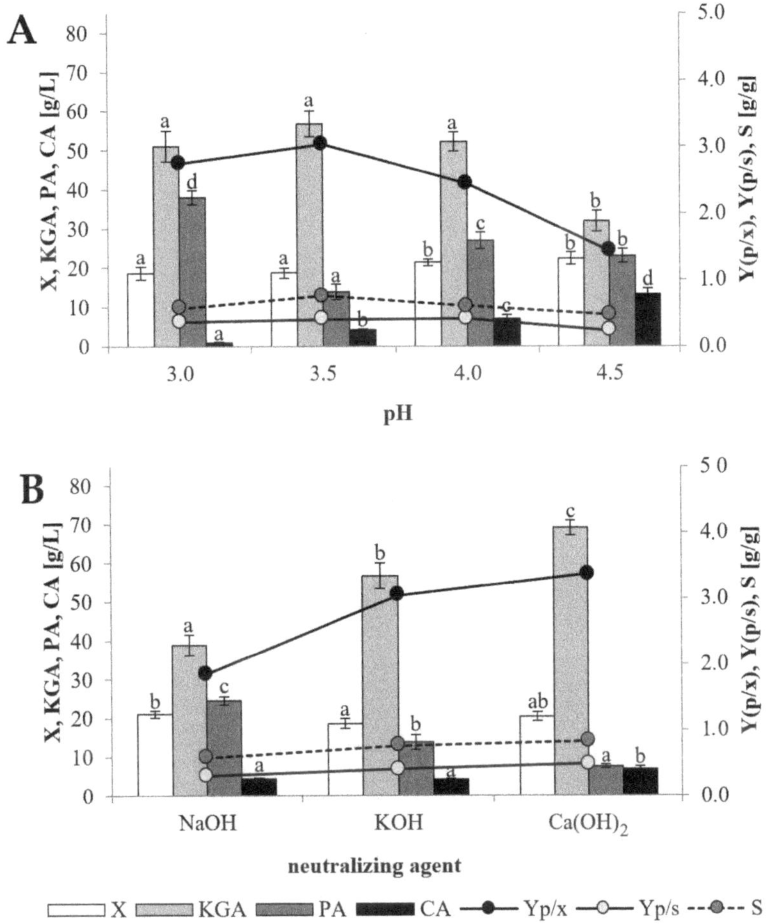

Figure 1. Impact of pH control by pH value (**A**) and type of neutralizing agent used (**B**) on yeast growth and acids formation during KGA biosynthesis performed by *Y. lipolytica* CBS 146773 in mixed glycerol/oil-based media. Culture conditions: 20% KOH (**A**), pH 3.5 (**B**), 800 rpm, 3 μg/L of thiamine. Abbreviations: X—biomass; KGA—α-ketoglutaric acid; PA—pyruvic acid; CA—citric acid; Y—yield of KGA with respect to biomass formed (p/x) and utilized substrates (p/s); S—selectivity of KGA relative to sum of acids formed (KGA/(KGA + PA + CA)). Mean values for a specific product concentration marked with different letters (a, b, c, ...) differ significantly at $p \leq 0.05$. Error bars indicate standard deviations.

In this study the second approach to the pH control strategy was to verify whether a different neutralizing agent may affect the process of KGA formation by the examined transformant yeast strain. For this purpose, biosynthesis conducted at pH 3.5 was maintained by automatic addition of a 20% solution of NaOH, KOH or $Ca(OH)_2$ (Figure 1B). In comparison to KOH, application of the two other neutralizing factors resulted in slightly higher biomass growth. The results of this experiment revealed that the type of factor used for neutralization has a significant influence on the efficiency of KGA biosynthesis. When NaOH was used, the yeast produced moderate concentrations of both KGA (39.0 g/L) and PA (24.6 g/L), which, despite the relatively low amount of CA, led to low selectivity of the process (57%). A significant increase in KGA biosynthesis was noted in the culture in which $Ca(OH)_2$ was applied, as yeast produced 69.1 g/L of KGA with the yield (Yp/s) of 0.49 g/g and selectivity of 83%. These results were in agreement with an earlier investigation that revealed a positive impact of Ca^{2+} ions on pyruvate carboxylase [28]. In the cultures of *Torulopsis glabrata* CCTCC M202019 growing on glucose, a lower concentration of KGA (1.3 g/L) was obtained when NaOH was used as a neutralizing agent, whereas application of $CaCO_3$ increased KGA synthesis (11.5–12.0 g/L) [28,29]. This relationship was also confirmed in the culture of *Y. lipolytica* WSK-Z06 performed in glycerol media, where the change of the neutralizing agent from NaOH to $CaCO_3$ resulted in an increase in the KGA:PA ratio from 22.0:36.9 g/L to 40.3:31.8 g/L [27]. As a consequence of the above described results, all subsequent experiments were carried out at pH 3.5 maintained by addition of $Ca(OH)_2$.

2.2. Manipulation of Aeration Level

In order to evaluate the impact of dissolved oxygen level cultures were performed at various agitation rates (400–900 rpm) while maintaining a constant air flow of 0.8 vvm. A significant effect of agitation rate on growth of the yeast was observed (Figure 2). The biomass level reached a maximum of 25.1 g/L at 900 rpm and decreased depending on the decrease in the agitation rate to 15.7 g/L at 400 rpm. In the range of 400–700 rpm biosynthesis of KGA (38.1–54.5 g/L) was accompanied by comparatively high production of PA (25.5–38 g/L). As a result, KGA production yield (Yp/s) and selectivity were at the level of 0.27–0.39 g/g and 53–59%, respectively. The highest concentration of KGA and the best parameters of its biosynthesis were obtained when the agitation rate reached 800 rpm, and a further increase to 900 rpm resulted in a decrease in the KGA biosynthesis efficiency. In these cultures, the selectivity of the process was found to be significantly higher (72–83%) than in the process that was conducted at a lower agitation speed, i.e., 400–700 rpm. The agitation rate is a parameter affecting the amount of oxygen dissolved in the culture broth. In this study, the application of an agitation rate in the range of 400–900 rpm corresponded to 20–60% pO_2, measured in the KGA production phase of the performed cultures. The aeration level has been identified as an important factor influencing KGA biosynthesis by *Y. lipolytica* growing on ethanol, rapeseed oil and biodiesel waste [23–25]. Similar to the results obtained in our study with the use of glycerol/oil media, the process of KGA production by *Y. lipolytica* VKM Y-2412 conducted on biodiesel waste (a substrate containing 70.8% glycerol and 23.9% fatty acids) was also promoted by high aeration [24]. The increase in aeration from 5% pO_2 to 50% pO_2 enabled an increase in the production of KGA from 56.8 to 80.4 g/L. In contrast, high aeration was not necessary for KGA biosynthesis from ethanol. In comparison to the culture with high aeration (50% pO_2), 1.3-times higher KGA formation (49.0 g/L) was observed when low aeration was applied (5% pO_2) [25]. As high aeration was found to stimulate KGA biosynthesis by the examined yeast strain, an agitation rate of 800 rpm was applied in all further experiments.

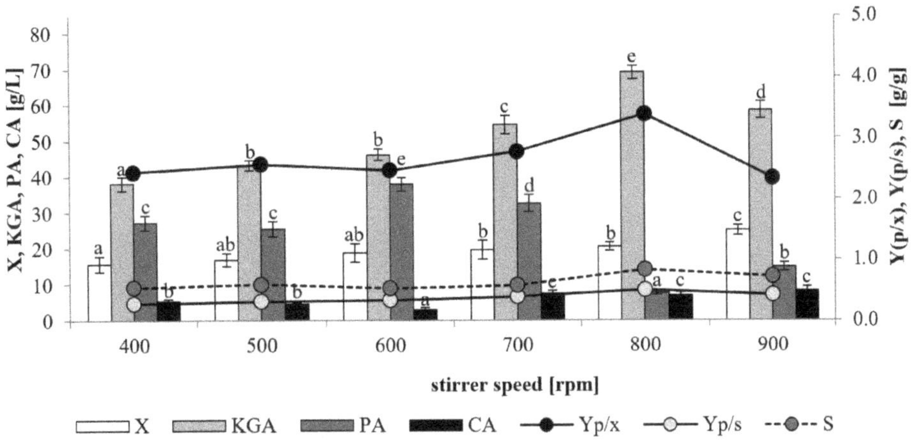

Figure 2. Impact of agitation on yeast growth and acids formation during KGA biosynthesis process performed by *Y. lipolytica* CBS 146773 in mixed glycerol/oil-based media. Culture conditions: 20% $Ca(OH)_2$, pH 3.5, 3 µg/L of thiamine. Abbreviations: X—biomass; KGA—α-ketoglutaric acid; PA—pyruvic acid; CA—citric acid; Y—yield of KGA with respect to biomass formed (p/x) and utilized substrates (p/s); S—selectivity of KGA relative to sum of acids formed (KGA/(KGA + PA + CA)). Mean values for a specific product concentration marked with different letters (a, b, c, ...) differ significantly at $p \leq 0.05$. Error bars indicate standard deviations.

2.3. Availability of Exogenous Vitamins

Yeast reported as producers of KGA have been characterized as auxotrophic for one, two or several vitamins. These vitamins are co-factors of enzymes in the Krebs cycle, and their exogenous level is one of the crucial factors affecting accumulation of KGA in auxotrophic cells [19]. *Y. lipolytica* is auxotrophic only to thiamine, limitation of which is known to reduce the activity of α-ketoglutarate dehydrogenase and therefore determines KGA oversynthesis [24]. It should be noted that auxotrophy only for one vitamin gives an advantage to the process performed by *Y. lipolytica* because it requires very precise control of only one vitamin. In order to obtain thiamine limitation in the cultures with *Y. lipolytica* CBS146773, the vitamin concentration was applied in the very low range of 1–4 µg/L (Figure 3A). It was clearly apparent, that thiamine had a significant impact on yeast growth, as the biomass level increased from 5.1 to 26.2 g/L with increased thiamine concentration. Increasing the vitamin addition from 1 to 3 µg/L resulted in rapid changes in both KGA and PA concentrations, but the opposite trend was observed for these acids—an increase from 11.0 to 69.1 g/L in the case of KGA and a decrease from 43.3 to 7.6 g/L in the case of PA. Thus, the selectivity of the process increased from 20% to 83% with the change of thiamine availability from 1 to 3 µg/L. No further improvement in KGA concentration or parameters of its biosynthesis was observed after the addition of 4 µg/L of thiamine. Because of the big differences in yeast growth observed between all the cultures it is worth paying attention to the parameter of yield of KGA calculated with respect to biomass formed (Yp/x). This parameter was the highest (3.37 g/g) when application of 3 µg/L of thiamine resulted in the highest amount of KGA produced. However, its value was very similar in the cultures with thiamine supplementation of 1 and 4 µg/L, where it reached 2.16 and 2.35 g/g, respectively, despite the amount of produced KGA (11.0—61.5 g/L, respectively) differing significantly between these processes. As mentioned above, the appropriate thiamine concentration is a crucial factor for effective KGA biosynthesis by the yeast belonging to the species *Y. lipolytica*. The optimal concentration for KGA biosynthesis requires a balance between the amount necessary for growth and the amount determining the decreased activity of α-ketoglutarate dehydrogenase and is a strain-dependent feature. It is reported in the literature that increasing availability of thiamine (up to 200 µg/L) stimulates the growth of the yeast, whereas for KGA synthesis a "peak" is observed at a certain low vitamin concentration (0.15–4 µg/L) specific to the kind of substrate, substrate feeding method and yeast strain applied for the process [22–26]. Moreover, it should be noted that by-product formation of PA also might be affected by thiamine concentration when yeast is grown on glycolytic carbon sources (glucose, fructose, glycerol, etc.), which are utilized via pyruvate because of modulation of thiamine-dependent pyruvate dehydrogenase activity [30].

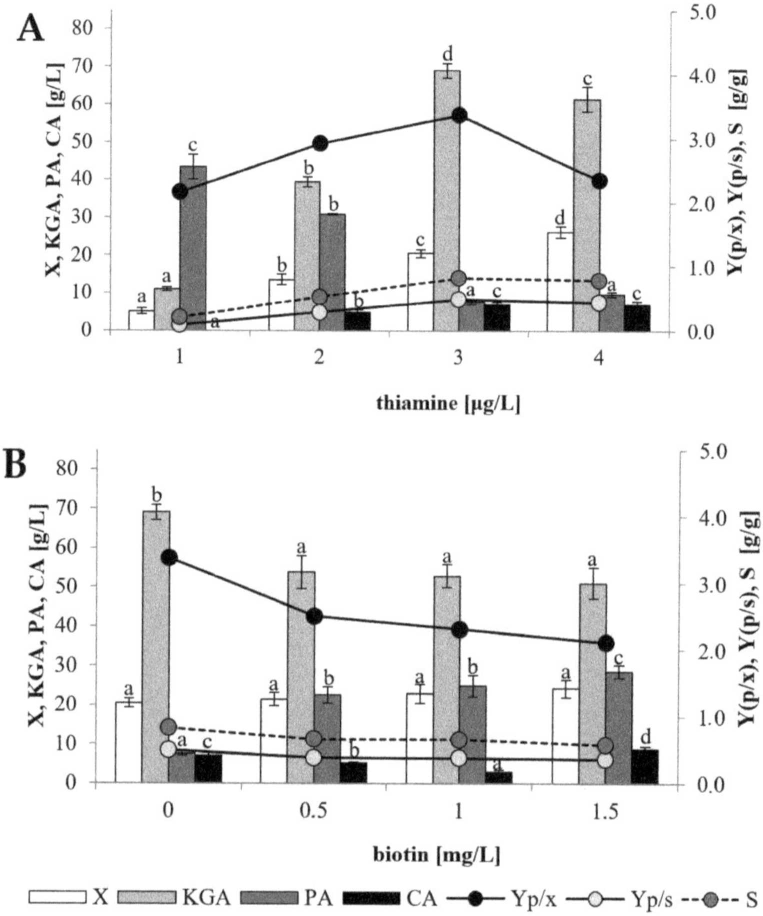

Figure 3. Impact of thiamine (**A**) and biotin (**B**) on the yeast growth and acids formation during KGA biosynthesis performed by *Y. lipolytica* CBS 146773 in mixed glycerol/oil-based media. Culture conditions: 20% $Ca(OH)_2$, pH 3.5, 800 rpm, 3 µg/L of thiamine (**B**). Abbreviations: X—biomass; KGA—α-ketoglutaric acid; PA—pyruvic acid; CA—citric acid; Y—yield of KGA with respect to biomass formed (p/x) and utilized substrates (p/s); S—selectivity of KGA relative to sum of acids formed (KGA/(KGA + PA + CA)). Mean values for a specific product concentration marked with different letters (a, b, c, ...) differ significantly at $p \leq 0.05$. Error bars indicate standard deviations.

The impact of exogenous biotin addition (0–1.5 mg/L) and all subsequent experiments were performed in media supplemented with 3 µg/L of thiamine. Biotin is another vitamin which may induce accumulation of KGA by affecting the activity of pyruvate carboxylase [30–32]. In this study, biotin addition to the culture stimulated yeast growth—in the processes supplemented with the vitamin, biomass was at the level of 21.5–24.2 g/L, whereas in the control culture its concentration reached 20.5 g/L (Figure 3B). However, no positive effect of biotin addition on KGA production was observed. The amount of KGA in the post-culture broth decreased from 69.1 to 51.1 g/L after culture supplementation with 1.5 mg/L of biotin. Simultaneously, PA concentration increased from 7.6 g/L in the culture not supplemented with biotin to 28.6 g/L in the process where 1.5 mg/L of the vitamin was used. Theoretically, biotin presence increases the activity of pyruvate carboxylase, which catalyzes the conversion of pyruvate to oxaloacetate. Therefore, biosynthesis of KGA from substrates metabolized by the glycolysis pathway (e.g., glycerol) should be enhanced by biotin supplementation. In the shake-flask culture of *Y. lipolytica* WSH-Z06, addition of 0.8 mg/L of biotin had only small positive effect on KGA biosynthesis from glycerol whereas PA production was unaffected [26]. Interesting observations were reported by Otto et al. [30], who studied the changes in the by-product spectrum during KGA biosynthesis from glycerol by *Y. lipolytica* H355A(PYC1) T3—a strain that overexpressed pyruvate carboxylase. In comparison to the mother strain H355, higher activity of the enzyme in the transformant strain resulted in a higher biomass level, a decrease in KGA production from 133.0 to 126.9 g/L, and a simultaneous slight increase in the formation of PA and other by-products. The positive effect on yeast growth was explained by accumulation of precursor molecules (oxaloacetic, malic, succinic and fumaric acids) caused by an imbalance between enhanced activity of pyruvate carboxylase and inhibited activity of pyruvate dehydrogenase (due to thiamine limitation). Assuming that, in the present study, the addition of biotin increased the activity of pyruvate carboxylase, the same tendency was noticeable—stimulation of yeast growth and PA production with a decrease in KGA formation. Similarly, the growth of *Y. lipolytica* VKM Y-2412 was slightly increased whereas KGA production was decreased and no effect on PA formation was noted when the process was conducted on biodiesel waste media upon supplementation with biotin (10–40 µg/L) [24].

2.4. The Effect of Iron

The presence of iron ions is another factor that may possibly affect KGA biosynthesis by modulating the activity of iron-dependent enzymes in the Krebs cycle, i.e., aconitase and succinate dehydrogenase [33,34]. In the present study the impact of iron was evaluated in the media supplemented with ammonium iron sulfate hexahydrate in an amount corresponding to 1–4 mg/L of iron ions. The presence of iron had no effect on growth of the examined yeast strain (Figure 4). The biomass level in the cultures supplemented with iron ions was in the range of 19.0–20.4 g/L, which was comparable to the level obtained for the control culture (20.5 g/L). Addition of iron ions resulted in lower production of KGA and enhanced accumulation of PA. In the supplemented cultures yeast produced 53.7–59.0 g/L and 17.0–29.2 g/L of these acids, respectively. Compared to the non-iron-supplemented process, formation of CA was found to be slightly decreased upon the addition of iron ions (2.7–5.7 g/L). The effect of iron on KGA biosynthesis by *Y. lipolytica* was investigated during cultivation on ethanol [35,36]. It was found that iron concentration of 0.5–2.0 mg/L stimulated KGA synthesis, whereas a further increase in iron concentration from 2–3 to 10.0 mg/L caused a gradual decrease in KGA accumulation. The positive impact of iron presence was also reported for KGA production on biodiesel waste by *Y. lipolytica* [24]. Effective biosynthesis of isocitric acid by *Y. lipolytica* from rapeseed oil was noted also upon iron supplementation [37]. In turn, the ability of *Y. lipolytica* to produce erythritol was unaffected by iron ions when yeast was cultivated in bioreactor cultures on glycerol media [38].

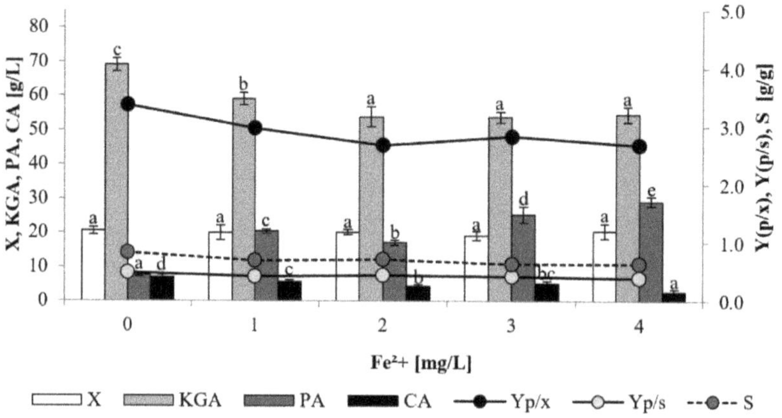

Figure 4. Impact of iron on yeast growth and acids formation during KGA biosynthesis performed by *Y. lipolytica* CBS 146773 in mixed glycerol/oil-based media. Culture conditions: 20% $Ca(OH)_2$, pH 3.5, 800 rpm, 3 μg/L of thiamine. Abbreviations: X—biomass; KGA—α-ketoglutaric acid; PA—pyruvic acid; CA—citric acid; Y—yield of KGA with respect to biomass formed (p/x) and utilized substrates (p/s); S—selectivity of KGA relative to sum of acids formed (KGA/(KGA + PA + CA)). Mean values for a specific product concentration marked with different letters (a, b, c, ...) differ significantly at $p \leq 0.05$. Error bars indicate standard deviations.

2.5. The Impact of Sorbitan Monolaurate

Sorbitan monolaurate, known under the trade name Span 20, is a non-ionic, water-insoluble, lipophilic emulsifier [39]. This surfactant is approved for use as a food additive (E493) and the acceptable daily intake is 10 mg/kg [40]. Literature reports mention a positive effect of Span 20 on the growth and excretion of some metabolites by microorganisms [41–44]. This surfactant, in addition to increasing the dispersion of oil substrates in the culture media, may also increase the permeability of microorganisms' cell membranes. Thus, it facilitates the excretion of metabolites and extracellular proteins from the cell. In this study, Span 20 was applied to the culture media in the concentration ranging from 0.25 to 1 g/L (Figure 5). In comparison to the control, the growth of the yeast was slightly lower in the presence of Span 20 and ranged 18.5–19.4 g/L. Nevertheless, addition of Span 20 had a significant positive impact on KGA biosynthesis. The increase in its addition up to 1 g/L resulted in a gradual increase in KGA production up to 82.4 g/L, corresponding to the yield (Yp/s) of 0.88 g/g and selectivity of 88% (Figures 5 and 6). Comparatively high PA amounts (11.0–18.1 g/L) were obtained in the cultures with Span 20 addition of 0.25–0.75 g/L. In turn, production of CA was slightly decreased by the presence of Span 20 and was in the range 3.5–5.0 g/L. The positive effect of Span 20 addition was reported previously for production of oxalic acid from fatty acid waste by *Aspergillus niger* [41]. In the process with the addition of 0.75 g/L of the surfactant, the production of oxalic acid increased from 34.7 to 48.4 g/L, compared to the control culture. The addition of 0.25 g/L of Span 20 to the culture of *Y. lipolytica* Wratislavia K1 cultivated on raw glycerol enhanced erythritol production from 149.6 g/L to 165.7 g/L [42]. Moreover, although very low by-product formation of CA and KGA was noted in this process, synthesis of both acids was enhanced by the presence of the surfactant. Application of Span 20 was also proved to have a positive effect on biomass formation and β-carotene production by *Blakeslea trispora* [43,44].

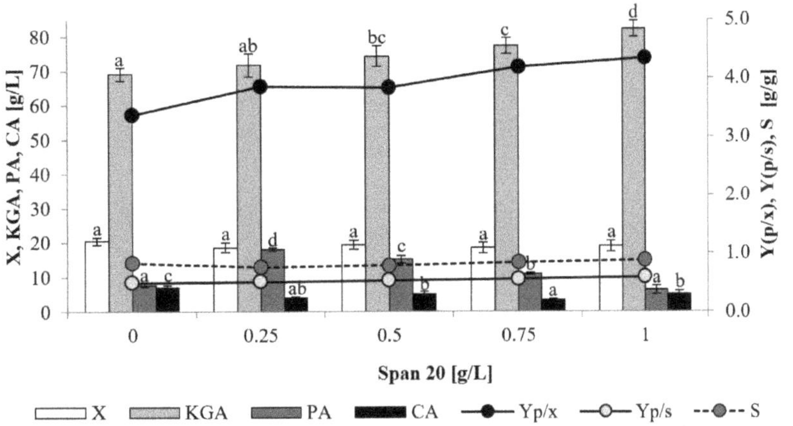

Figure 5. Impact of Span 20 on yeast growth and acids formation during KGA biosynthesis performed by *Y. lipolytica* CBS 146773 in mixed glycerol/oil-based media. Culture conditions: 20% $Ca(OH)_2$, pH 3.5, 800 rpm, 3 μg/L of thiamine. Abbreviations: X—biomass; KGA—α-ketoglutaric acid; PA—pyruvic acid; CA—citric acid; Y—yield of KGA with respect to biomass formed (p/x) and utilized substrates (p/s); S—selectivity of KGA relative to sum of acids formed (KGA/(KGA + PA + CA)). Mean values for a specific product concentration marked with different letters (a, b, c, ...) differ significantly at $p \leq 0.05$. Error bars indicate standard deviations.

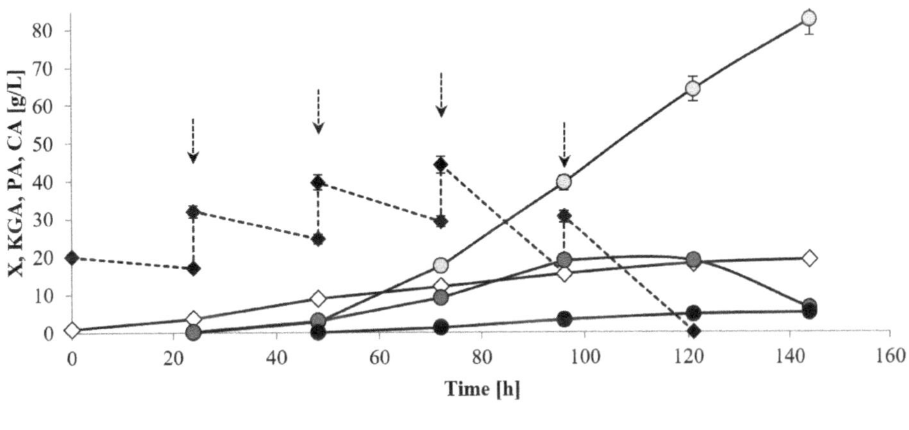

Figure 6. Time-course of yeast growth and acid formation during KGA biosynthesis performed by *Y. lipolytica* CBS 146773 in mixed glycerol/oil-based media supplemented with 1 g/L of Span 20. Culture conditions: 20% $Ca(OH)_2$, pH 3.5, 800 rpm, 3 μg/L of thiamine, 1 g/L of Span 20. Abbreviations: GLY—glycerol; for other abbreviations, see Figure 1. Each arrow indicates addition of one portion of 30 g/L of mixed substrates (see: Materials and Methods, Section 3.2).

3. Materials and Methods

3.1. Microorganisms

The strain *Yarrowia lipolytica* CBS146773, formerly named 1.31.GUT1/6.CIT1/3.E34672 [21], was the subject of this study. The strain was stored on YM agar slants at 4 °C in the culture collection belonging to the Department of Biotechnology and Food Microbiology at Wrocław University of Environmental and Life Sciences (Poland).

3.2. Media Composition and Culture Conditions

Seed cultures were performed in 300 mL baffled flasks containing 50 mL of an inoculation medium that consisted of (g/L): edible rapeseed oil—20.0; NH_4Cl—9.0; KH_2PO_4—2.0; $MgSO_4 \cdot 7 H_2O$—1.4; $CaCO_3$—10.0; and thiamine—3 μg/L dissolved in distilled water. The pH was 3.5. The medium was inoculated from agar slants and next the culture was grown for 72 h at 29 °C and 140 rpm on a rotary shaker (CERTOMAT IS; Sartorius, Germany).

Batch fermentation was conducted in a 5 L bioreactor (BIOSTAT B Plus; Sartorius, Germany) with a working final volume of 2 L. Bioreactor production medium composition was (g/L): pure glycerol (98%; Wratislavia-Bio; Poland)—20.0; NH_4Cl—9.0; KH_2PO_4—2.0; $MgSO_4 \cdot 7 H_2O$—1.4; and thiamine—3 μg/L, dissolved in tap water. The pH was 3.5. In some cultures, the medium was supplemented with thiamine (1–4 μg/L), biotin (0.5–1.5 mg/L), $(NH_4)_2Fe(SO_4)_2 \cdot 6 H_2O$ (7.03–28.12 mg/L) or Span 20 (0.25–1 g/L). The volume of 150 mL of seed culture was used to inoculate production medium in fermenter and the process was conducted for 144 h at 29 °C, with agitation rate of 800 rpm, aeration rate of 0.8 vvm and pH 3.5 maintained by the automatic addition of a 20% solution of $Ca(OH)_2$ (or NaOH/KOH, when indicated). During the cultivation, the process was fed at 24 h intervals (i.e., 24, 48, 72, 96 h, as indicated in Figure 6) with 4 sterile portions of 30 g/L of mixed substrates (glycerol and rapeseed oil in equal amounts of 15 g/L). Any changes in the culture parameters and composition of the medium are indicated for specific experiments in the Results section. All chemicals used in the investigation were of analytical purity (Sigma-Aldrich, Steinheim, Germany). Prior to inoculation, all media were sterilized at 121 °C for 30 min.

3.3. Analytical Methods

The samples collected from the bioreactor fermentation process were analyzed in terms of biomass level (dry matter) and concentration of glycerol and acids—ketoglutaric, pyruvic and citric. The sample preparation and analytical methods were described previously by Rywińska et al. [22]. The results are presented as mean values of the process performed in duplicate. Statistical analysis was performed using one-way analysis of variance (Statistica 13.0 software; StatSoft, Tulsa, OK, USA). The significant differences in the data (X, KGA, PA, CA) were compared by Duncan's multiple range test. ($p \leq 0.05$).

4. Conclusions

In this study the effects of selected media components and culture conditions were evaluated in order to enhance the biosynthesis of KGA by *Y. lipolytica* CBS146773. The possible impact of evaluated factors is presented in Figure 7. In the research, the source of carbon and energy was a mixture of glycerol and rapeseed oil. The addition of Span 20 was used to increase the dispersion of oil droplets and increase the permeability of cell membranes, which may facilitate the secretion of the produced metabolites [41,42]. Notably, substrates applied in this investigation are utilized by the yeast cell in different metabolic pathways. Glycerol is first transformed by the action of glycerol kinase to glycerol-3-phosphate, which, after being converted to glyceraldehyde, undergoes further transformations in the glycolytic pathway, resulting in the formation of pyruvate. In the mitochondrion, pyruvate dehydrogenase catalyzes the conversion of pyruvate into acetyl-CoA, which is incorporated into the tricarboxylic acid cycle with the action of citrate synthase. The transformant strain used in this study was characterized by overexpression of genes encoding glycerol kinase and citrate synthase, which was aimed at increasing the efficiency of the above-described metabolic pathway [21]. Moreover, a gene encoding previously uncharacterized mitochondrial organic acid transporter was overexpressed in the yeast strain to investigate whether it might facilitate secretion of organic acids from mitochondrion and increase the extracellular concentration of KGA as a final product.

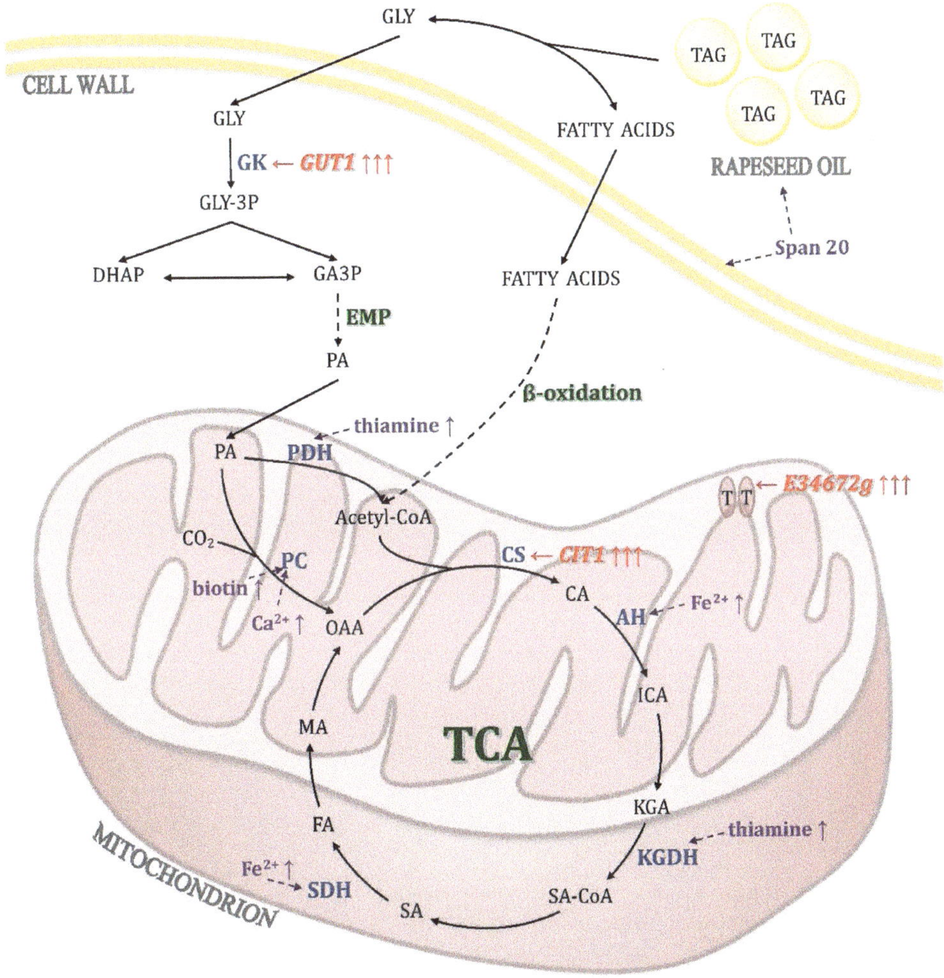

Figure 7. Scheme of metabolic pathways involved in conversion of glycerol and rapeseed oil to KGA by *Yarrowia lipolytica* CBS146773 and the putative effect of selected factors on the efficiency of the process. Abbreviations: ↑: stimulatory effect; AH: aconitate dehydrogenase; CA: citric acid; *CIT1*: gene encoding citrate synthase; CS: citrate synthase; DHAP: dihydroxyacetone phosphate; *E34672g*: gene *YALI0E34672g* encoding mitochondrial acid transporter; EMP: Embden–Meyerhof–Parnas glycolytic pathway; FA: fumaric acid; GA3P: glyceraldehyde-3-phosphate; GK: glycerol kinase; GLY: glycerol; GLY-3P: glycerol-3-phosphate; *GUT1*: gene encoding glycerol kinase; ICA: isocitric acid; KGA: α-ketoglutaric acid; KGDH: α-ketoglutarate dehydrogenase; MA: malic acid; OAA: oxaloacetic acid; PA: pyruvic acid; PC: pyruvate carboxylase; PDH: pyruvate dehydrogenase; SA: succinic acid; SA-CoA: succinyl-CoA; SDH: succinate dehydrogenase; T: mitochondrial acid transporter; TAG: triacylglyceride; TCA: tricarboxylic acid cycle.

The second substrate—rapeseed oil—is hydrolyzed by extracellular lipases to glycerol and fatty acids. In the cell the latter undergo β-oxidation, resulting in the formation of acetyl-CoA—a compound that connects the pathways of glycerol and oil utilization. Formed from acetyl-CoA and oxaloacetate, citrate is further converted to isocitric acid by aconitase, and the subsequent transformation leads to the formation of KGA. Theoretically, at this stage, metabolism can be stimulated by the presence of iron ions, since aconitase is an iron-dependent enzyme [33,34]. In the presented study, however, we did not observe a positive effect of iron supplementation on KGA biosynthesis. Overproduction of KGA requires inhibition of thiamine-dependent ketoglutarate dehydrogenase, catalyzing KGA conversion to succinyl-CoA. In *Y. lipolytica*, this is possible by limiting exogenous thiamine, as this yeast is unable to synthesize the pyrimidine structure of this vitamin [24]. In accordance with literature reports, in the presented study, the concentration of thiamine was noted as the key factor determining the effective production of KGA by *Y. lipolytica* [22–26]. However, it should be noted that the differences in the amount of thiamine reported as necessary for KGA production might be dependent not only on the strain but also on the kind of substrate applied in the process. In comparison to glycolytic substrates, the use of fatty substrates omits the reaction catalyzed by the second thiamine-dependent enzyme—pyruvate dehydrogenase; hence, the cell's need for this vitamin may be lower in such a process. A side effect of using thiamine limitation in cultures conducted on glycerol media is the inhibition of pyruvate dehydrogenase resulting in accumulation of pyruvate. This effect can be counteracted by increasing the activity of pyruvate carboxylase, which converts pyruvate to oxaloacetate [31]. In this study, two factors that might stimulate the activity of this enzyme were examined: biotin and calcium ions. The presence of calcium ions was found to significantly enhance the biosynthesis of KGA, whereas a positive effect of biotin supplementation was not observed.

The results of the experiments performed in this study to identify the best conditions for effective KGA biosynthesis indicated the following: maintenance of pH at 3.5 by neutralization with the use of $Ca(OH)_2$, an agitation rate of 800 rpm and the addition of 3 µg/L of thiamine and 1 g/L of Span 20. In our earlier investigation, this transformant strain was identified as a good producer of KGA from mixed media which, after preliminary optimization of the process conditions (C:N:P ratio), was able to biosynthesize 53.1 g/L of KGA with productivity of 0.35 g/L h and yield (Yp/s) of 0.53 g/g [21]. In the present work selection of culture conditions enabled the increase in KGA biosynthesis to 82.4 g/L. Moreover, the parameters of KGA biosynthesis were significantly improved—the productivity increased to 0.57 g/L h and the yield (Yp/s) reached 0.59 g/g. A similar amount of the acid was obtained after optimization of the KGA production process performed by *Y. lipolytica* VKM Y-2412 on biodiesel waste containing glycerol and fatty acids [24]. In the optimal conditions, yeast produced 80.4 g/L of KGA with the selectivity of 96.7%. However, the cultivation process was significantly longer (192 h) than in the present research (144 h). In the literature data the highest KGA production was reported for *Y. lipolytica* H355A(PYC1-IDP1) T5 engineered for overexpression of isocitrate dehydrogenase and pyruvate carboxylase, which was able to synthesize 186.0 g/L of KGA from raw glycerol with productivity of 1.75 g/L h [45]. However, the yield of KGA production, which is one of the factors determining the process attractiveness in terms of industrial production, obtained in the process with strain H355A(PYC1-IDP1) T5 reached only 0.36 g/g and was significantly lower than the yield obtained in this research with strain CBS146773.

Author Contributions: Conceptualization, L.T.-H. and A.R.; methodology, L.T.-H., A.R. and W.R.; software, L.T.-H.; validation, L.T.-H. and A.R.; formal analysis, L.T.-H.; investigation, W.R., L.T.-H., Z.L. and A.R.; resources, W.R.; data curation, L.T.-H. and W.R.; writing—original draft preparation, L.T.-H.; writing—review and editing, L.T.-H.; visualization, L.T.-H.; supervision, A.R. and W.R.; project administration, W.R.; funding acquisition, A.R., L.T.-H. and W.R. All authors have read and agreed to the published version of the manuscript.

Funding: This research was funded by The National Centre for Research and Development under the Project No. POIR.04.01.02-00-0028/18 entitled "Development of an innovative technology for the production of dietary supplements based on alpha-ketoglutaric acid obtained on the biological way with *Yarrowia lipolytica* yeast".

Data Availability Statement: Not applicable.

Conflicts of Interest: The authors declare no conflict of interest. The funders had no role in the design of the study; in the collection, analyses, or interpretation of data; in the writing of the manuscript; or in the decision to publish the results.

References

1. Otto, C.; Yovkova, V.; Barth, G. Overproduction and secretion of α-ketoglutaric acid by microorganisms. *Appl. Microbiol. Biotechnol.* **2011**, *92*, 689–695. [CrossRef] [PubMed]
2. Barrett, D.G.; Yousaf, M.N. Poly(triol α-ketoglutarate) as biodegradable, chemoselective, and mechanically tunable elastomers. *Macromolecules* **2008**, *41*, 6347–6352. [CrossRef]
3. Stottmeister, U.; Aurich, A.; Wilde, H.; Andersch, J.; Schmidt, S.; Sicker, D. White biotechnology for green chemistry: Fermentative 2-oxocarboxylic acids as novel building blocks for subsequent chemical syntheses. *J. Ind. Microbiol. Biotechnol.* **2005**, *32*, 651–664. [CrossRef] [PubMed]
4. Luo, Z.; Yu, S.; Zeng, W.; Zhou, J. Comparative analysis of the chemical and biochemical synthesis of keto acids. *Biotechnol. Adv.* **2021**, *47*, 107706. [CrossRef] [PubMed]
5. Filip, R.; Pierzynowski, S.G. The role of glutamine and α-ketoglutarate in gut metabolism and the potential application in medicine and nutrition. *J. Clin. Res.* **2007**, *1*, 9–15.
6. Zdzisińska, B.; Żurek, A.; Kandefer-Szerszeń, M. Alpha-ketoglutarate as a molecule with pleiotropic activity: Well-known and novel possibilities of therapeutic use. *Arch. Immunol. Ther. Exp.* **2017**, *65*, 21–36. [CrossRef]
7. Cynober, L.A. The use of alpha-ketoglutarate salts in clinical nutrition and metabolic care. *Curr. Opin. Clin. Nutr. Meta. Care* **1999**, *2*, 33–37. [CrossRef]
8. Wernerman, J.; Hammarqvist, F.; Vinnars, E. Alpha-ketoglutarate and postoperative muscle catabolism. *Lancet* **1990**, *335*, 701–703. [CrossRef]
9. Riedel, E.; Nündel, M.; Hampl, H. alpha-Ketoglutarate application in hemodialysis patients improves amino acid metabolism. *Nephron* **1996**, *74*, 261–265. [CrossRef]
10. Filip, R.S.; Pierzynowski, S.G.; Lindegard, B.; Wernerman, J.; Haratym-Maj, A.; Podgurniak, M. Alpha-ketoglutarate decreases serum levels of C-terminal cross-linking telopeptide of type I collagen (CTX) in postmenopausal women with osteopenia: Six-month study. *Int. J. Vitam. Nutr. Res.* **2007**, *77*, 89–97. [CrossRef]
11. Zimmermann, E.; Wassmer, S.; Steudle, V. Long-term treatment with calcium-alpha-ketoglutarate corrects secondary hyperparathyroidism. *Miner. Electrol. Metab.* **1996**, *22*, 196–199.
12. Shahmirzadi, A.A.; Edgar, D.; Liao, C.Y.; Hsu, Y.M.; Lucanic, M.; Shahmirzadi, A.A.; Wiley, C.D.; Gan, G.; Kim, D.E.; Kasler, H.G.; et al. Alpha-ketoglutarate, an endogenous metabolite, extends lifespan and compresses morbidity in aging mice. *Cell Metab.* **2020**, *32*, 447–456. [CrossRef] [PubMed]
13. Kowalik, S.; Śliwa, E.; Tatara, M.R.; Krupski, W.; Majcher, P.; Studziński, T. Influence of alpha-ketoglutarate on mineral density and geometrical and mechanical parameters of femora during postnatal life in piglets. *Bull. Vet. Inst. Pulawy.* **2005**, *49*, 107–111.
14. Welborn, J.R.; Shpun, S.; Dantzler, W.H.; Wright, S.H. Effect of alpha-ketoglutarate on organic anion transport in single rabbit renal proximal tubules. *Am. J. Physiol. Ren. Physiol.* **1998**, *274*, 165–174. [CrossRef] [PubMed]
15. Velvizhi, S.; Dakshayani, K.B.; Subramanian, P. Effects of alpha-ketoglutarate on antioxidants and lipid peroxidation products in rats treated with ammonium acetate. *Nutrition* **2002**, *18*, 747–750. [CrossRef]
16. Szczygiełda, M.; Prochaska, K. Effective separation of bio-based alpha-ketoglutaric acid from post-fermentation broth using bipolar membrane electrodialysis (EDBM) and fouling analysis. *Biochem. Eng. J.* **2021**, *166*, 107883. [CrossRef]
17. Zeng, W.; Du, G.; Chen, J.; Li, J.; Zhou, J. A high-throughput screening procedure for enhancing α-ketoglutaric acid production in *Yarrowia lipolytica* by random mutagenesis. *Process Biochem.* **2015**, *50*, 1516–1522. [CrossRef]
18. Yin, X.; Li, J.H.; Shin, H.D.; Du, G.C.; Liu, L.; Chen, J. Metabolic engineering in the biotechnological production of organic acids in the tricarboxylic acid cycle of microorganisms: Advances and prospects. *Biotechnol. Adv.* **2015**, *33*, 830–841. [CrossRef]
19. Guo, H.; Su, S.; Madzak, C.; Zhou, J.; Chen, H.; Chen, G. Applying pathway engineering to enhance production of alpha-ketoglutarate in *Yarrowia lipolytica*. *Appl. Microbiol. Biotechnol.* **2016**, *100*, 9875–9884. [CrossRef]
20. Kamzolova, S.V.; Morgunov, I.G. Selection of Producer of α-Ketoglutaric Acid from Ethanol-Containing Wastes and Impact of Cultivation Conditions. *Fermentation* **2022**, *8*, 362. [CrossRef]
21. Tomaszewska-Hetman, L.; Rywińska, A.; Lazar, Z.; Juszczyk, P.; Rakicka-Pustułka, M.; Janek, T.; Kuźmińska-Bajor, M.; Rymowicz, W. Application of a new engineered strain of *Yarrowia lipolytica* for effective production of calcium ketoglutarate dietary supplements. *Int. J. Mol. Sci.* **2021**, *22*, 7577. [CrossRef]
22. Rywińska, A.; Tomaszewska-Hetman, L.; Rakicka-Pustułka, M.; Juszczyk, P.; Rymowicz, W. Alpha-ketoglutaric acid production from a mixture of glycerol and rapeseed oil by *Yarrowia lipolytica* using different substrate feeding strategies. *Sustainability* **2020**, *12*, 6109. [CrossRef]
23. Kamzolova, S.V.; Morgunov, I.G. α-Ketoglutaric acid production from rapeseed oil by *Yarrowia lipolytica* yeast. *Appl. Microbiol. Biot.* **2013**, *97*, 5517–5525. [CrossRef] [PubMed]
24. Kamzolova, S.V.; Morgunov, I.G. Optimization of medium composition and fermentation conditions for α-ketoglutaric acid production from biodiesel waste by *Yarrowia lipolytica*. *Appl. Microbiol. Biot.* **2020**, *104*, 7979–7989. [CrossRef] [PubMed]

25. Chernyavskaya, O.G.; Shishkanova, N.V.; Il'chenko, A.P.; Finogenova, T.V. Synthesis of alpha-ketoglutaric acid by *Yarrowia lipolytica* yeast grown on ethanol. *Appl. Microbiol. Biotechnol.* **2000**, *53*, 152–158. [CrossRef]
26. Zhou, J.; Zhou, J.; Du, G.; Liu, L.; Chen, J. Screening of thiamine-auxotrophic yeast for α-ketoglutaric acid overproduction. *Lett. Appl. Microbiol.* **2010**, *51*, 264–271. [CrossRef]
27. Yu, Z.; Du, G.; Zhou, J.; Chen, J. Enhanced α-ketoglutaric acid production in *Yarrowia lipolytica* WSH-Z06 by an improved integrated fed-batch strategy. *Bioresour. Technol.* **2012**, *114*, 597–602. [CrossRef]
28. Liu, L.M.; Li, Y.; Du, G.C.; Chen, J. $CaCO_3$ stimulates alpha-ketoglutarate accumulation during pyruvate fermentation by *Torulopsis glabrata*. *Chin. J. Biotechnol.* **2003**, *19*, 745–749.
29. Huang, H.J.; Liu, L.M.; Li, Y.; Du, G.C.; Chen, J. Redirecting Carbon Flux in *Torulopsis glabrata* from pyruvate to α-ketoglutaric acid by changing metabolic co-factors. *Biotechnol. Lett.* **2006**, *28*, 95–98. [CrossRef]
30. Otto, C.; Yovkova, V.; Aurich, A.; Mauersberger, S.; Barth, G. Variation of the by-product spectrum during α-ketoglutaric acid production from raw glycerol by overexpression of fumarase and pyruvate car-boxylase genes in *Yarrowia lipolytica*. *Appl. Microbiol. Biotechnol.* **2012**, *95*, 905–917. [CrossRef]
31. Liu, L.; Li, Y.; Zhu, Y.; Du, G.; Chen, J. Redistribution of carbon flux in *Torulopsis glabrata* by altering vitamin and calcium level. *Metab. Eng.* **2007**, *9*, 21–29. [CrossRef]
32. Verseck, S.; Karau, A.; Weber, M. Fermentative Production of Alpha-Ketoglutaric Acid. Evonik Degussa. GmbH. Patent WO2009053489A1, 30 April 2009.
33. Fickers, P.; Cheng, H.; Lin, C.S.K. Sugar alcohols and organic acids synthesis in *Yarrowia lipolytica*: Where are we? *Microorganisms* **2020**, *8*, 574. [CrossRef] [PubMed]
34. Babaei, M.; Kildegaard, K.R.; Niaei, A.; Hosseini, M.; Ebrahimi, S.; Sudarsan, S.; Angelidaki, I.; Borodina, I. Engineering olaginous yeast as the host for fermentative succinic acid production from glucose. *Front. Bioeng. Biotechnol.* **2019**, *7*, 361. [CrossRef] [PubMed]
35. Kamzolova, S.V.; Chiglintseva, M.N.; Lunina, J.N.; Morgunow, I.G. α-Ketoglutaric acid production by *Yarrowia lipolytica* and its regulation. *Appl. Microbiol. Biotechnol.* **2012**, *96*, 783–791. [CrossRef] [PubMed]
36. Kamzolova, S.V.; Chiglintseva, M.N.; Yusupova, A.I.; Vinokurova, N.G.; Lysanskaya, V.Y.; Morgunov, I.G. Biotechnological potential of *Yarrowia lipolytica* grown under thiamine limitation. *Food Technol. Biotechnol.* **2012**, *50*, 412–419.
37. Kamzolova, S.V.; Dedyukhina, E.G.; Samoilenko, V.A.; Lunina, J.N.; Puntus, I.F.; Allayarov, R.L.; Chiglintseva, M.N.; Mironov, A.A.; Morgunov, I.G. Isocitric acid production from rapeseed oil by *Yarrowia lipolytica* yeast. *Appl. Microbiol. Biotechnol.* **2013**, *97*, 9133–9144. [CrossRef] [PubMed]
38. Tomaszewska, L.; Rymowicz, W.; Rywińska, A. Mineral supplementation increases erythrose reductase activity in erythritol biosynthesis from glycerol by *Yarrowia lipolytica*. *Appl. Biochem. Biotechnol.* **2014**, *172*, 3069–3078. [CrossRef]
39. Cottrell, T.; Peij, J. Sorbitan esters and polysorbates. In *Emulsifiers in Food Technology*, 2nd ed.; Norn, V., Ed.; Wiley-Blackwell: Hoboken, NJ, USA, 2015; pp. 271–295.
40. Mortensen, A.; Aguilar, F.; Crebelli, R.; Di Domenico, A.; Dusemund, B.; Frutos, M.J.; Galtier, P.; Gott, D.; Gundert-Remy, U.; Leblanc, J.C.; et al. Re-evaluation of sorbitan monostearate (E 491), sorbitan tristearate (E 492), sorbitan monolaurate (E 493), sorbitan monooleate (E 494) and sorbitan monopalmitate (E 495) when used as food additives. *EFSA J.* **2017**, *15*, e04788. [CrossRef]
41. Musiał, I.; Rymowicz, W.; Witkowska, D. Effect of Span 20 concentration on oxalic acid production from post-refining fatty acids by *Aspergillus niger* XP. *Chem. Pap.* **2006**, *60*, 388–390. [CrossRef]
42. Rakicka, M. The effect of Span 20 addition on erythritol production from crude glycerol by *Yarrowia lipolytica*. *Acta Sci. Pol. Biotechnol.* **2013**, *12*, 15–24.
43. Xu, F.; Yuan, Q.P.; Zhu, Y. Improved production of lycopene and β-carotene by *Blakeslea trispora* with oxygen-vectors. *Process Biochem.* **2007**, *42*, 289–293. [CrossRef]
44. Hu, W.L.; Dai, D.H. Effect of non-ionic surfactant and oxygen-vector on the production of β-carotene and biomass from *Blakeslea trispora*. *Adv. Mater. Res.* **2013**, *726–731*, 315–319. [CrossRef]
45. Yovkova, V.; Otto, C.; Aurich, A.; Mauersberger, S.; Barth, G. Engineering the α-ketoglutarate over production from raw glycerol by overexpression of the genes encoding NADP+-dependent isocitrate dehydrogenase and pyruvate carboxylase in *Yarrowia lipolytica*. *Appl. Microbiol. Biotechnol.* **2014**, *98*, 2003–2013. [CrossRef] [PubMed]

catalysts

Article

An *O*-Demethylation Metabolite of Rabeprazole Sulfide by *Cunninghamella blakesleeana* 3.970 Biotransformation

Ming Song [1], Hongxiang Zhu [2], Jian Wang [3], Weizhuo Xu [2,*] and Wei Xu [2,*]

1 School of Functional Food and Wine, Shenyang Pharmaceutical University, Shenyang 110016, China
2 School of Life Sciences and Biopharmaceuticals, Shenyang Pharmaceutical University, Shenyang 110016, China
3 School of Pharmaceutical Engineering, Shenyang Pharmaceutical University, Shenyang 110016, China
* Correspondence: weizhuo.xu@syphu.edu.cn (W.X.); shxuwei8720@163.com (W.X.); Tel./Fax: +86-024-43520301 (Weizhuo Xu); Tel.: +86-024-43520307 (Wei Xu)

Abstract: To explore the potential metabolites from rabeprazole sulfide, seven strains of filamentous fungi were screened for their biotransformation abilities. Among these strains, *Cunninghamella blakesleeana* 3.970 exhibited the best result. Four different culture media were screened in order to identify the most optimal for subsequent research. Single factors such as the initial pH of culture media, culture time, inoculation volume, and media volume were individually investigated to provide the optimum biotransformation conditions. Then, an orthogonal optimization process using a five-factor, four-level $L_{16}(4^5)$ experiment was designed and performed. Finally, when the substrate concentration is 3 g/L, one major metabolite was detected with a transformation rate of 72.4%. Isolated by semipreparative HPLC, this metabolite was further detected by ESI-MS and NMR. The final data analysis indicated that the metabolite is *O*-demethylation rabeprazole sulfide.

Keywords: rabeprazole; sulfide; biotransformation; *Cunninghamella blakesleeana* 3.970; *O*-demethylation rabeprazole sulfide

Citation: Song, M.; Zhu, H.; Wang, J.; Xu, W.; Xu, W. An *O*-Demethylation Metabolite of Rabeprazole Sulfide by *Cunninghamella blakesleeana* 3.970 Biotransformation. *Catalysts* **2023**, *13*, 15. https://doi.org/10.3390/catal13010015

Academic Editors: Zhilong Wang and Tao Pan

Received: 14 November 2022
Revised: 16 December 2022
Accepted: 17 December 2022
Published: 22 December 2022

1. Introduction

Rabeprazole is a kind of proton pump inhibitor, which has been widely used in the treatment of gastric acid secretion, and can also be applied to eradicate *Helicobacter pylori* [1]. As a benzimidazole derivative, rabeprazole has a heterocyclic molecule consisting of a pyridine and benzimidazole moiety linked by a methylsulfinyl group. This general structure guarantees the similar pharmacological properties of other proton pump inhibitors. Rabeprazole sulfide (rabeprazole thioether) is an essential intermediate for the production of rabeprazole [2,3]. Meanwhile, sulfide is also one of the metabolites produced by the rabeprazole metabolites in vivo [4–8].

Rabeprazole sulfide is an important imidazole compound, which is not only a precursor of rabeprazole synthesis, but is also one of the main metabolites of rabeprazole in vivo. As the primary metabolite of rabeprazole, rabeprazole sulfide can undergo further metabolic reactions such as demethylation and sulfuric acid binding. These demethylated metabolites often have pharmacological activity. Pharmacologically active metabolites can contribute significantly to the overall therapeutic and adverse effects of drugs. Therefore, to fully understand the mechanism of action of drugs, it is important to recognize the role of active metabolites [9–12].

Some microorganisms, especially the fungi belonging to *Cunninghamella* species, possess cytochrome P-450 mono-oxygenase systems analogous to those in mammals [13]. Hence, microbial transformation has been proposed as a complementary in vitro model for mammalian drug metabolism. *Cunnihamella* is a genus of fungi in the family *Cunninghamellaceae*, which has been implemented for in vitro biotransformations for a long time [14–20]. Various reaction types have been reported for these kinds of fungi. Previous studies have demonstrated their hydroxylation [21–24], epoxidation [25–27], amination [28], demethylation [29], and oxidation from sulfide to sulfoxide, which is usually the final step for the production of prazole-type molecules [30–32].

Demethylated metabolites often have some unexpected pharmacological effects. Metamizole (dipyrone) is an analgesic with antipyretic and spasmolytic properties, which has been in use for almost 100 years. The primary metabolite of metamizole, 4-methylaminoantipyrine, can be *N*-demethylated to 4-aminoantipyrine. 4-aminoantipyrine plays a role as a non-steroidal anti-inflammatory drug, a non-narcotic analgesic, an antirheumatic drug, a peripheral nervous system drug, and an EC 1.14.99.1 (prostaglandin-endoperoxide synthase) inhibitor. It is used as a reagent for biochemical reactions, such as the production of peroxides or phenols [33,34]. Papaverine plays a role as a vasodilator agent and an anti-spasmodic drug. The metabolism of papaverine with *Cunninghamella echinulate* results in *O*-demethylation. 4′-*O*-demethylated papaverine, 3′-*O*-demethylated papaverine, and 6-*O*-demethylated papaverine have been isolated. In silico docking studies of these metabolites using crystals of human phosphodiesterase 10a (hPDE10a) revealed that the compounds 4′-*O*-demethylated papaverine and 6-*O*-demethylated papaverine possess better docking scores and binding poses with favorable interactions than the native ligand papaverine [35].

In this report, seven strains of filamentous fungi are investigated for their ability to metabolize the rabeprazole sulfide, in order to explore the reaction types of the whole cell transformation. Instead of hunting for the sulfoxide metabolites obtained, an *O*-demethyl rabeprazole sulfide, a new compound with potential pharmacological activity, was finally isolated and identified (Figure 1), which is the first report of the *O*-demethyl reaction in rabeprazole sulfide to date.

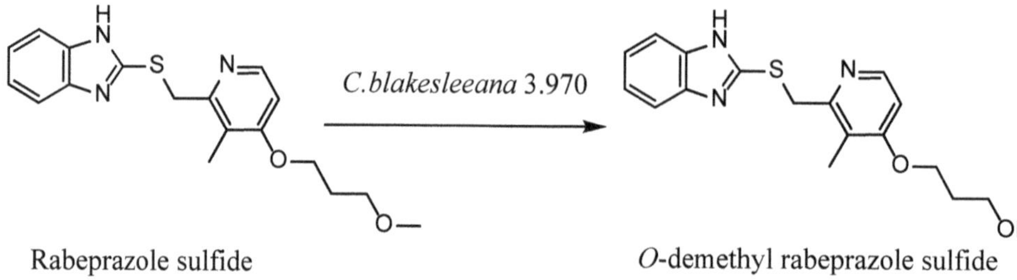

Figure 1. Rabeprazole sulfide and its *O*-demethyl metabolite obtained by the biotransformation of *Cunninghamella blakesleeana* 3.970.

2. Results and Discussion

2.1. Whole-Cell Biotransformation Results of Rabeprazole Sulfide

Among the seven strains, *Cunninghamella elegans* 3.910, *Cunninghamella echimnulata* 3.967, *Cunninghamella blakesleeana* 3.970, and *Absidia coerulea* 41,050 displayed different transformation results for rabeprazole sulfide. *Cunninghamella elegans* 3.910 and *Cunninghamella blakesleeana* 3.970 demonstrated a rather significant major metabolite spectrum, whereas *Gibberella fujiluroi* 40,272, *Gibberella* sp. 2498, and *Caldariomyces fumago* 16,373 did not indicate any positive transformation result. According to the transformation TLC results (Figure 2), the fungus *Cunninghamella blakesleeana* 3.970 converted rabeprazole sulfide to rabeprazole and another, novel compound, and the conversion rate of this new compound was high. Therefore, *Cunninghamella blakesleeana* 3.970 was selected for further study.

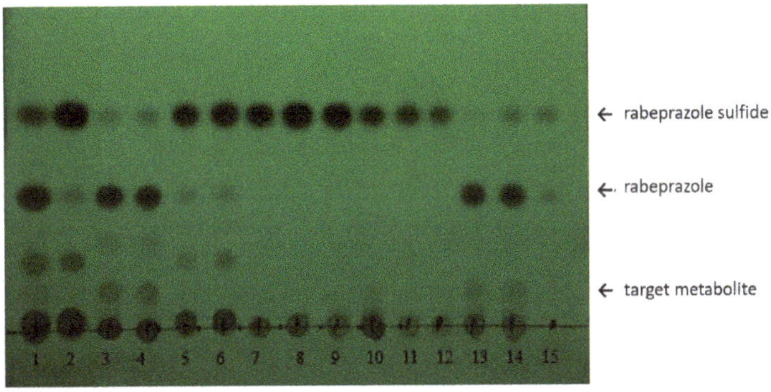

Figure 2. TLC screening results for seven strains of filamentous fungi for rabeprazole sulfide. Lanes 1 and 2, *Cunninghamella elegans* 3.910, 3 and 4, *Cunninghamella echimnulata* 3.967, 5 and 6, *Absidia coerulea* 41,050, 7 and 8, *Gibberella* sp. 2498, 9 and 10, *Caldariomyces fumago* 16,373, 11 and 12, *Gibberella fujiluroi* 40,272, 13 and 14, *Cunninghamella blakesleeana* 3.970, and 15, standard mixture of rabeprazole sulfide and rabeprazole.

2.2. *Optimization of Transformation Media*

In order to obtain the transformation products of rabeprazole sulfide, four microbial media were screened. The evaluation criteria were set according to both the growth of the strain and the HPLC detection diagram for the transformation solution. According to Figure S1, fewer mycelium balls were observed in medium 1 and medium 2, and the color of the mycelium ball was a little pink in medium 3, which is not its usual appearance. Only in medium 4 were the numbers of the mycelium balls ambient and the growth of mycelium good.

It can be seen from the HPLC results in Figure S1 that under the same substrate concentration, the ability of transformation media 1, 2, and 3 to transform rabeprazole sulfide is low, and the transformation products are relatively miscellaneous. However, transformation medium 4 had a higher ability to transform rabeprazole sulfide, and only one transformation product was produced. Therefore, based on the results in Figures S1 and S2, medium 4 was selected as the medium for subsequent fermentation.

2.3. *Single-Factor Evaluation for Rabeprazole Sulfide Biotransformation*

2.3.1. Standard Curve of Rabeprazole Sulfide

Figure 3 indicates that the linear correlation coefficient $R^2 = 0.9995$ and the correlation equation $y = 18{,}285x + 66.794$ for the concentration and peak area of rabeprazole sulfide, with a good linear correlation. This standard curve can be used to calculate the conversion rate of rabeprazole sulfide.

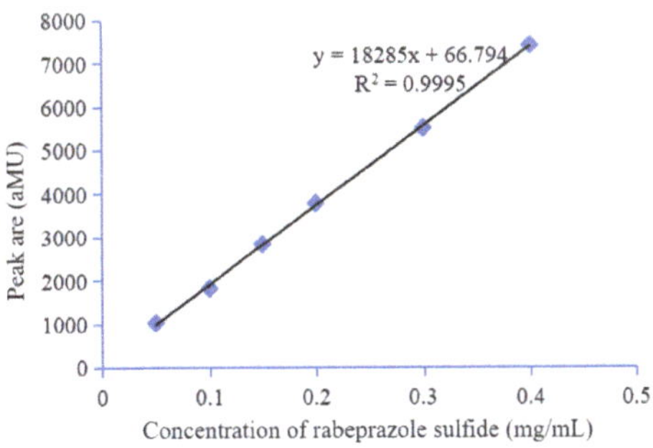

Figure 3. Standard curve of rabeprazole sulfide.

2.3.2. Evaluation of Culture Time

The culture duration of whole-cell catalysis has an important effect on the reaction. If the reaction time is too short, the reaction may not be accomplished. If the reaction time is too long, product inhibition may occur and the total transformation rate may decrease. In this section, we investigate the effect of conversion time on the production of metabolites; the results are shown in Figure 4.

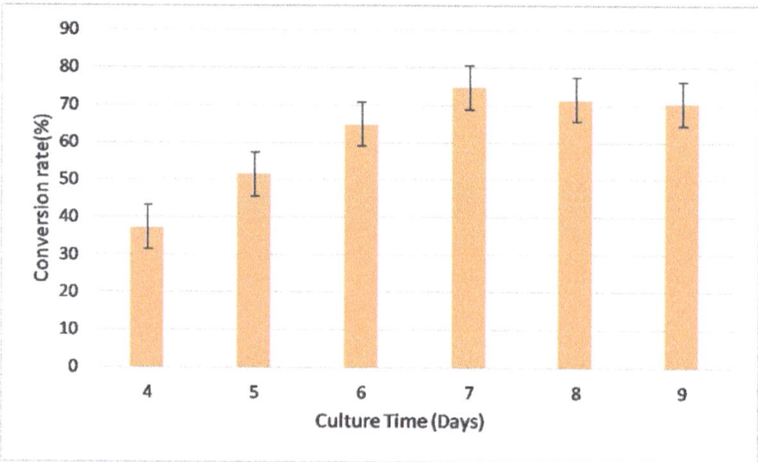

Figure 4. Effects of culture time on metabolite conversion rate (initial pH = 6.5, media volume 50 mL, inoculation volume 10%, substrate concentration 1 mg/mL).

When the transformation was performed at 28 °C, 200 rpm, it could be seen that the culture time ranged between 6 and 9 days. A culture time of 7 days resulted in the optimal conversion rate of 74.71%.

2.3.3. Evaluation of Initial pH

The initial pH of the media is vital for strain growth and transformation. A pH that is too low or high will cause the strains to undergo non-ambient growth and affect their metabolism capability. This is why different strains desire specific initial pH values in their media. In this section, a pH range of 5.5 to 8.0 was selected for the investigation. From Figure 5, it can be seen that both the acid and alkaline conditions are not fit for biotransformation, so the best initial pH is set to 6.0.

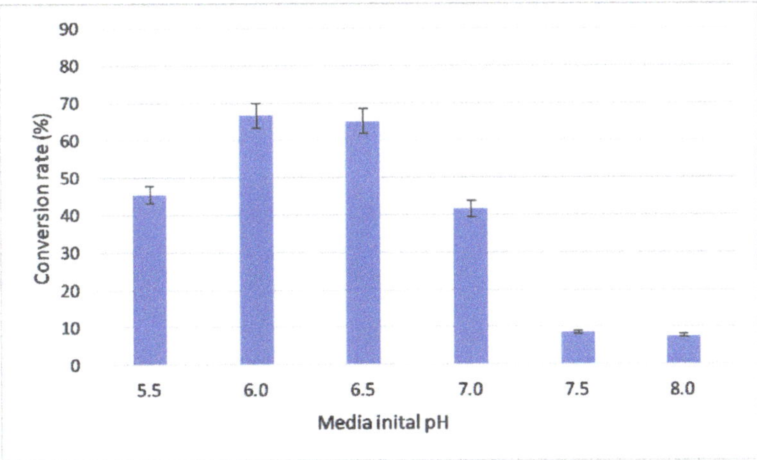

Figure 5. Effects of the initial pH of the media on the metabolite conversion rate (culture time 5 d, media volume 50 mL, inoculation volume 10%, substrate concentration 1 mg/mL).

2.3.4. Evaluation of Media Volume

The media volume is crucial for the transformation effects, which are related to the dissolved oxygen when the conical flask volume is constant. Media volumes from 30 mL to 120 mL were selected for the investigation. From Figure 6, it can be seen that a media volume of 50 mL displayed the best transformation effect.

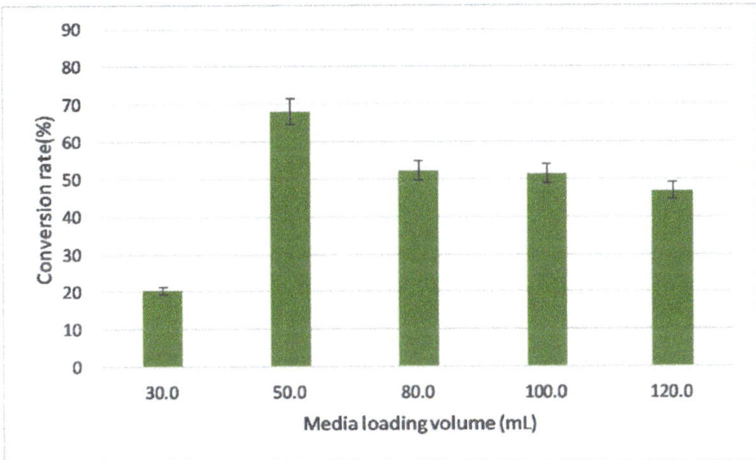

Figure 6. Effects of media volume on metabolite conversion rate (initial pH = 6.5, culture time 5 d, inoculation volume 10%, substrate concentration 1 mg/mL).

2.3.5. Evaluation of Inoculation Volume

Inoculation volume is a critical factor for the biotransformation process. If the inoculation volume is high, the strains will grow to the exponential phase too early and the nutrients will be depleted quickly, such that the enzymes in the strains may not sufficiently be induced and expressed, and the biotransformation may not be adequate. If the inoculation volume is low, the strains may take too long a time to grow and induce enzymes, which may decrease their metabolism capabilities. Based our previous experiences, inoculation volumes ranging from 6% to 14% were selected for the investigation.

From Figure 7, it can be seen that an inoculation volume of 10% is the best for biotransformation.

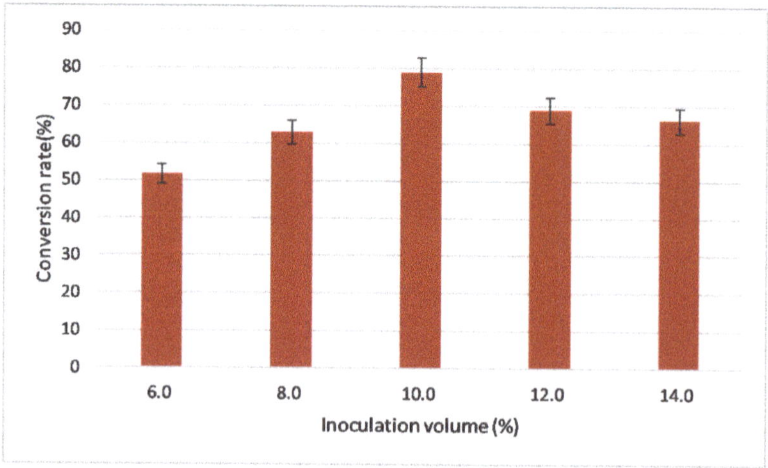

Figure 7. Effects of inoculation volume on metabolite conversion rate (initial pH = 6.5, culture time 5 d, media volume 50 mL, substrate concentration 1 mg/mL).

2.3.6. Evaluation of Substrate Concentration

In this study, rabeprazole sulfide is not a natural substrate, which may cause severe inhibition and toxic effects for the strains. Hence, a proper substrate concentration is essential for successful biotransformation. A substrate concentration range from 1 mg/mL to 5 mg/mL was set for the investigation.

From Figure 8, it can be seen from the decreasing conversion rate tendency that 1 mg/mL substrate concentration is the best for biotransformation.

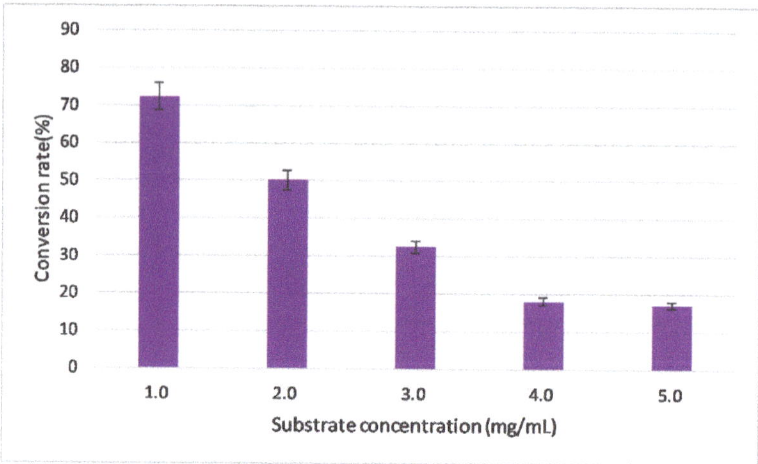

Figure 8. Effects of substrate concentration on metabolite conversion rate (initial pH = 6.5, culture time 5 d, media volume 50 mL, substrate concentration 1 mg/mL).

2.4. Orthogonal Optimization

According to the above single-factor investigation results, a 5-factor, 4-level, orthogonal $L_{16}(4^5)$ optimization process was designed, as shown in Table 1.

Table 1. Orthogonal experimental factors and level assignments.

Factor / Level	A/Culture Time (Days)	B/Inoculation Volume (%)	C/Media Volume (mL)	D/Media Starting pH	E/Substrate Concentration (g/L)
1	5	8	30	5.5	1
2	6	10	50	6.0	2
3	7	12	80	6.5	3
4	8	14	100	7.0	4

After the design and execution of the experiments, it could be seen that the strength of impact of the various factors on biotransformation was in the order D > B > E > C > A, i.e., the most significant influence was that of the initial pH value of the media, as shown in Table 2. When the range analysis was performed, it was found that the best combination was $D_2B_1E_3C_4A_4$, so an initial pH of 6.0, inoculation volume of 8%, substrate concentration of 3 mg/mL, media volume of 100 mL, and culture time of 8 days were the best values. These values are consistent between the best results from the orthogonal design and those from the No.13 experimental group (total 16 groups), with a conversion rate of 72.4%. This is almost a double increase in the conversion rate compared with the unoptimized results.

Table 2. Orthogonal experimental results and analysis.

Factors / Experiment No.	A	B	C	D	E	Conversion Rate
1	1	1	1	1	1	0.352
2	1	2	2	2	2	0.315
3	1	3	3	3	3	0.348
4	1	4	4	4	4	0.201
5	2	1	2	3	4	0.213
6	2	2	1	4	3	0.124
7	2	3	4	1	2	0.336
8	2	4	3	2	1	0.566
9	3	1	3	4	2	0.347
10	3	2	4	3	1	0.171
11	3	3	1	2	4	0.135
12	3	4	2	1	3	0.343
13	4	1	4	2	3	0.724
14	4	2	3	1	4	0.053
15	4	3	2	4	1	0.094
16	4	4	1	2	3	0.033
K1	1.216	1.636	0.644	1.084	1.184	
K2	1.24	0.664	0.964	1.812	1.032	
K3	0.996	0.912	1.312	0.764	1.54	
K4	1.304	1.144	1.432	0.764	0.604	
k1	0.304	0.409	0.161	0.271	0.296	
k2	0.310	0.166	0.241	0.453	0.258	
k3	0.249	0.228	0.328	0.191	0.385	
k4	0.326	0.286	0.358	0.191	0.151	
$R_{(\text{Range analysis})}$	0.084	0.243	0.197	0.244	0.234	
Best combination	A2	B1	C4	D2	E1	

2.5. Structure Identification for the Rabeprazole Sulfide Biotransformation

The isolated metabolite was resolved in DMSO for the MS and NMR analysis. According to the ESI-MS results, showing that the m/z $[M+H]^+$ is 330.25 (Figure S3) and the m/z

[M-H]$^-$ is 328.27 (Figure S4), it could be inferred that the target metabolite has a molecular weight of 329.12, which is only a Dalton loss of 14 compared with the original rabeprazole sulfide with a molecular weight of 343.44. Hence, the loss of a methyl group is assumed.

According to the comparison of the NMR data between the substrate and metabolite, it could be found that a methyl and a carbon signal are absent in the H spectrum and C spectrum. Combined with the MS and NOESY data (Figure S7) indicating a methyl loss, it can be concluded that the target metabolite is the *O*-dimethyl rabeprazole sulfide (Figure 1). The analysis is given below.

^{1}H-NMR (600 MHz, DMSO-d_6) δH: 8.23 (1H, d, *J* = 5.6 Hz), 7.45 (2H, dd, *J*1 = 5.8, *J*2 = 3.2 Hz), 7.11 (2H, dd, *J*1 = 5.9, *J*2 = 3.1 Hz), 6.95 (1H, d, *J* = 5.6 Hz), 4.69 (2H, s), 4.12 (2H, t, *J* = 6.2 Hz), 3.58 (2H, t, *J* = 6.2 Hz), 2.21 (3H, s), 1.87–1.91 (2H, m).

^{13}C-NMR (150 MHz, DMSO-d_6) δC: 163.27, 155.16, 150.90, 148.24, 121.73, 120.20, 106.78, 65.50, 57.56, 36.74, 32.31, 10.88.

3. Materials and Methods

3.1. Materials and Strains

Rabeprazole sulfide was synthesized in the Department of Pharmaceutical Engineering, Shenyang Pharmaceutical University. Methanol and acetonitrile were purchased from Concord Technology Co., Ltd. (Tianjin, China). Peptone and yeast extract were purchased from HopeBio Co., Ltd. (Qingdao, China). All other chemical reagents were purchased from Yuwang Chemical Co., Ltd. (Shenyang, China).

Cunninghamella blakesleeana 3.970 and *Cunninghamella elegans* 3.910 were stored in our lab reservoir. *Cunninghamella echimnulata* 3.967, *Absidia coerulea* 41,050, *Gibberella fujiluroi* 40,272, *Gibberella* sp. 2498, and *Caldariomyces fumago* 16,373 were purchased from China Center of Industrial Culture Collection (CICC).

3.2. Media and Strain Cultivations

Slant and agar media consisted of potato (200 g), glucose (20 g), agar (20 g), and 1000 mL distilled water. Seed culture media consisted of potato starch (45 g), yeast extract (3 g), corn steep liquor (10 g), $CaCO_3$ (3 g), $MgSO_4$ (0.5 g), and $FeSO_4$ (0.05 g). Four kinds of biotransformation culture media ingredients are listed in Table 3. For all media, the pH was adjusted to 6.5 with 6 M HCl, autoclaved at 115 °C for 30 min, and cooled before use.

Table 3. Four different biotransformation media formulas.

Ingredients	No.1 (g)	No.2 (g)	No.3 (g)	No.4 (g)
Sucrose	30		15	
Yeast extracts	20	0.5		2
Corn steep liquor	5			
Peptone		0.5	5	5
Glucose		4	15	50
K_2HPO_4	16	0.5	1	4
KH_2PO_4	2			5
$MgSO_4$	0.5		0.5	0.2
$FeSO_4$	0.05		0.01	0.05
KCl		0.5	0.5	

Two loops of mycelium from a fresh growth slant were inoculated into the seed culture media and cultured at 28 °C, 200 rpm for 48 h. Then, 5 mL of seed culture was inoculated into the transformation media (50 mL in a 250 mL conical flask, rabeprazole sulfide as substrate) and cultured at 28 °C, 200 rpm for 5 days. Substrate controls were set without inoculating the fungi into the media and strain controls were set without adding the substrate into the media, with all other conditions remaining the same.

3.3. *Thin-Layer Chromatography (TLC)*

The biotransformation results of the fungal extracts were analyzed by TLC performed on a Merck Silica gel 60 F254 plate (Merck, Darmstadt, Germany) with ethyl acetate:methanol:ammonia water (90:9:1 $v/v/v$) as the mobile phase, and a UV detection wavelength of 254 nm. A mixture of rabeprazole sulfide and rabeprazole (10 mg) was dissolved in 100 mL of isopropyl alcohol used as a standard solution. For TLC detection, 10 μL of standard solution and fungal extracts was spotted onto TLC plates.

3.4. *Extraction of Metabolites and HPLC Detection*

Biotransformation cultures were collected, ultrasonically treated for 30 min, and centrifugated at 3000 rpm for 10 min to retrieve the supernatant. Pellets were washed and extracted three times using ethyl acetate. The combined water phase and organic phase were extracted three times with an equal volume of ethyl acetate. The organic phase was then isolated and evaporated by reduced pressure distillation to obtain the raw metabolites. The raw metabolites were dissolved in methanol for further use.

The HPLC detection process for the metabolites was derived from the Pharmacopeia of the People's Republic of China (2020 Volume II) and optimized according to our practical work. For High-performance liquid chromatography (HPLC), a Shimadzu WondaSil C18 Superb (250 × 4.6 mm × 5 μm) column (Shimadzu Corp., Kyoto, Japan) was used. The mobile phase consisted of 0.015 mol/L Na_2HPO_4:acetonitrile = 60:40, the detection wavelength was 290 nm, the column temperature was 30 °C, the loading speed was 1 mL/min, and the loading volume was 10 μL.

3.5. *Isolation and Identification of Major Metabolite*

The isolation of the target metabolite was performed by semipreparative HPLC using an Elite SinoChrom ODS-BP (250 × 10 mm × 5 μm, Dalian, China) consisting of a quaternary pump, a vacuum degasser, a variable wavelength detector, and an autosampler. The mobile phase consisted of acetonitrile and water in different percentages, and the proper fractions were collected.

ESI-MS and NMR analysis were performed under standard conditions.

3.6. *Determination of the Standard Curve*

Rabeprazole sulfide was accurately weighed, and a solution of 0.40 mg/mL was prepared, diluted step by step with methanol. Standard samples of 0.40, 0.30, 0.20, 0.15, 0.10, and 0.05 mg/mL were prepared and detected with HPLC, and a standard curve was drawn with rabeprazole sulfide concentration as the abscissa and peak area as the ordinate. This standard curve was used to calculate the conversion of rabeprazole sulfide. The total conversion rate was calculated according to the following equation:

$$\text{Conversion rate } \% = \frac{1 - \text{residue amount of raberprazole sulfide}}{\text{Starting amount of rabeprazole sulfide}}$$

4. Conclusions and Discussion

By screening seven filamentous fungi, rabeprazole sulfide was the most efficiently decomposed into *O*-dimethyl rabeprazole sulfide by *Cunninghamella blakesleeana* 3.970, with a conversion rate of 72.4%. This is not a common biotransformation reaction type, with only few studies having suggested the *N*-demethylation [36,37] and *O*-dimethyl reactions [38,39].

The original aim of this study was to explore the possibility of whether these fungi could exhibit some oxidase activity so that the rabeprazole sulfide could be transformed into rabeprazole, i.e., to determine whether the thioether structure could be oxidized into the sulfoxide. Despite *Cunninghamella* species having exhibited various kinds of oxidation in previous works, no oxidation metabolites were identified in this research. Several reports have shown that *Absidia* and *Gibberella* species could exhibit oxidase activity as their P450, but positive oxidation metabolites have yet to be generated. More importantly, *Caldariomyces fumago* was selected for its strong inner chloroperoxidase expression, which exhibited multiple oxidation types, such as epoxidation [40,41], hydroxylation [42–47], halogenation [48], etc. However, none of the above reactions were identified in this rabeprazole sulfide biotransformation.

Rabeprazole sulfide is actually not a natural substrate of the fungus, which was mentioned in the previous section on the influence of substrate concentration on the conversion rate. With the increase in substrate concentration, the conversion rate decreases (Figure 8), and a too-high concentration may produce some inhibition. It is speculated that the *O*-demethylation reaction of rabeprazole sulfide is not completed in one step. It may be that rabeprazole sulfide is first converted into one or several intermediate products and then further converted into *O*-demethyl rabeprazole sulfide.

Some microorganisms can transform drugs and other xenobiotic compounds in a manner similar to that in mammals, and the utilization of microbial systems as models for mimicking and predicting the metabolism of drugs in humans and animals has received considerable attention. Liu et al. reported that the main metabolic pathways of *Cunninghamella blakesleeana* to transform verapamil are *N*-dealkylation, *O*-demethylation, and sulfate coupling, and the *O*-demethylated metabolites of verapamil have the same potency as the parent drug [49]. Xie ZY et al. reported that approximately 92.5% of pantoprazole was metabolized to six metabolites by *Cunninghamella blakesleeana* AS 3.153, one of which was 4′-*O*-demethyl-pantoprazole thioether [30]. These studies did not deeply explore the mechanism of *O*-demethylation when *Cunninghamella blakesleeana* transformed the substrate. Therefore, it is of interest to further study the mechanism of the *O*-demethylation of *Cunninghamella blakesleeana* and the pharmacological or biological characteristics of these *O*-demethylated metabolites.

In addition, further investigation should be undertaken in the hunt for fungi that may oxidize thioether to sulfoxide. Meanwhile, other rational designs for the engineered enzyme to facilitate this reaction are also welcomed.

Supplementary Materials: The following are available online at https://www.mdpi.com/article/10.3390/catal13010015/s1, Figure S1. Growth of *Cunninghamella blakesleeana* 3.970 in 4 different transformation media. Figure S2. HPLC detection results of *Cunninghamella blakesleeana* 3.970 in 4 different transformation media. Figure S3. HPLC detection results of rabeprazole sulfide and its metabolites. Figure S4. HPLC detection results of rabeprazole sulfide metabolites for MS and NMR analysis. Figure S5. MS(ESI)m/z[M+H]$^+$ diagram for the metabolite. Figure S6. MS(ESI)m/z[M+H]$^-$ diagram for the metabolite. Figure S7. ^{1}H NMR diagram for the *O*-demethyl rabeprazole sulfide. Figure S8. ^{13}C NMR diagram for the *O*-demethyl rabeprazole sulfide. Figure S9. ^{1}H NOESY diagram for the *O*-demethyl rabeprazole sulfide.

Author Contributions: M.S. (data curation, investigation); H.Z. (investigation, methodology); J.W. (rabeprazole and its sulfide preparation); W.X. (Weizhuo Xu) and W.X. (Wei Xu) (resources, supervision, writing—review and editing). All authors have read and agreed to the published version of the manuscript.

Funding: This research received no external funding.

Data Availability Statement: Data are available upon reasonable request.

Conflicts of Interest: The authors declare no conflict of interest.

References

1. FDA. Approval for Aciphex (Rabeprazole Sodium) Tablets. 2002. Available online: https://www.accessdata.fda.gov/drugsatfda_docs/nda/2002/021456_aciphex.cfm (accessed on 25 April 2003).
2. Duan, Z.; Wang, Y.; Zhang, L.; Cao, X.; Fu, L.; Li, Z.; Zhang, J. An application of continuous flow microreactor in the synthesis and extraction of rabeprazole. *Int. J. Chem. React. Eng.* **2021**, *19*, 287–294. [CrossRef]
3. Nishiguchi, S.; Izumi, T.; Kouno, T.; Sukegawa, J.; Ilies, L.; Nakamura, E. Synthesis of Esomeprazole and Related Proton Pump Inhibitors through Iron-Catalyzed Enantioselective Sulfoxidation. *ACS Catal.* **2018**, *8*, 9738–9743. [CrossRef]

4. Lu, C.; Jia, Y.; Song, Y.; Li, X.; Sun, Y.; Zhao, J.; Wang, S.; Shi, L.; Wen, A.; Ding, L. Application of a liquid chromatographic/tandem mass spectrometric method to a urinary excretion study of rabeprazole and two of its metabolites in healthy human urine. *J. Chromatogr. B* **2015**, *988*, 75–80. [CrossRef] [PubMed]
5. Zannikos, P.N.; Doose, D.R.; Leitz, G.J.; Rusch, S.; Gonzalez, M.D.; Solanki, B.; Haddad, I.; Mulberg, A.E. Pharmacokinetics and Tolerability of Rabeprazole in Children 1 to 11 Years Old with Gastroesophageal Reflux Disease. *J. Pediatr. Gastroenterol. Nutr.* **2011**, *52*, 691–701. [CrossRef] [PubMed]
6. Ren, S.; Park, M.-J.; Kim, A.; Lee, B.-J. In vitro metabolic stability of moisture-sensitive rabeprazole in human liver microsomes and its modulation by pharmaceutical excipients. *Arch. Pharmacal Res.* **2008**, *31*, 406–413. [CrossRef]
7. Miura, M.; Satoh, S.; Tada, H.; Habuchi, T.; Suzuki, T. Stereoselective metabolism of rabeprazole-thioether to rabeprazole by human liver microsomes. *Eur. J. Clin. Pharmacol.* **2005**, *62*, 113–117. [CrossRef]
8. Sharara, A.I. Rabeprazole: The role of proton pump inhibitors in Helicobacter pylori eradication. *Expert Rev. Anti-Infect. Ther.* **2005**, *3*, 863–870. [CrossRef]
9. Lun, J.; Ma, S.; Xue, M.; Zhao, P.; Song, Y.; Guo, X. Simultaneous enantiomeric analysis of five proton-pump inhibitors in soil and sediment using a modified QuEChERS method and chiral high performance liquid chromatography coupled with tandem mass spectrometry. *Microchem. J.* **2020**, *160*, 105625. [CrossRef]
10. Marelli, S.; Pace, F. Rabeprazole for the treatment of acid-related disorders. *Expert Rev. Gastroenterol. Hepatol.* **2012**, *6*, 423–435. [CrossRef]
11. Yousuf, M.; Jamil, W.; Mammadova, M.Y.A.K. Microbial Bioconversion: A Regio-specific Method for Novel Drug Design and Toxicological Study of Metabolites. *Curr. Pharm. Biotechnol.* **2019**, *20*, 1156–1162. [CrossRef]
12. Asha, S.; Vidyavathi, M. *Cunninghamella*—A microbial model for drug metabolism studies—A review. *Biotechnol. Adv.* **2009**, *27*, 16–29. [CrossRef]
13. Murphy, C.D. Drug metabolism in microorganisms. *Biotechnol. Lett.* **2015**, *37*, 19–28. [CrossRef]
14. Keum, Y.-S.; Lee, Y.-H.; Kim, J.-H. Metabolism of methoxychlor by *Cunninghamella elegans* ATCC36112. *J. Agric Food Chem.* **2009**, *57*, 7931–7937. [CrossRef]
15. Choudhary, M.I.; Khan, N.T.; Musharraf, S.G.; Anjum, S.; Rahman, A.U. Biotransformation of adrenosterone by filamentous fungus, *Cunninghamella elegans*. *Steroids* **2007**, *72*, 923–929. [CrossRef]
16. Hanson, R.L.; Matson, J.A.; Brzozowski, D.B.; LaPorte, T.L.; Springer, D.M.; Patel, R.N. Hydroxylation of Mutilin by Streptomyces griseus and *Cunninghamella echinulata*. *Org. Process Res. Dev.* **2002**, *6*, 482–487. [CrossRef]
17. Ibrahim, A.; Khalifa, S.I.; Khafagi, I.; Youssef, D.T.; Khan, S.; Mesbah, M.; Khan, I. Microbial Metabolism of Biologically Active Secondary Metabolites from *Nerium oleander* L. *Chem. Pharm. Bull.* **2008**, *56*, 1253–1258. [CrossRef]
18. Kouzi, S.A.; Chatterjee, P.; Pezzuto, J.M.; Hamann, M.T. Microbial Transformations of the Antimelanoma Agent Betulinic Acid. *J. Nat. Prod.* **2000**, *63*, 1653–1657. [CrossRef]
19. Ning, L.; Zhan, J.; Qu, G.; Zhong, L.; Guo, H.; Bi, K.; Guo, D. Biotransformation of triptolide by *Cunninghamella blakesleana*. *Tetrahedron* **2003**, *59*, 4209–4213. [CrossRef]
20. Zhang, D.; Evans, F.E.; Freeman, J.P.; Duhart, B.; Cerniglia, C.E. Biotransformation of amitriptyline by *Cunninghamella elegans*. *Drug Metab. Dispos.* **1995**, *23*, 1417–1425.
21. Wu, Y.; Lu, Y.; Yi, Y.; Wang, A.; Wang, W.; Yang, M.; Fan, B.; Chen, G. Biotransformation of asiatic acid by *Cunninghamella echinulata* and *Circinella muscae* to discover anti-neuroinflammatory derivatives. *Nat. Prod. Res.* **2022**, *36*, 1–6. [CrossRef]
22. Khan, M.F.; Murphy, C.D. Cytochrome P450 5208A3 is a promiscuous xenobiotic biotransforming enzyme in *Cunninghamella elegans*. *Enzym. Microb. Technol.* **2022**, *161*, 110102. [CrossRef] [PubMed]
23. Bai, Y.; Zhao, Y.; Gao, X.; Zhang, D.; Ma, Y.; Yang, L.; Sun, P. A Novel Antimalarial Metabolite in Erythrocyte from the Hydroxylation of Dihydroartemisinin by *Cunninghamella elegans*. *Front. Chem.* **2022**, *10*, 850133. [CrossRef] [PubMed]
24. Sousa, I.P.; de SousaTeixeira, M.V.; Freitas, J.A.; Ferreira, A.G.; Pires, L.M.; Santos, R.A.; Heleno, V.C.G.; Furtado, N.A.J.C. Production of More Potent Anti- *Candida* Labdane Diterpenes by Biotransformation Using *Cunninghamella elegans*. *Chem. Biodivers.* **2022**, *19*, e202100757. [CrossRef] [PubMed]
25. Barrero, A.F.; Oltra, J.E.; Raslan, A.D.S.; Saúde, D.A. Microbial Transformation of Sesquiterpene Lactones by the Fungi *Cunninghamella echinulata* and *Rhizopus oryzae*. *J. Nat. Prod.* **1999**, *62*, 726–729. [CrossRef] [PubMed]
26. Zhang, D.; Hansen, E.B., Jr.; Deck, J.; Heinze, T.M.; Henderson, A.; Korfmacher, W.A.; Cerniglia, C.E. Fungal transformations of antihistamines: Metabolism of cyproheptadine hydrochloride by *Cunninghamella elegans*. *Xenobiotica* **1997**, *27*, 301–315. [CrossRef]
27. Zhou, L.; Xu, W.; Chen, Y.; Zhao, J.; Yu, N.; Fu, B.; You, S. Stereoselective epoxidation of curcumol and curdione by *Cunninghamella elegans* AS 3.2028. *Catal. Commun.* **2012**, *28*, 191–195. [CrossRef]
28. Grafinger, K.E.; Wilke, A.; König, S.; Weinmann, W. Investigating the ability of the microbial model *Cunninghamella elegans* for the metabolism of synthetic tryptamines. *Drug Test. Anal.* **2018**, *11*, 721–729. [CrossRef]
29. Nykodemová, J.; Šuláková, A.; Palivec, P.; Češková, H.; Rimpelová, S.; Šíchová, K.; Leonhardt, T.; Jurásek, B.; Hájková, K.; Páleníček, T.; et al. 2C-B-Fly-NBOMe Metabolites in Rat Urine, Human Liver Microsomes and *C. elegans*: Confirmation with Synthesized Analytical Standards. *Metabolites* **2021**, *11*, 775. [CrossRef]
30. Xie, Z.Y.; Huang, H.H.; Zhong, D.F. Biotransformation of pantoprazole by the fungus *Cunninghamella blakesleeana*. *Xenobiotica* **2005**, *35*, 467–477. [CrossRef]

31. Sutherland, J.B.; Freeman, J.P.; Heinze, T.M.; Moody, J.D.; Parshikov, I.A.; Williams, A.J.; Zhang, D. Oxidation of phenothiazine and phenoxazine by *Cunninghamella elegans*. *Xenobiotica* **2001**, *31*, 799–809. [CrossRef]
32. Zhang, D.; Freeman, J.P.; Sutherland, J.B.; Walker, A.E.; Yang, Y.; Cerniglia, C.E. Biotransformation of chlorpromazine and methdilazine by *Cunninghamella elegans*. *Appl. Environ. Microbiol.* **1996**, *62*, 798–803. [CrossRef]
33. Bachmann, F.; zu Schwabedissen, H.E.M.; Duthaler, U.; Krähenbühl, S. Cytochrome P450 1A2 is the most important enzyme for hepatic metabolism of the metamizole metabolite 4-methylaminoantipyrine. *Br. J. Clin. Pharmacol.* **2021**, *88*, 1885–1896. [CrossRef]
34. Ren, H.; Dhanaraj, P.; Enoch, I.V.; Paulraj, M.S.; Paulraj, M.S. Synthesis and Biological Evaluation of 4-Aminoantipyrine Analogues. *Med. Chem.* **2022**, *18*, 26–35. [CrossRef]
35. Eliwa, D.; Albadry, M.A.; Ibrahim AR, S.; Kabbash, A.; Meepagala, K.; Khan, I.A.; El-Aasr, M.; Ross, S.A. Biotransformation of papaverine and in silico docking studies of the metabolites on human phosphodiesterase 10a. *Phytochemistry* **2021**, *183*, 112598. [CrossRef]
36. Yang, W.; Jiang, T.; Acosta, D.; Davis, P.J. Microbial models of mammalian metabolism: Involvement of cytochrome P450 in the N-demethylation of N-methylcarbazole by *Cunninghamella echinulata*. *Xenobiotica* **1993**, *23*, 973–982. [CrossRef]
37. Hansen, E.B., Jr.; Cho, B.P.; Korfmacher, W.A.; Cerniglia, C.E. Fungal transformations of antihistamines: Metabolism of brompheniramine, chlorpheniramine, and pheniramine to N-oxide and N-demethylated metabolites by the fungus *Cunninghamella elegans*. *Xenobiotica* **1995**, *25*, 1081–1092. [CrossRef]
38. Fan, H.X.; Zhou, Z.Q.; Peng, J.; Wu, B.J.; Chen, H.R.; Bao, X.F.; Mu, Z.Q.; Jiao, W.H.; Yao, X.S.; Gao, H. A microbial model of mammalian metabolism: Biotransformation of 4,5-dimethoxyl-canthin-6-one using *Cunninghamella blakesleeana* CGMCC 3.970. *Xenobiotica* **2017**, *47*, 284–289. [CrossRef]
39. Ibrahim, A.-R.S.; Galal, A.M.; Ahmed, M.S.; Mossa, G.S. O-Demethylation and Sulfation of 7-Methoxylated Flavanones by *Cunninghamella elegans*. *Chem. Pharm. Bull.* **2003**, *51*, 203–206. [CrossRef]
40. Liu, Y.; Wang, Y.; Jiang, Y.; Hu, M.; Li, S.; Zhai, Q. Biocatalytic synthesis of C3 chiral building blocks by chloroperoxidase-catalyzed enantioselective halo-hydroxylation and epoxidation in the presence of ionic liquids. *Biotechnol. Prog.* **2015**, *31*, 724–729. [CrossRef]
41. Morozov, A.N.; Chatfield, D.C. Chloroperoxidase-catalyzed epoxidation of cis-beta-methylstyrene: Distal pocket flexibility tunes catalytic reactivity. *J. Phys. Chem. B* **2012**, *116*, 12905–12914. [CrossRef]
42. de Hoog, H.M.; Nallani, M.; Cornelissen, J.J.L.M.; Rowan, A.E.; Nolte, R.J.M.; Arends, I.W.C.E. Biocatalytic oxidation by chloroperoxidase from Caldariomyces fumago in polymersome nanoreactors. *Org. Biomol. Chem.* **2009**, *7*, 4604–4610. [CrossRef] [PubMed]
43. Kellner, H.; Pecyna, M.J.; Buchhaupt, M.; Ullrich, R.; Hofrichter, M. Draft Genome Sequence of the Chloroperoxidase-Producing Fungus *Caldariomyces fumago* Woronichin DSM1256. *Genome Announc.* **2016**, *4*, e00774-16. [CrossRef] [PubMed]
44. Dong, X.; Li, H.; Jiang, Y.; Hu, M.; Li, S.; Zhai, Q. Rapid and efficient degradation of bisphenol A by chloroperoxidase from Caldariomyces fumago: Product analysis and ecotoxicity evaluation of the degraded solution. *Biotechnol. Lett.* **2016**, *38*, 1483–1491. [CrossRef] [PubMed]
45. Buchhaupt, M.; Hüttmann, S.; Sachs, C.C.; Bormann, S.; Hannappel, A.; Schrader, J. *Caldariomyces fumago* DSM1256 Contains Two Chloroperoxidase Genes, Both Encoding Secreted and Active Enzymes. *J. Mol. Microbiol. Biotechnol.* **2015**, *25*, 237–243. [CrossRef] [PubMed]
46. Buchhaupt, M.; Hüttmann, S.; Schrader, J. White Mutants of Chloroperoxidase-Secreting *Caldariomyces fumago* as Superior Production Strains, Revealing an Interaction between Pigmentation and Enzyme Secretion. *Appl. Environ. Microbiol.* **2012**, *78*, 5923–5925. [CrossRef]
47. Buchhaupt, M.; Ehrich, K.; Hüttmann, S.; Guder, J.; Schrader, J. Over-expression of chloroperoxidase in *Caldariomyces fumago*. *Biotechnol. Lett.* **2011**, *33*, 2225–2231. [CrossRef]
48. Höfler, G.T.; But, A.; Hollmann, F. Haloperoxidases as catalysts in organic synthesis. *Org. Biomol. Chem.* **2019**, *17*, 9267–9274. [CrossRef]
49. Sun, L.; Huang, H.-H.; Liu, L.; Zhong, D.-F. Transformation of Verapamil by *Cunninghamella blakesleeana*. *Appl. Environ. Microbiol.* **2004**, *70*, 2722–2727. [CrossRef]

catalysts

MDPI

Article

Biodegradation of Crystalline and Nonaqueous Phase Liquid-Dissolved ATRAZINE by *Arthrobacter* sp. ST11 with Cd^{2+} Resistance

Jiameng Zhang, Zhiliang Yu, Yaling Gao, Meini Wang, Kai Wang and Tao Pan *

Jiangxi Province Key Laboratory of Mining and Metallurgy Environmental Pollution Control, School of Resource and Environmental Engineering, Jiangxi University of Science and Technology, Ganzhou 341000, China
* Correspondence: t.pan@jxust.edu.cn; Tel.: +86-13698057756

Abstract: A newly isolated cadmium (Cd)-resistant bacterial strain from herbicides-polluted soil in China could use atrazine as the sole carbon, nitrogen, and energy source for growth in a mineral salt medium (MSM). Based on 16S rRNA gene sequence analysis and physiochemical tests, the bacterium was identified as *Arthrobacter* sp. and named ST11. The biodegradation of atrazine by ST11 was investigated in experiments, with the compound present either as crystals or dissolved in di(2-ethylhexyl) phthalate (DEHP) as a non-aqueous phase liquid (NAPL). After 48 h, ST11 consumed 68% of the crystalline atrazine in MSM. After being dissolved in DEHP, the degradation ratio of atrazine was reduced to 55% under the same conditions. Obviously, the NAPL-dissolved atrazine has lower bioavailability than the crystalline atrazine. Cd^{2+} at concentrations of 0.05–1.5 mmol/L either had no effect (<0.3 mmol/L), slight effects (0.5–1.0 mmol/L), or significantly (1.5 mmol/L) inhibited the growth of ST11 in Luria-Bertani medium. Correspondingly, in the whole concentration range (0.05–1.5 mmol/L), Cd^{2+} promoted ST11 to degrade atrazine, whether crystalline or dissolved in DEHP. Refusal to adsorb Cd^{2+} may be the main mechanism of high Cd resistance in ST11 cells. These results may provide valuable insights for the microbial treatment of arable soil co-polluted by atrazine and Cd.

Keywords: atrazine; *Arthrobacter* sp. ST11; biodegradation; nonaqueous-phase liquid; cadmium

Citation: Zhang, J.; Yu, Z.; Gao, Y.; Wang, M.; Wang, K.; Pan, T. Biodegradation of Crystalline and Nonaqueous Phase Liquid-Dissolved ATRAZINE by *Arthrobacter* sp. ST11 with Cd^{2+} Resistance. *Catalysts* **2022**, *12*, 1653. https://doi.org/10.3390/catal12121653

Academic Editor: Jasmina Nikodinović-Runić

Received: 25 September 2022
Accepted: 14 December 2022
Published: 15 December 2022

1. Introduction

Arable soil is often polluted with herbicides [1]. Atrazine (6-chloro-4-N-ethyl-2-N-propan-2-yl-1,3,5-triazine-2,4-diamine) is one of the most widely-used persistent chlorine herbicides that often remain in agricultural fields and water bodies for several years at concentrations of hundreds of μg/kg [2]. Atrazine has lower bio-accessibility when present in an unavailable phase, such as crystalline form, soil and sediment solids, or non-aqueous phase liquid (NAPL). The environmental persistence of atrazine has been shown to be a vastly significant problem [2]. No strong evidence could prove that atrazine causes cancer; however, it affects the endocrine response and thus has a potential effect on human reproduction and development [3]. Although it was banned for use by the European Union in 2004, atrazine remains legal in China [2]. Atrazine is expected to persist in arable soil sources for decades [4], thereby calling for appropriate remediation measures.

Cadmium (Cd), a potentially toxic heavy metal with no known biological function, occurs widely in nature [5]. A national-scale study of soil Cd pollution in China reported that the average and maximum concentrations of Cd in arable soil were 0.0024 and 1.36 mmol/kg, respectively [6]. As one of the most toxic trace elements in the environment, Cd could cause serious health problems to microorganisms, plants, animals, and humans [5,7].

Bioremediation is cost-effective and environmentally friendly and, thus, has become one of the most popular approaches for removing atrazine [2]. Biodegradation by indigenous microbial populations is considered an important process that affects the fate of atrazine in contaminated sites [2]. Since the 1990s, different atrazine-degrading bacteria and fungi have been isolated from contaminated sites [8,9], including *Arthrobacter* [10], *Nocardioides* [11], *Shewanella* [12], *Rhodococcus* [13], *Stenotrophomonas* [14], *Pseudomonas* [15], *Paenarthrobacter* [16], and *Trametes* [17]. Bacteria are the foundation of microbial bioremediation, which can outperform fungi in the potential for atrazine-specific bioaugmentation [18]. However, the low bioavailability of atrazine in the environment hindered the degradation efficiency of these strains. Although many enhancement methods have been applied [19–21], little is known about the mode of acquisition of atrazine when it is present in an unavailable phase. Heavy metals and synthetic pesticides often co-occur in soil, although their hazards are usually evaluated separately and in bulk soil [22]. Coexisting heavy metals may stimulate or inhibit the biodegradation of herbicides [23,24]. When Cd pollution coexists, the fate of atrazine in the soil will be difficult to predict. The potential ecological risk of combined pollution of atrazine and Cd in waters and soils still exists, and it cannot be ignored, even when said risk is lower than that of atrazine or Cd alone [25,26]. Overall, the search and isolation of specific bacterial strains that could degrade atrazine efficiently with Cd resistance are of great interest.

Arthrobacter is prevalent in the agricultural soil environment, and it could degrade many kinds of environmental pollutants [10,27]. Many *Arthrobacter* strains have been isolated to degrade atrazine, such as *Arthrobacter* Sp. DNS10, *Arthrobacter* sp. LY-1, and *Arthrobacter Aurescens* TC1 [28–30]. In addition, some *Arthrobacter* strains were reported to have metal resistance [31,32]. However, atrazine-degrading *Arthrobacter* strains with Cd resistance have not yet been found.

The three primary goals of this study were designed to address gaps that currently exist in the research. The first goal was to isolate high-efficiency atrazine-degrading bacteria with Cd resistance and characterize them through 16S rRNA gene sequencing analysis and physiochemical tests. The second goal was to investigate the bioavailability of crystalline and NAPL-dissolved atrazine to the strain. The third goal was to explore the mechanisms of the effect of Cd^{2+} on atrazine biodegradation. Finally, some valuable insights into the treatment of atrazine in soil with Cd pollution were provided.

2. Results and Discussion

2.1. Identification and Characterization of Test Strain

An efficient atrazine-degrading bacterium ST11 was isolated from herbicide-polluted soil. Under scanning electron microscopy (SEM), ST11 appeared as rods when rapidly dividing and cocci when in the stationary phase (Figure S1), showing the typical characteristics of *Arthrobacter* cells [33]. The strain was further identified as *Arthrobacter* sp. by physiochemical tests (Table S1) and 16S rDNA analysis (GenBank OP435654). A phylogenetic tree was constructed using an approximate maximum-likelihood analysis (Figure 1).

2.2. Characterization of Cd^{2+} Resistance

The ST11 strain was assessed for its ability to grow in the presence of Cd^{2+}. As shown in Figure 2, no difference was found from the control samples when ST11 was incubated for 48 h in a culture medium containing <0.3 mmol/L Cd^{2+}. The growth of ST11 was slightly suppressed in the presence of 0.5–1.0 mmol/L Cd^{2+} and significantly suppressed in the presence of 1.5 mmol/L Cd^{2+}. The effective concentration-25 (EC_{25}) and effective concentration-50 (EC_{50}) were defined as the Cd^{2+} concentration required to obtain 25% and 50% cell growth inhibition effects, respectively. The EC_{25} and EC_{50} of ST11 toward Cd^{2+} were 1.21 and 2.01 mmol/L, respectively.

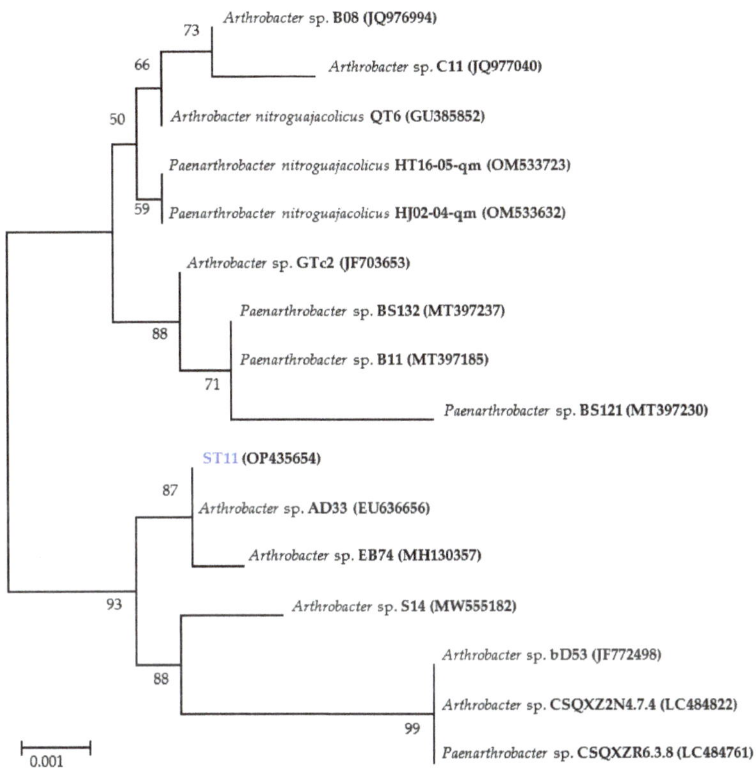

Figure 1. Phylogenetic analysis of strain ST11 and related species by the Neighbor-Joining method based on 16S rRNA gene sequences. Bootstrap values (%) are indicated at the nodes, and the scale bars represent 0.001 substitutions per site.

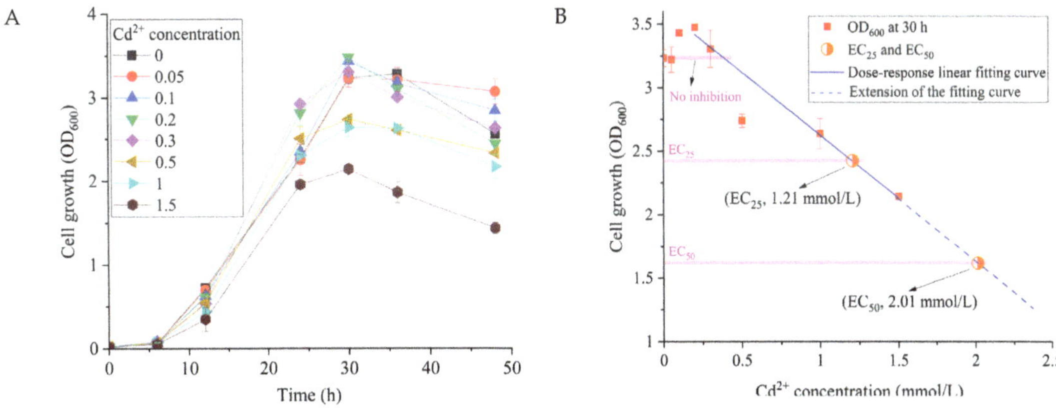

Figure 2. Growth curves (**A**) and inhibition fitting curve (**B**) of ST11 in the presence of Cd^{2+}. The cells were cultured with the Luria-Bertani (LB) medium in the presence of Cd^{2+} (0–1.5 mmol/L) at 30 °C at 150 rpm in darkness for 48 h. At the timed interval, flasks were taken out and 1 mL culture was withdrawn for OD_{600} measurements.

The adsorption capacities of ST11 for Cd^{2+} were shown in Figure 3. Before and after cell adsorption, the difference of Cd^{2+} detected was negligible. The result shows that ST11 cells have limit adsorption capacity for Cd^{2+}.

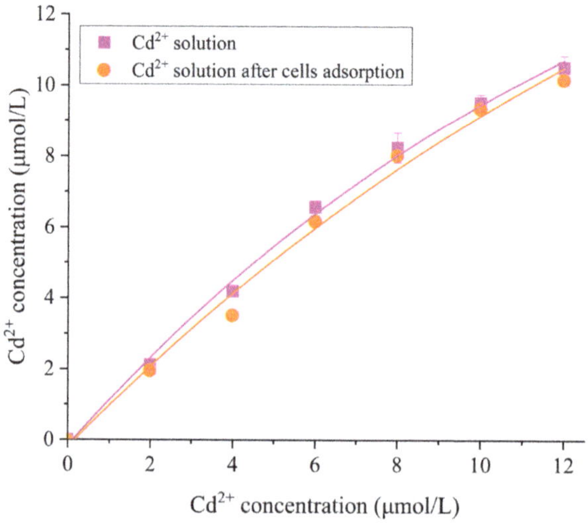

Figure 3. Adsorption isotherm of Cd^{2+} on ST11 cells as a function of Cd^{2+} concentration. The cells were harvested at 6380× *g* centrifugation for 10 min after ST11 was cultured in LB for 48 h. The cells were washed three times with deionized water to remove the residual medium. The obtained wet cells were used directly for Cd^{2+} adsorption without further treatment. The adsorption of Cd^{2+} by ST11 cells was carried out at 25 °C and pH 7.

Arthrobacter is prevalent in the agricultural soil environment, and it could degrade many kinds of environmental pollutants, but evidence of its resistance to heavy metals is limited [10]. For instance, a quinaldine-degrading *Arthrobacter* sp. Rue61a was suggested to have the potential to tolerate heavy metals, such as Zn^{2+}, Pb^{2+}, and Co^{2+} [31]. However, for atrazine-degrading *Arthrobacter*, a strain with the ability to resist heavy metals has not been reported. In this study, *Arthrobacter* sp. ST11 was shown to survive in the presence of up to 1.5 mmol/L Cd^{2+}. The microbial metabolism of heavy metals includes extracellular complexation, extracellular precipitation, cation outflow, in-vivo detoxification, and in vivo complexation [32]. Living ST11 cells do not adsorb Cd^{2+}, so the resistance of ST11 to heavy metals may be based on cation outflow or the absence of a Cd^{2+} binding site on the cell membrane of ST11.

2.3. *Biodegradation of Crystalline and NAPL-Dissolved Atrazine*

The cell growth and biodegradation of crystalline and NAPL-dissolved atrazine by ST11 are shown in Figure 4A,B, respectively. Di(2-ethylhexyl) phthalate (DEHP) is a commonly used typical NAPL [34] that is biocompatible and could not be used by ST11. Meanwhile, atrazine could be degraded by physical, chemical, and biological methods [35]. In the absence of microorganisms, they could attenuate naturally under some physical and chemical factors [18]. A non-inoculated control experiment was set up in the present study to exclude the effects of temperature, pH, dissolved oxygen, ionic strength, and other physical and chemical factors. For the whole culture cycle, significant bacterial growth of non-investigated samples was not detected under these conditions. For inoculated samples, during the first 16 h, the numbers of cells produced at the two systems were not statistically (Student's *t*-test) significant. The bacterial growth rates during this phase were 0.0171 ± 0.0013 and 0.0185 ± 0.0005 OD_{600}/h with 1.85 mmol/L crystalline and DEHP-dissolved atrazine, respectively. Subsequent bacterial growth led to an increased number of cells until growth stopped after 40 h. At that time, the average final OD_{600} values of cells were 0.61 ± 0.01 and 0.44 ± 0.03 with crystalline and DEHP-dissolved atrazine, respectively. In the following 8 h, the cells almost stopped growing. Most of the bioavailable atrazine at that point was assumed to be consumed. Therefore, residual atrazine was detected after 48 h of culture (Figure 4B). Physical and chemical factors were found to degrade atrazine by no more than 15%. With crystalline atrazine as the substrate, the degradation ratio of atrazine increased significantly and reached 68% after 48 h. When atrazine was dissolved in DEHP, the degradation ratio of atrazine decreased to less than 55%, which is consistent with the low number of cells.

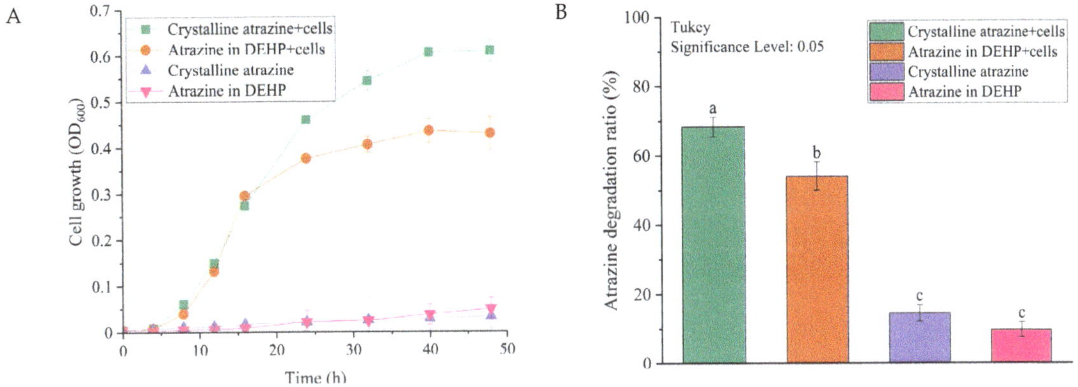

Figure 4. Cell growth (**A**) and biodegradation of atrazine (**B**) in crystalline form or in DEHP by ST11. The initial concentration of atrazine was 1.85 mmol/L. For biological samples, 1 mL of bacterial suspension (OD_{600} 1.0) was inoculated. For blank tests, no inoculation was prepared to control the non-biological effects. (**A**) The growth was monitored spectrophotometrically by measuring OD_{600} at an interval of 4–8 h. (**B**) At the end of culture (48 h), the biodegradation ratio of atrazine was detected. Data points represent the mean of three replicates and error bars show the standard deviation. Different lower-case letters (a, b, and c) over the bars indicate significant differences at $p < 0.05$.

The atrazine degradation pathway includes hydrolysis, deamination, dealkylation, and ring cleavage [2]. Liquid Chromatography-Ultraviolet (LC-UV) analysis was performed at 225 nm to determine whether a new metabolite was generated after atrazine degradation (Figure S2). A new absorption peak (Figure S2c) appeared at the retention time of 4.145 min, which was not found in the atrazine standard (Figure S2a) and the culture medium before atrazine was degraded (Figure S2b). In the process of atrazine biodegradation by the *Bacillus licheniformis* ATLJ-5 strain, intermediate metabolites hydroxyatrazine and n-isopropylammelide were detected by LC-UV [8]. Similarly, during the process of ST11 using atrazine, a UV detectable metabolite was found. This finding suggested that the new compound was a metabolite of the degraded atrazine.

Microscopic observation explained that atrazine dissolved in DEHP had decreased bioavailability (Figure 5). For the microbial adhesion to hydrocarbons (MATH) experiment of ST11, the solution was vigorously homogenized and then allowed to stand for 2 h to ensure complete phase separation. The OD_{600} of cells in the water phase increased from 0.6 to 0.68 (Figure 5B), possibly because some DEHP micro-droplets suspended in the water phase hindered the transmission of light. The DEHP oil phase (Figure 5B) was as clear and transparent as the sterile sample (Figure 5A), and no obvious emulsification was observed. The liquid at the oil-water interface was sampled for microscopic observation (Figure 5C). The results showed that the ST11 cells were evenly scattered in the water phase and not observed on the oil-water interface. Therefore, ST11 is a water-soluble bacterium that could only use atrazine dissolved in the water phase in the oil-water two-phase system.

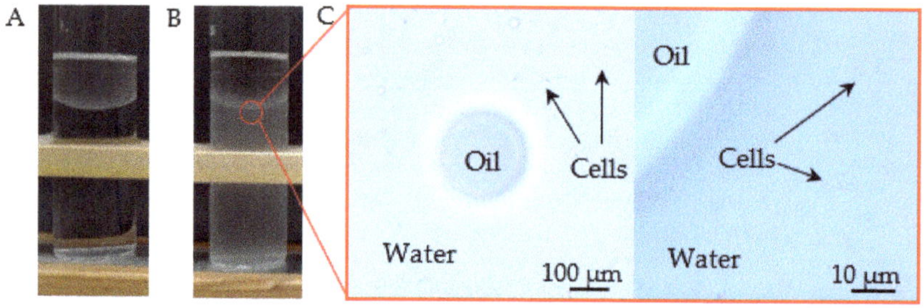

Figure 5. Visual aspect of two-phase systems with DEHP as a NAPL phase. (**A**) 5 mL of DEHP, 20 mL of MSM; (**B**) 5 mL of DEHP, 20 mL of cells suspension in MSM with OD_{600} 0.6; (**C**) Microscopic images of oil-water interface. The arrow indicates ST11 cells.

Atrazine has low bioavailability because of its low aqueous solubility (log P_{ow}, 2.28; water solubility, 0.153 mmol/L at 20 °C) [36]. However, when atrazine is used as a substrate in crystalline and DEHP-soluble forms, it could be speculated to have different bioavailability. The mass transfer of atrazine is not limited by the mass transfer rate, and ST11 cells could only use the substrate dissolved in water. When atrazine is a crystal, the mass transfer between the solid and water is only limited by the saturated solubility of atrazine. At this time, the concentration of atrazine in the aqueous phase could be approximately regarded as the saturated solubility (0.153 mmol/L). When atrazine is dissolved in a NALP, the mass transfer of atrazine between the NAPL and water is affected by the octanol-water partition coefficient (log P_{ow}, 2.28) of atrazine. In the experiments, the additional amount of atrazine in DEHP was 11.13 mmol/L. If the difference between DEHP and octanol was not considered, the concentration of atrazine in the aqueous phase was only 0.058 mmol/L. Therefore, when atrazine was used as a substrate in crystal form or dissolved in DEHP, the bioavailability of the former was 2.8 times that of the latter. The low bioavailability of atrazine in DEHP delayed its degradation by ST11 (Figure 4B).

For other hydrophobic organic compounds in soil, such as polycyclic aromatic hydrocarbons, the bioavailability mechanism of substrates in the non-aqueous phase has been deeply studied [34,37]. Although the bioremediation of atrazine has been extensively studied [38], little is known about the mode of its acquisition when it is present in an unavailable phase, such as soil and sediment solids or NAPL. For a naphthalene-degrading *Arthrobacter* sp., the attachment of this strain to the NAPL-water interface is necessary, and the cells at the interface could degrade organic compounds at rates higher than those of abiotic partitioning [39]. However, for ST11 cells that could not adhere to the oil-water interface (Figure 5), the biodegradation of atrazine in DEHP was limited by the mass transfer of the substrate between the two oil-water phases. Therefore, the findings of this study indicated that when atrazine is dissolved in NAPL in soil, its biodegradation is likely to be restricted, which may be one of the reasons why atrazine is a persistent contaminant in arable soil [4].

2.4. *Effect of Cd^{2+} on the Growth of Strain ST11 and Atrazine Biodegradation*

The effects of Cd^{2+} on ST11 cell density and degradation of atrazine in crystal form or dissolved in DEHP were determined (Figure 6). At Cd^{2+} concentrations below 0.3 mmol/L, cell growth and atrazine degradation were stimulated with Cd^{2+} concentration. The highest removal of atrazine (almost 100% at 48 h) in crystal form or dissolved in DEHP occurred in the case of Cd^{2+} concentration of 0.1 and 0.2 mmol/L, respectively. The DEHP-dissolved atrazine with decreased bioavailability, which may cause more Cd^{2+}, is needed to obtain high degradation ratio. When the concentration of Cd^{2+} was increased from 0.5 to 1.5 mmol/L, the effect of stimulating cell growth and atrazine degradation gradually weakened. However, in this Cd^{2+} concentration range, the growth was partially inhibited when ST11 was used soluble substrates (Figure 2A). A notable detail that at the same Cd^{2+}

concentration (such as 0.1 mmol/L), the biodegradation enhancement effect of Cd^{2+} on crystalline atrazine was stronger than that of the DEHP-dissolved atrazine. As mentioned, the main reason could be the lower bioavailability of atrazine in DEHP. Correspondingly, the promotion of atrazine biodegradation in the presence of the Cd^{2+} concentration of 0.05–1.5 mmol/L could be attributed to the enhanced atrazine catabolism (Figure 6). Low concentrations of heavy metal ions could stimulate the catabolism of organic substances [40]. Cu^{2+} concentrations of 15 and 2 mg/L stimulated the degradation of decabromodiphenyl ether and benzo[a]pyrene, respectively [41,42]. An assessment of Cd pollution in arable soil in China showed that the average and maximum Cd concentrations were 0.0024 and 1.36 mmol/kg, respectively [6], both lower than the 1.5 mmol/L shown in the present study. Therefore, ST11 is not affected by soil Cd when it degrades atrazine in arable soils.

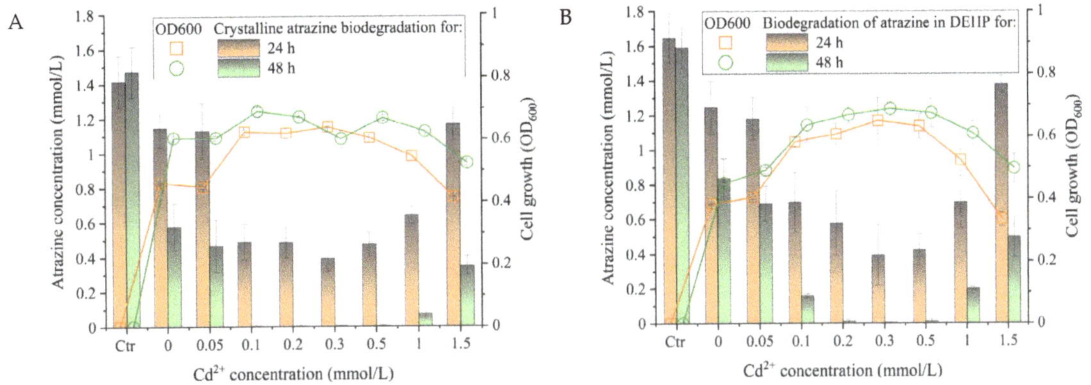

Figure 6. Effect of Cd^{2+} of different concentrations on bacterial growth and atrazine biodegradation. (**A**) crystalline atrazine; (**B**) DEHP dissolved atrazine. The initial concentration of atrazine was 1.85 mmol/L. Ctr: control samples with no cells inoculated. For each biological flask, 1 mL of bacterial suspension (OD_{600} 1.0) was inoculated. At an interval of 24 h, the growth was monitored spectrophotometrically by measuring OD_{600} and the residual atrazine was detected. Data points represent the mean of three replicates and error bars show the standard deviation.

As is well known, Cd, similar to other heavy metals, such as copper, zinc, and lead, has biological toxicity [5]. Cd could attach to proteins with sulfhydryl functional groups or glutathione, thus interfering with the synthesis of cysteine or directly damaging DNA [43]. In addition, Cd could easily penetrate the cell membrane of bacteria, causing the emission of substances in cells and affecting the metabolic process of cells [44]. The results of the present study showed that Cd inhibited and stimulated the growth of ST11 using soluble substrates (Cd^{2+} concentration of 0.5–1.5 mmol/L) and hydrophobic substrate atrazine (Cd^{2+} concentration of $\leq$1.5 mmol/L), respectively. ST11 cells have negligible Cd adsorption capacity, which may be one of the reasons why they could resist high Cd concentration. In addition, the antagonistic effect of hydrophobic organics on Cd could be used to mitigate Cd effect on soil living organisms [45]. The results of the present study are consistent with that of the previous study. The biodegradation of crystalline atrazine and NAPL-solubilized atrazine by ST11 could not be affected by Cd^{2+} with concentrations up to 1.5 mmol/L.

3. Materials and Methods

3.1. Chemicals

Atrazine (product number TCI-A1650, 97.0%), $CdCl_2 \cdot 2.5H_2O$ ($\geq$99.0%), ethyl acetate (>99%), and methanol (>99%) were purchased from Sinopharm (Shanghai, China). DEHP ($\geq$99.5%) was obtained from Aladdin (Shanghai, China). 2-(5-Bromo-2-pyridylazo)-5-(diethylamino) phenol (5-Br-PADAP, 97%) and Triton X-114 (laboratory grade) were received from Sigma-Aldrich (Shanghai, China). The other reagents and solvents were of analytical grade and used directly. The stock solution of atrazine was prepared by dissolving atrazine in dichloromethane (92.73 mmol/L).

3.2. *Microorganism and Culture Conditions*

Soil samples were collected by hand-picking in a 5–10 cm soil layer with a modified sampling technique [46]. Each sample was placed in a portable icebox and then transferred to the laboratory. An efficient atrazine-degrading bacterium *Arthrobacter* sp. ST11 was isolated from herbicide-polluted soil. The bacterium was cultivated on LB medium (10 g tryptone, 5 g yeast extract, and 10 g NaCl per liter of Milli-Q deionized water; pH 6.8; stored at room temperature after sterilization) at 30 °C, with shaking at 150 rpm for 18 h. The cells were harvested by centrifugation at 6380× *g* for 10 min at 4 °C and washed twice with mineral salt medium (MSM, 5.8 g Na_2HPO_4, 0.9 g KH_2PO_4, 0.2 g $MgSO_4{\cdot}7H_2O$, and 1 mL of trace element solution per liter of Milli-Q deionized water; pH 6.5). The composition of the trace element solution was as follows: 0.4 g $Na_2B_4O_7{\cdot}10H_2O$, 0.5 g $Na_2MoO_4{\cdot}2H_2O$, 0.8 g $CuSO_4{\cdot}5H_2O$, 2 g $FeSO_4{\cdot}7H_2O$, 2 g $MnSO_4{\cdot}H_2O$, 10 g $ZnSO_4{\cdot}7H_2O$, and 5 g EDTA disodium per liter of Milli-Q deionized water; pH 6.5. The cells were resuspended with MSM, and the optical density at 600 nm (OD_{600}) was adjusted to 1.0.

3.3. *Cd^{2+} Resistance and Growth Curve of ST11*

The resistance of ST11 to Cd^{2+} was investigated. The Cd^{2+} concentrations ranged from 0 to 1.5 mmol/L for ST11. The cells were cultured with an LB medium without Cd^{2+} at 30 °C at 150 rpm in darkness for 24 h. After the cells were washed three times and resuspended with Milli-Q deionized water, 1 mL of cell culture (OD_{600} = 1.0) was transferred into 150 mL sterilized flasks containing 30 mL of LB medium with various Cd^{2+} concentrations. The flasks were incubated at 150 rpm in darkness at 30 °C for 48 h. Cell growth was recorded by measuring the OD_{600} of the medium. A survival curve was generated using the growth changes in cultures supplemented with various concentrations of Cd^{2+} compared with no-Cd^{2+}-treated control samples. The regular interval for the biodegradation experiment was 48 h.

3.4. *Cd^{2+} Adsorption Experiment*

The adsorption of ST11 cells for Cd^{2+} was examined. After culturing in an LB medium for 60 h, ST11 cells were harvested through centrifugation (H2050R, Xiangyi, China) at 6380× *g* and 4 °C for 10 min. The cells were then washed three times with Milli-Q deionized water to remove the residual medium. The obtained wet cells were used directly for Cd^{2+} adsorption without further treatment. An aqueous solution of Cd^{2+} with a concentration of 0–12 μmol/L was prepared, and 10 mL of the solution was added to a 15 mL glass test tube. Furthermore, the wet cells were added to the tube, and the cell concentration in the aqueous solution was set as OD_{600} = 1.0. The above solutions were shaken at 25 °C and 150 rpm for 1 h. After centrifugation, the residual amount of Cd^{2+} in the supernatant was detected.

3.5. *Biodegradation of Crystalline Atrazine*

For experiments with crystalline atrazine, the bacterium was grown at 30 °C and 150 rpm in 150 mL-triangular flasks containing 30 mL of MSM supplied with different concentrations of atrazine as the sole source of carbon, nitrogen, and energy. A total of 1 mL of resuspended cells (OD600 1.0) was added into MSM. A certain volume of atrazine stock solution was evenly spread at the bottom of the pre-dried triangular flasks to avoid the influence of insoluble atrazine crystals on the absorbance detection of cells in the culture medium. A layer of homogeneous atrazine crystals was fixed at the bottom of the flask after the volatilization of dichloromethane for 12 h. In the process of cell culture, these crystals do not fall off from the bottom of the flask [47]. The regular interval for the biodegradation experiment was 48 h.

The biodegradation of atrazine was estimated indirectly as bacterial growth and atrazine content reduction. The bacteria were collected by centrifuging the culture medium at 6380× *g* and 4 °C for 10 min. The cells were washed three times with deionized water to remove the residual medium. The obtained wet cells were resuspended with MSM and the OD_{600} was determined. The complete biodegradation of atrazine was confirmed microscopically by the disappearance of atrazine crystals at the bottom of triangular flasks and by ethyl acetate extraction of the crystals after biodegradation and analysis of the extracts with an Agilent high-performance liquid chromatography (HPLC) system.

3.6. Biodegradation of Atrazine in NAPL

DEHP was used as the NAPL for determining the ST11 biodegradability of NAPL-dissolved atrazine. Experiments were performed in triangular flasks containing 24 mL of MSM and 6 mL of DEHP containing 9.25 mmol/L atrazine. The concentration of atrazine in the system was 1.85 mmol/L. For the inoculated sample, 1 mL of resuspended cells (OD600 1.0) was added to MSM. For the non-inoculated control experiment, no ST11 cells were added. The samples measured in triplicate were incubated at 30 °C on a rotary shaker at 150 rpm for 48 h.

The adherence of suspended bacteria to the liquid organic phase was tested. The method used was a modified assay of MATH [48]. DEHP (5 mL) and 20 mL of the cell suspension (OD_{600} 0.6) in MSM were vigorously homogenized in a test tube for 1 min by a vortex agitator (SA8, Thomas Scientific, Swedesboro, NJ. USA). After 2 h of equilibration, the cells' adhesion to DEHP was estimated from the loss of aqueous phase absorbance at 600 nm. Furthermore, optical imaging of the droplets at the oil-water interface was carried out to detect the location of bacterial cells by phase-contrast microscopy (Eclipse E200, Nikon, Tokyo, Japan).

3.7. Effect of Cd^{2+} Ions

The effect of Cd^{2+} ions on the metabolic activity of ST11 was investigated when atrazine was used as the substrate in crystal form or dissolved in DEHP. The concentrations of Cd^{2+} in MSM were set to 0–1.5 mmol/L. MSM separately supplemented with different concentrations of Cd^{2+} and 1.85 mmol/L atrazine was cultured at 30 °C, with shaking at 150 rpm for 48 h. Subsequently, the cell density of ST11 was monitored, and the residual amounts of atrazine were detected at 48 h. A sterile control culture without ST11 was included to assess the abiotic loss of atrazine. All measurements were carried out in triplicate.

3.8. Analytical Methods

The method for atrazine extraction was as follows: each triangular flask was added with 30 mL of ethyl acetate and shaken at 200 rpm for 1 h. Atrazine was extracted into the ethyl acetate layer. Furthermore, the aqueous and organic phases were separated using a separatory funnel. The organic phase was centrifuged at 6380× *g* for 10 min. A total of 500 µL of the supernatant was placed in a glass tube, and ethyl acetate was evaporated in an oven at 70 °C. The residue was extracted with 1 mL of methanol-water solution (85%, *v*/*v*). The extracts of the same sample were merged and filled into a 2 mL sample bottle after filtration through a 0.22 µm organic filter membrane, and then sealed and stored in a 4 °C refrigerator. Atrazine was detected using an HPLC system (Agilent 1260, Tokyo, Japan) with a UV detector. An Agilent HC-C18 (5 µm, 150 mm × 4.6 mm) column was used for the analysis of atrazine samples. Deionized water and methanol at a ratio of 85:15 were used as the mobile phase with a flow rate of 1 mL/min. The detection wavelength was 225 nm, and the retention time was 5.11 min. The detection limit of this method was 2 mg/L.

Cd^{2+} was determined by spectrophotometry after cloud point extraction [49]. In a 10 mL polystyrene tube, 0.2 mL of Cd^{2+} sample, 1 mL of NH_3-NH_4Cl buffer (pH 9.5), 0.53 mL of 5-Br-PADAP ethanol solution (0.2 g/L), and 0.6 mL of Triton X-114 solution (2%, *v*/*v*) were added successively. Furthermore, the liquid in the tube was diluted to 10 mL with deionized water and mixed with a vortex agitator for 30 s. The mixed solution was heated at 45 °C for 15 min and then centrifuged at 1595× *g* for 5 min to accelerate the phase separation of the cloud point system. The solution after phase separation was quickly placed into an ice water mixture for quick freezing for 30 min to make the coacervate phase viscous. After the water phase was discarded, the coacervate phase was dissolved in 2 mL of absolute ethanol by shaking in a vortex agitator. Subsequently, these samples were detected at 560 nm via spectrophotometry (UV2700, Shimadzu, Tokyo, Japan). Each sample was measured at least three times.

4. Conclusions

In this study, an atrazine-degrading bacterial strain with Cd resistance from herbicide-polluted soil was isolated. Based on 16S rRNA gene sequence analysis and physiochemical tests, the bacterium was identified as *Arthrobacter* sp., and designated as strain ST11. The ST11 cells were grown in MSM culture with atrazine as the sole source of carbon, nitrogen, and energy. The strain could degrade atrazine in crystal form or present in DEHP as NAPL. To the authors' knowledge, this study was the first to report an *Arthrobacter* strain actively degrading crystalline and NAPL-dissolved atrazine with Cd resistance. Cd^{2+} with concentrations < 0.5 and 0.5–1.5 mmol/L did not affect or slightly inhibited the growth of ST11 in LB, respectively. Correspondingly, in the whole concentration range (0.05–1.5 mmol/L), Cd^{2+} promoted ST11 to degrade atrazine, whether in crystalline form or dissolved in DEHP. Refusal to adsorb Cd^{2+} may be the main mechanism of high Cd resistance in ST11 cells. Therefore, ST11 may be a potential candidate for the industrial elimination of atrazine from contaminated arable soils with Cd pollution. However, in-situ studies that examine the mechanisms of atrazine degradation and Cd resistance are necessary before this strain could be used in practice.

Supplementary Materials: The following supporting information can be downloaded at: https://www.mdpi.com/article/10.3390/catal12121653/s1, Table S1: Morphological, physiological, and biochemical properties of *Arthrobacter* sp. ST11; Figure S1: Photographs of colonies and cells of strain ST11: (A) photograph of ST11 colonies on LB solid media plate; (B) SEM photograph of ST11 cultured for 12 h; (C) SEM photograph of ST11 cultured for 36 h; Figure S2: The LC-UV chromatograms at 225 nm of atrazine standard (a), sample extract before atrazine degradation (b), and sample extract after atrazine degradation (c). The retention time of 5.081, 5.211, and 5.251 min is the absorption peak of atrazine. The retention time of 4.145 min was the absorption peak of a new metabolite.

Author Contributions: Conceptualization, J.Z.; Funding acquisition, T.P.; Investigation, J.Z.; Methodology, J.Z., Z.Y., Y.G. and M.W.; Resources, K.W.; Software, J.Z., Z.Y., Y.G. and M.W.; Supervision, T.P.; Validation, T.P.; Writing–original draft, J.Z.; Writing–review & editing, M.W., K.W. and T.P. All authors have read and agreed to the published version of the manuscript.

Funding: This work was funded by the National Natural Science Foundation of China (21866015); National Key Research and Development Program Project (2019YFC1805100); and the Program of Qingjiang Excellent Young Talents of JXUST (JXUSTQJYX2019011).

Data Availability Statement: The datasets are available from the corresponding author on reasonable request.

Conflicts of Interest: The authors declare that they have no known competing financial interests or personal relationships that could have appeared to influence the work reported in this paper.

References

1. Wołejko, E.; Wydro, U.; Odziejewicz, J.I.; Koronkiewicz, A.; Jabłońska-Trypuć, A. Biomonitoring of Soil Contaminated with Herbicides. *Water* **2022**, *14*, 1534. [CrossRef]
2. Rostami, S.; Jafari, S.; Moeini, Z.; Jaskulak, M.; Keshtgar, L.; Badeenezhad, A.; Azhdarpoor, A.; Rostami, M.; Zorena, K.; Dehghani, M. Current Methods and Technologies for Degradation of Atrazine in Contaminated Soil and Water: A Review. *Environ. Technol. Innov.* **2021**, *24*, 102019. [CrossRef]
3. Jowa, L.; Howd, R. Should Atrazine and Related Chlorotriazines Be Considered Carcinogenic for Human Health Risk Assessment? *J. Environ. Sci. Health Part C Environ. Carcinog. Ecotoxicol. Rev.* **2011**, *29*, 91–144. [CrossRef] [PubMed]

4. Mudhoo, A.; Garg, V.K. Sorption, Transport and Transformation of Atrazine in Soils, Minerals and Composts: A Review. *Pedosphere* **2011**, *21*, 11–25. [CrossRef]
5. Vig, K.; Megharaj, M.; Sethunathan, N.; Naidu, R. Bioavailability and Toxicity of Cadmium to Microorganisms and Their Activities in Soil: A Review. *Adv. Environ. Res.* **2003**, *8*, 121–135. [CrossRef]
6. Zhang, X.; Chen, D.; Zhong, T.; Zhang, X.; Cheng, M.; Li, X. Assessment of Cadmium (Cd) Concentration in Arable Soil in China. *Environ. Sci. Pollut. Res.* **2015**, *22*, 4932–4941. [CrossRef]
7. Genchi, G.; Sinicropi, M.S.; Lauria, G.; Carocci, A.; Catalano, A. The Effects of Cadmium Toxicity. *Int. J. Environ. Res. Public. Health* **2020**, *17*, 3782. [CrossRef]
8. Zhu, J.; Fu, L.; Jin, C.; Meng, Z.; Yang, N. Study on the Isolation of Two Atrazine-Degrading Bacteria and the Development of a Microbial Agent. *Microorganisms* **2019**, *7*, 80. [CrossRef]
9. Zhang, J.; Liang, S.; Wang, X.; Lu, Z.; Sun, P.; Zhang, H.; Sun, F. Biodegradation of Atrazine by the Novel *Klebsiella Variicola* Strain FH-1. *Biomed Res. Int.* **2019**, *2019*, 4756579. [CrossRef]
10. Guo, X.; Xie, C.; Wang, L.; Li, Q.; Wang, Y. Biodegradation of Persistent Environmental Pollutants by *Arthrobacter* sp. *Environ. Sci. Pollut. Res.* **2019**, *26*, 8429–8443. [CrossRef]
11. Fang, H.; Lian, J.; Wang, H.; Cai, L.; Yu, Y. Exploring Bacterial Community Structure and Function Associated with Atrazine Biodegradation in Repeatedly Treated Soils. *J. Hazard. Mater.* **2015**, *286*, 457–465. [CrossRef] [PubMed]
12. Ye, J.; Zhang, J.; Gao, J.; Li, H.; Liang, D.; Liu, R. Isolation and Characterization of Atrazine-Degrading Strain *Shewanella* sp. YJY4 from Cornfield Soil. *Lett. Appl. Microbiol.* **2016**, *63*, 45–52. [CrossRef] [PubMed]
13. Kolekar, P.D.; Patil, S.M.; Suryavanshi, M.; Suryawanshi, S.S.; Khandare, R.; Govindwar, S.P.; Jadhav, J.P. Microcosm Study of Atrazine Bioremediation by Indigenous Microorganisms and Cytotoxicity of Biodegraded Metabolites. *J. Hazard. Mater.* **2019**, *374*, 66–73. [CrossRef] [PubMed]
14. Radwan, E.K.; El Sebai, T.N.M.; Ghafar, H.H.A.; Khattab, A.E.-N.A. Atrazine Mineralization by *Stenotrophomonas Maltophilia* and *Agrobacterium Tumefaciens* Egyptian Soil Isolates. *Desalination Water Treat.* **2019**, *171*, 325–330. [CrossRef]
15. Sharma, A.; Kalyani, P.; Trivedi, V.D.; Kapley, A.; Phale, P.S. Nitrogen-Dependent Induction of Atrazine Degradation Pathway in *Pseudomonas* sp. Strain AKN5. *FEMS Microbiol. Lett.* **2019**, *366*, fny277. [CrossRef]
16. Chen, S.; Li, Y.; Fan, Z.; Liu, F.; Liu, H.; Wang, L.; Wu, H. Soil Bacterial Community Dynamics Following Bioaugmentation with *Paenarthrobacter* sp. W11 in Atrazine-Contaminated Soil. *Chemosphere* **2021**, *282*, 130976. [CrossRef]
17. Castro-Gutiérrez, V.; Masís-Mora, M.; Carazo-Rojas, E.; Mora-López, M.; Rodríguez-Rodríguez, C.E. Fungal and Bacterial Co-Bioaugmentation of a Pesticide-Degrading Biomixture: Pesticide Removal and Community Structure Variations during Different Treatments. *Water. Air. Soil Pollut.* **2019**, *230*, 247. [CrossRef]
18. Lopes, P.R.M.; Cruz, V.H.; de Menezes, A.B.; Gadanhoto, B.P.; Moreira, B.R.D.A.; Mendes, C.R.; Mazzeo, D.E.C.; Dilarri, G.; Montagnolli, R.N. Microbial Bioremediation of Pesticides in Agricultural Soils: An Integrative Review on Natural Attenuation, Bioaugmentation and Biostimulation. *Rev. Environ. Sci. Biotechnol.* **2022**, *21*, 851–876. [CrossRef]
19. Benson, J.J.; Sakkos, J.K.; Radian, A.; Wackett, L.P.; Aksan, A. Enhanced Biodegradation of Atrazine by Bacteria Encapsulated in Organically Modified Silica Gels. *J. Colloid Interface Sci.* **2018**, *510*, 57–68. [CrossRef]
20. Jiang, Z.; Zhang, X.; Wang, Z.; Cao, B.; Deng, S.; Bi, M.; Zhang, Y. Enhanced Biodegradation of Atrazine by *Arthrobacter* sp. DNS10 during Co-Culture with a Phosphorus Solubilizing Bacteria: *Enterobacter* sp. P1. *Ecotoxicol. Environ. Saf.* **2019**, *172*, 159–166. [CrossRef]
21. Jakinala, P.; Lingampally, N.; Kyama, A.; Hameeda, B. Enhancement of Atrazine Biodegradation by Marine Isolate *Bacillus Velezensis* MHNK1 in Presence of Surfactin Lipopeptide. *Ecotoxicol. Environ. Saf.* **2019**, *182*, 109372. [CrossRef] [PubMed]
22. Meite, F.; Granet, M.; Imfeld, G. Ageing of Copper, Zinc and Synthetic Pesticides in Particle-Size and Chemical Fractions of Agricultural Soils. *Sci. Total Environ.* **2022**, *824*, 153860. [CrossRef] [PubMed]
23. Sun, S.; Fan, Z.; Zhao, J.; Dai, Z.; Zhao, Y.; Dai, Y. Copper Stimulates Neonicotinoid Insecticide Thiacloprid Degradation by *Ensifer Adhaerens* TMX-23. *J. Appl. Microbiol.* **2021**, *131*, 2838–2848. [CrossRef] [PubMed]
24. Jiang, W.; Yao, G.; Jing, X.; Liu, X.; Liu, D.; Zhou, Z. Effects of Cd^{2+} and Pb^{2+} on Enantioselective Degradation Behavior of α-Cypermethrin in Soils and Their Combined Effect on Activities of Soil Enzymes. *Environ. Sci. Pollut. Res.* **2021**, *28*, 47099–47106. [CrossRef] [PubMed]
25. Wang, J.-H.; Zhu, L.-S.; Meng, Y.; Wang, J.; Xie, H.; Zhang, Q.-M. The Combined Stress Effects of Atrazine and Cadmium on the Earthworm *Eisenia Fetida*. *Environ. Toxicol. Chem.* **2012**, *31*, 2035–2040. [CrossRef] [PubMed]
26. Wang, Q.; Xie, D.; Peng, L.; Chen, C.; Li, C.; Que, X. Phytotoxicity of Atrazine Combined with Cadmium on Photosynthetic Apparatus of the Emergent Plant Species *Iris Pseudacorus*. *Environ. Sci. Pollut. Res.* **2022**, *29*, 34798–34812. [CrossRef]
27. Lee, G.L.Y.; Zakaria, N.N.; Futamata, H.; Suzuki, K.; Zulkharnain, A.; Shaharuddin, N.A.; Convey, P.; Zahri, K.N.M.; Ahmad, S.A. Metabolic Pathway of Phenol Degradation of a Cold-Adapted Antarctic Bacteria, Arthrobacter sp. *Catalysts* **2022**, *12*, 1422. [CrossRef]
28. Zhang, Y.; Ge, S.; Jiang, M.; Jiang, Z.; Wang, Z.; Ma, B. Combined Bioremediation of Atrazine-Contaminated Soil by *Pennisetum* and *Arthrobacter* sp. Strain DNS10. *Environ. Sci. Pollut. Res.* **2014**, *21*, 6234–6238. [CrossRef]
29. Li, Y.; Liang, D.; Sha, J.; Zhang, J.; Gao, J.; Li, H.; Liu, R. Isolating and Identifying the Atrazine-Degrading Strain *Arthrobacter* sp. LY-1 and Applying It for the Bioremediation of Atrazine-Contaminated Soil. *Pol. J. Environ. Stud.* **2019**, *28*, 1267–1275. [CrossRef]

30. Kundu, K.; Marozava, S.; Ehrl, B.; Merl-Pham, J.; Griebler, C.; Elsner, M. Defining Lower Limits of Biodegradation: Atrazine Degradation Regulated by Mass Transfer and Maintenance Demand in *Arthrobacter Aurescens* TC1. *Isme J.* **2019**, *13*, 2236–2251. [CrossRef]
31. Niewerth, H.; Schuldes, J.; Parschat, K.; Kiefer, P.; Vorholt, J.A.; Daniel, R.; Fetzner, S. Complete Genome Sequence and Metabolic Potential of the Quinaldine-Degrading Bacterium *Arthrobacter* sp. Rue61a. *BMC Genomics* **2012**, *13*, 534. [CrossRef] [PubMed]
32. Xu, L.; Shi, W.; Zeng, X.-C.; Yang, Y.; Zhou, L.; Mu, Y.; Liu, Y. Draft Genome Sequence of *Arthrobacter* sp. Strain B6 Isolated from the High-Arsenic Sediments in Datong Basin, China. *Stand. Genomic Sci.* **2017**, *12*, 11. [CrossRef] [PubMed]
33. Luscombe, B.M.; Gray, T.R.G. Characteristics of Arthrobacter Grown in Continuous Culture. *Microbiology* **1974**, *82*, 213–222. [CrossRef]
34. Garcia-Junco, M.; De Olmedo, E.; Ortega-Calvo, J.J. Bioavailability of Solid and Non-Aqueous Phase Liquid (NAPL)-Dissolved Phenanthrene to the Biosurfactant-Producing Bacterium *Pseudomonas Aeruginosa* 19SJ. *Environ. Microbiol.* **2001**, *3*, 561–569. [CrossRef] [PubMed]
35. He, H.; Liu, Y.; You, S.; Liu, J.; Xiao, H.; Tu, Z. A Review on Recent Treatment Technology for Herbicide Atrazine in Contaminated Environment. *Int. J. Environ. Res. Public. Health* **2019**, *16*, 5129. [CrossRef]
36. Nemeth-Konda, L.; Füleky, G.; Morovjan, G.; Csokan, P. Sorption Behaviour of Acetochlor, Atrazine, Carbendazim, Diazinon, Imidacloprid and Isoproturon on Hungarian Agricultural Soil. *Chemosphere* **2002**, *48*, 545–552. [CrossRef]
37. Pan, T.; Liu, C.; Wang, M.; Zhang, J. Interfacial Biodegradation of Phenanthrene in Bacteria-Carboxymethyl Cellulose-Stabilized Pickering Emulsions. *Appl. Microbiol. Biotechnol.* **2022**, *106*, 3829–3836. [CrossRef]
38. Xu, X.; Zarecki, R.; Medina, S.; Ofaim, S.; Liu, X.; Chen, C.; Hu, S.; Brom, D.; Gat, D.; Porob, S.; et al. Modeling Microbial Communities from Atrazine Contaminated Soils Promotes the Development of Biostimulation Solutions. *ISME J.* **2019**, *13*, 494–508. [CrossRef]
39. Ortega-Calvo, J.J.; Alexander, M. Roles of Bacterial Attachment and Spontaneous Partitioning in the Biodegradation of Naphthalene Initially Present in Nonaqueous-Phase Liquids. *Appl. Environ. Microbiol.* **1994**, *60*, 2643–2646. [CrossRef]
40. Pal, A.; Bhattacharjee, S.; Saha, J.; Sarkar, M.; Mandal, P. Bacterial Survival Strategies and Responses under Heavy Metal Stress: A Comprehensive Overview. *Crit. Rev. Microbiol.* **2022**, *48*, 327–355. [CrossRef]
41. Chen, S.; Yin, H.; Ye, J.; Peng, H.; Zhang, N.; He, B. Effect of Copper(II) on Biodegradation of Benzo[a]Pyrene by Stenotrophomonas Maltophilia. *Chemosphere* **2013**, *90*, 1811–1820. [CrossRef] [PubMed]
42. Lu, M.; Zhang, Z.-Z.; Wu, X.-J.; Xu, Y.-X.; Su, X.-L.; Zhang, M.; Wang, J.-X. Biodegradation of Decabromodiphenyl Ether (BDE-209) by a Metal Resistant Strain, *Bacillus Cereus* JP12. *Bioresour. Technol.* **2013**, *149*, 8–15. [CrossRef] [PubMed]
43. Helbig, K.; Grosse, C.; Nies, D.H. Cadmium Toxicity in Glutathione Mutants of *Escherichia coli*. *J. Bacteriol.* **2008**, *190*, 5439–5454. [CrossRef] [PubMed]
44. Bruins, M.R.; Kapil, S.; Oehme, F.W. Microbial Resistance to Metals in the Environment. *Ecotoxicol. Environ. Saf.* **2000**, *45*, 198–207. [CrossRef] [PubMed]
45. Elyamine, A.M.; Afzal, J.; Rana, M.S.; Imran, M.; Cai, M.; Hu, C. Phenanthrene Mitigates Cadmium Toxicity in Earthworms *Eisenia Fetida* (Epigeic Specie) and *Aporrectodea Caliginosa* (Endogeic Specie) in Soil. *Int. J. Environ. Res. Public. Health* **2018**, *15*, 2384. [CrossRef]
46. Liu, C.; Wang, M.; Zhang, J.; Qian, Y.; Xiao, K.; Wang, R.; Dong, W.; Pan, T. A Polycyclic Aromatic Hydrocarbon Degrading Strain and Its Potential to Degrade Phenanthrene in Various Enhanced Systems. *Chin. J. Biotechnol.* **2021**, *37*, 3696–3707.
47. Pan, T.; Deng, T.; Zeng, X.; Dong, W.; Yu, S. Extractive Biodegradation and Bioavailability Assessment of Phenanthrene in the Cloud Point System by *Sphingomonas Polyaromaticivorans*. *Appl. Microbiol. Biotechnol.* **2016**, *100*, 431–437. [CrossRef]
48. Rosenberg, M. Microbial Adhesion to Hydrocarbons: Twenty-Five Years of Doing MATH. *FEMS Microbiol. Lett.* **2006**, *262*, 129–134. [CrossRef]
49. Wang, J.; Ren, Z.; Wu, L.; Li, M. Determination of Trace Cd in Water by Spectrophotometry after Cloud Point Extraction. *Phys. Test. Chem. Anal. Part B Chem. Anal.* **2012**, *48*, 735–736, 739.

catalysts

MDPI

Article

Metabolic Pathway of Phenol Degradation of a Cold-Adapted Antarctic Bacteria, *Arthrobacter* sp.

Gillian Li Yin Lee [1], Nur Nadhirah Zakaria [1], Hiroyuki Futamata [2,3], Kenshi Suzuki [3], Azham Zulkharnain [4], Noor Azmi Shaharuddin [1], Peter Convey [5,6], Khadijah Nabilah Mohd Zahri [1] and Siti Aqlima Ahmad [1,*]

1 Department of Biochemistry, Faculty of Biotechnology and Biomolecular Sciences, Universiti Putra Malaysia, Serdang 43400, Selangor, Malaysia
2 Graduate School of Science and Technology, Shizuoka University, Naka-ku, Hamamatsu 432-8561, Japan
3 Research Institute of Green Science and Technology, Shizuoka University, Suruga-ku, Shizuoka 422-8529, Japan
4 Department of Bioscience and Engineering, College of Systems Engineering and Science, Shibaura Institute of Technology, 307 Fukasaku, Minuma-ku, Saitama 337-8570, Japan
5 British Antarctic Survey, NERC, High Cross, Madingley Road, Cambridge CB3 0ET, UK
6 Department of Zoology, University of Johannesburg, P.O. Box 524, Auckland Park 2006, South Africa
* Correspondence: aqlima@upm.edu.my

Abstract: Phenol is an important pollutant widely discharged as a component of hydrocarbon fuels, but its degradation in cold regions is challenging due to the harsh environmental conditions. To date, there is little information available concerning the capability for phenol biodegradation by indigenous Antarctic bacteria. In this study, enzyme activities and genes encoding phenol degradative enzymes identified using whole genome sequencing (WGS) were investigated to determine the pathway(s) of phenol degradation of *Arthrobacter* sp. strains AQ5-05 and AQ5-06, originally isolated from Antarctica. Complete phenol degradative genes involved only in the ortho-cleavage were detected in both strains. This was validated using assays of the enzymes catechol 1,2-dioxygenase and catechol 2,3-dioxygenase, which indicated the activity of only catechol 1,2-dioxygenase in both strains, in agreement with the results from the WGS. Both strains were psychrotolerant with the optimum temperature for phenol degradation, being between 10 and 15 °C. This study suggests the potential use of cold-adapted bacteria in the bioremediation of phenol pollution in cold environments.

Keywords: phenol; metabolites; whole genome sequencing; xenobiotics

Citation: Lee, G.L.Y.; Zakaria, N.N.; Futamata, H.; Suzuki, K.; Zulkharnain, A.; Shaharuddin, N.A.; Convey, P.; Zahri, K.N.M.; Ahmad, S.A. Metabolic Pathway of Phenol Degradation of a Cold-Adapted Antarctic Bacteria, *Arthrobacter* sp. *Catalysts* **2022**, *12*, 1422. https://doi.org/10.3390/catal12111422

Academic Editors: Zhilong Wang and Tao Pan

Received: 6 September 2022
Accepted: 7 November 2022
Published: 12 November 2022

1. Introduction

Antarctica has long been considered one of the last pristine and remote wilderness areas on Earth. However, in recent years, pollution by hydrocarbons such as polycyclic aromatic hydrocarbons, chlorinated biphenyls, and phenols has become an increasing concern in Antarctica [1–4]. A previous study on the impacts of lubricant oil on Antarctic infauna reported the presence of not readily biodegradable phenol additives in both "conventional" and "biodegradable" lubricants [5]. Hydrocarbons from oil, diesel, and lubricant spills tend to persist due to the harsh conditions in Antarctica [6–8]. Furthermore, aromatic petroleum components such as phenol and phenolic compounds (PCs) are harmful to Antarctic terrestrial and aquatic ecosystems [9,10]. In addition, persistent organic pollutants (POPs), including polychlorinated biphenyls (PCBs), are associated with human activities and are subject to long-range atmospheric transport [11]. The direct discharge of phenol into the environment causes severe damage and threats to terrestrial and aquatic ecosystems and, even at low concentrations, it is extremely toxic to aquatic life [10,12–14].

Phenol is a hazardous pollutant that must be removed from the environment [15–17]. Several methods have been published to treat phenol, including the application of physical, chemical, and biological processes. Many studies have shown that bioremediation is more effective in degrading toxic compounds compared to physicochemical treatments and leads to the complete degradation of these compounds [16,18]. Among the various microorganisms reported, phenol-degrading bacteria are the most extensively studied [1,10,19]. Several reports have indicated that members of the genera *Pseudomonas, Rhodococcus, Acinetobacter*, and *Arthrobacter* are capable of degrading phenol [20–22]. In cold environments, the bioremediation of polluted sites requires cold-adapted microorganisms [23–25]. A number of phenol-degrading microorganisms exhibiting cold-adapted features have been isolated from cold regions, such as the alpine areas and Antarctica. Certain bacteria are able to utilize phenol as a carbon and energy source under aerobic or anaerobic conditions [26]. The application of next-generation sequencing (NGS) has catalysed a revolution in biodegradation by predicting the genes involved in biodegradation and other metabolic processes of bacteria [27].

Bioremediation is increasingly viewed as an appropriate remediation technology for cleaning industrial wastewater and contaminated sites [3,28,29]. Aromatic pollutants, such as phenol, are recalcitrant and exhibit anti-microbial properties. The main objectives of this study were to identify the pathway(s) of phenol degradation using whole genome sequencing (WGS) and enzyme assays in two phenol-degrading strains of bacteria originally isolated from Antarctica.

2. Results and Discussion

Detection of Phenol Degradative Genes via WGS

The key structural features of the draft genomes of strains AQ5-05 and AQ5-06 are summarized in Table 1. The draft genome size of strain AQ5-05 was 4647352 base pairs (bp) in 116 contigs with an average G + C content of 65.7%, while strain AQ5-06 was 4,139,264 bp in 70 contigs with an average G + C content of 65.7%. The annotated draft genome of strain AQ5-05 contained a total of 4294 coding sequences (CDSs), 9 rRNAs, and 57 tRNAs and that of strain AQ5-06 contained a total of 3981 CDSs, 3 rRNAs, and 50 tRNAs. In comparison, the total CDSs of completely sequenced cold-adapted *Arthrobacter alpinus* ERGS4:06 and *Arthrobacter* sp. strain FB24 were 3538 and 3279, respectively [30,31]. Members of the genus *Arthrobacter* are known as high-GC-content Actinobacteria, consistent with the findings here and on the previously reported *Arthrobacter* spp. [32–35]. According to Kim et al. [36], members of the genus *Arthrobacter* typically have a DNA GC content of 59 to 66%, in agreement with the genome of strains AQ5-05 and AQ5-06. Similar to the genomic size of strain AQ5-06, a 4.12 Mb draft genome sequence for Arthrobacter sp. strain 7749 isolated from marine sediment samples from the Antarctic Peninsula was reported by Sastre et al. [35].

Table 1. Genomic features of *Arthrobacter* sp. strains AQ5-05 and AQ5-06.

Feature	Count/Value	
	AQ5-05	**AQ5-06**
Genome size (bp)	4,647,352	4,139,264
GC content (%)	65.7	64.9
Number of contigs	116	70
Length of the longest contig (bp)	554,363	406,582
Number of Subsystems	415	411
Number of CDSs	4294	3981
Number of rRNAs	9	3
Number of tRNAs	57	50

Figure 1 displays the number of genes in each subsystem and the subsystem coverage. According to RAST annotation, the genome sequences for strain AQ5-05 are classified into 415 subsystems (Table 1). Based on Figure 1A, 48% of the total CDSs are classified into subsystems, while 68% of the CDSs are not included in the subsystems. The subsystem of amino acid and its derivatives contained the highest number of coding sequences (495 counts), followed by 461 counts in the subsystem of carbohydrates and 301 counts in the subsystem of the cofactors, vitamins, prosthetic groups, and pigments. Figure 1B presents the overview of each subsystem feature and the subsystem coverage. Forty-two percent of a total of 3981 CDSs were categorised into subsystems, while fifty-eight percent of the CDSs were not categorised into subsystems. The subsystem with the highest number of counts was carbohydrate (507 counts), followed by the subsystem of amino acids and derivatives (453 counts) and 287 counts in the subsystem of cofactors, vitamins, prosthetic groups, and pigments that are essential for maintaining the life of bacterial cells [37].

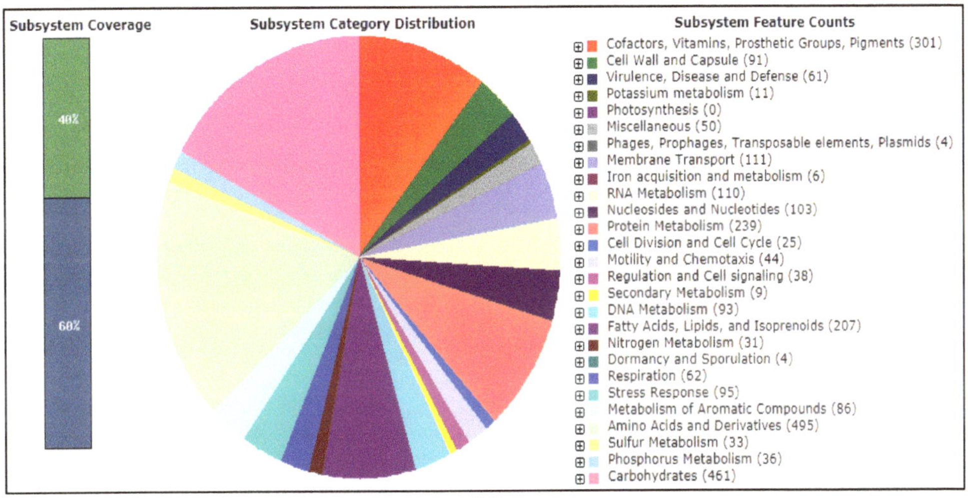

(**A**)

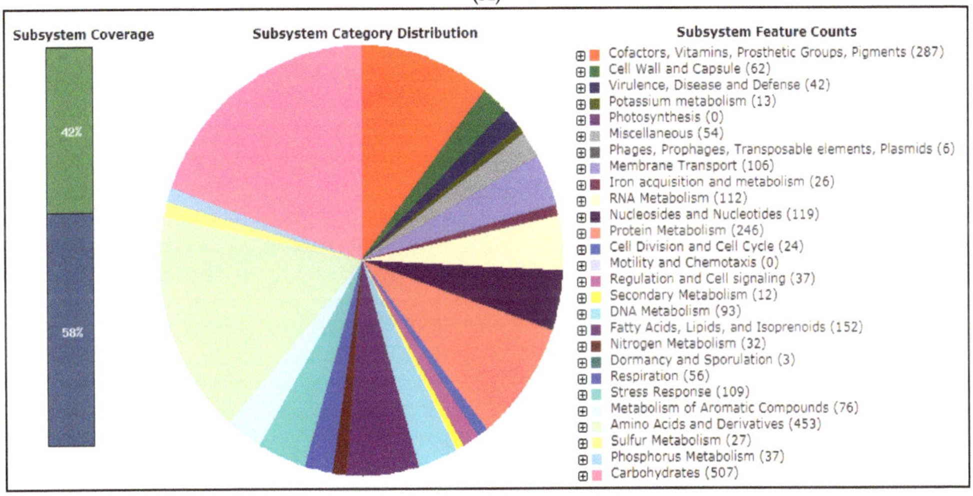

(**B**)

Figure 1. Subsystem category distribution for *Arthrobacter* sp. Strains AQ5-06 (**A**) and AQ5-06 (**B**) were annotated using the RAST server. The red box marks the genes involved in the metabolism of aromatic compounds.

In order to survive in extremely cold environments, polar bacteria synthesise cold-shock proteins (Csps) in response to rapid temperature declines [38]. In strain AQ5-05,

95 counts of genes were responsible for stress responses with three classified as Csps. Rapid temperature reduction decreases the cell membrane fluidity and enzyme activity, and reduces the efficiency of transcription and translation due to the stabilisation of nucleic acid secondary structures and the inhibition of protein folding and ribosome function [38,39]. The genome of strain AQ5-06 contained 109 counts of the genes responsible for stress responses, including five Csps. Of these, four were classified as cold-shock protein A (CspA) and one as cold-shock protein C (CspC). Many bacteria produce Csps, including psychrophiles, mesophiles, and even thermophiles [38–40]. Dsouza et al. [41] noted that Csps were identified in seven temperate and seven Antarctic *Arthrobacter* genomes examined. Previous studies have noted that the production of CspA is induced by cold shock [38,42]. For instance, the overexpression of CspA was reported in *Psychromonas arctica* KOPRI 22215 isolated from Arctic marine sediment [43]. CspC, responsible for the regulation of the growth and expression of the stress response proteins RpoS and UspA, has also been reported in previous studies [44–46].

Seventy-six counts were categorised in the subsystem for the metabolism of aromatic compounds, with seven counts involved in the central aromatic intermediate metabolism, such as catechol, benzoate, and phenol. Similar findings have been reported in the genomes of several *Arthrobacter* spp., such as *Arthrobacter* sp. Edens01 [32], *Arthrobacter* sp. W1 [47], and *A. antarcticus* strain W2 [48]. Genome analysis of *A. alpinus* R3.8 isolated from a soil sample obtained at Rothera Point, Adelaide Island revealed genes involved in the bioremediation of xenobiotic compounds, including naphthalene [49].

Two phenol biodegradation pathways initiated by the oxygenation of phenol followed by the oxidation of catechol at the 1,2-(ortho) or 2,3-(meta) positions have been reported as being typical aerobic phenol degradation pathways for bacteria [19,50]. According to the gene annotation using RAST, genes encoding phenol degradation genes, including phenol 2-monooxgenase, and enzymes involved in the ortho-cleavage of catechol were present in strain AQ5-05.

The physical maps of the gene clusters for both phenol 2-monooxygenase and catechol 1,2 dioxygenase are shown in Figure 2A, with the genes found in the phenol and catechol degradation gene clusters summarised in Table 2. The results from BLASTP searches revealed that the amino acid sequence of ORF6 shared a 90.1% identity with the phenol 2-monooxygenase of *Arthrobacter* sp. UCD-GKA (accession no. WP_071214692) and 76.9% with the PheA/TfdB family putative FAD monooxygenase from *PaenArthrobacter aurescens* strain TC1 (accession no. WP_011777158). According to Figure 2A, genes encoding the enzymes phosphoribosylformylglycinamidine synthase (ORF1-3), 3-oxoacyl-[acyl-carrier protein] (ORF7), and antibiotic biosynthesis monooxygenase (ORF8) were found within the gene cluster of the phenol 2-monooxygenase of strain AQ5-05. The enzyme of fatty acid biosynthesis, 3-oxoacyl-[acyl-carrier protein] catalyses the first of the two reduction steps in the elongation cycle of fatty acid synthesis, while phosphoribosylformylglycinamidine synthases are involved in the purine biosynthetic pathway [51,52]. The antibiotic biosynthesis monooxygenase of bacteria catalyses oxygen-dependent and cofactor-independent oxidations or monooxygenations involved in the biosynthetic pathways of polyketide antibiotics [53]. Antibiotic coding genes, such as antibiotic biosynthesis monooxygenase, are also present in the genome of *A. crystallopoietes* strain BAB-32, which may have industrial and medical applications [54]. A previous study has reported the presence of antibiotic activities for seven antibiotic-producing *Arthrobacter* spp. isolated from the Arctic Ocean under different culture conditions [55]. However, no other phenol catabolic genes were found within the gene cluster.

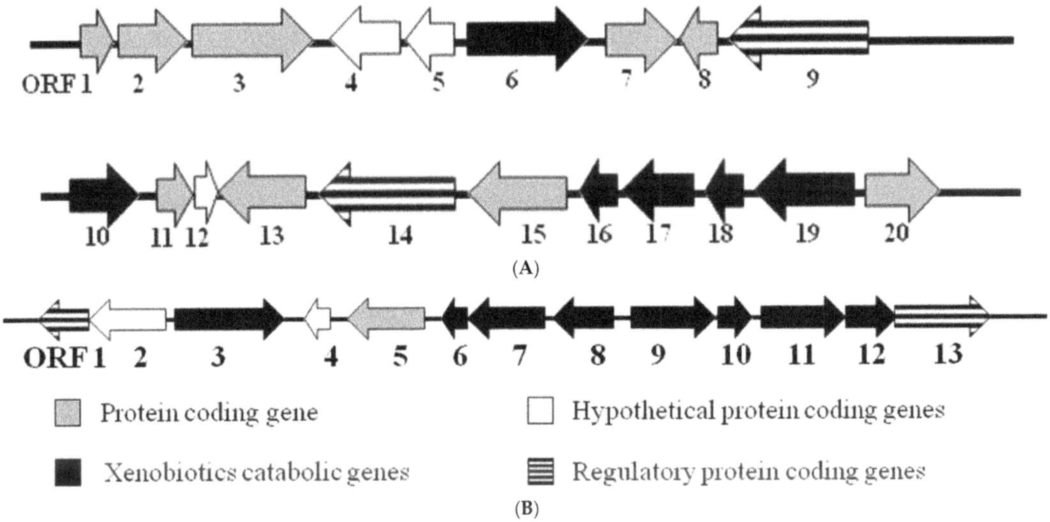

Figure 2. Physical map of gene cluster containing phenol 2-monooxygenase and catechol 1,2-dioxygenase genes from *Arthrobacter* sp. strains AQ5-05 (**A**) and AQ5-06 (**B**).

Table 2. The putative functions of the genes in phenol and catechol degradation gene clusters from *Arthrobacter* sp. strain AQ5-05.

ORF	Putative Functions
1	Phosphoribosylformylglycinamidine synthase subunit, purS
2	Phosphoribosylformylglycinamidine synthase subunit, purQ
3	Phosphoribosylformylglycinamidine synthase subunit, purl
4	Hypothetical protein
5	Hypothetical protein
6	Phenol 2-monooxygenase, PheA/TdfB family
7	3-oxoacyl-[acyl-carrier protein] reductase
8	Antibiotic biosynthesis monooxygenase
9	LuxR family transcriptional regulator
10	Catechol 1,2-dioxygenase, catA
11	Flavin reductase
12	Hypothetical protein
13	Pimeloyl-ACP methyl ester carboxylesterase
14	LuxR family transcriptional regulator
15	AAHS family benzoate transporter-like MFS transporter, benK
16	1,6-dihydroxycyclohexa-2,4-diene-1-carboxylate dehydrogenase, benD
17	Benzoate 1,2-dioxygenase reductase subunit, benC
18	Benzoate 1,2-dioxygenase beta subunit, benB
19	Benzoate 1,2-dioxygenase alpha subunit, benA
20	Benzoate membrane transport protein, benE

In the gene-encoding for enzyme C12O, the catA of strain AQ5-05 was shown as ORF10, sharing a 96.5% amino acid sequence identity with the C12O of *Arthrobacter* sp. UCD-GKA (accession no. WP_071214119). Apart from catA, benzoate catabolic genes, benABCD encoding for benzoate 1,2-dioxygenases and dihydroxycyclohexadiene carboxylate dehydrogenase, and benEK encoding for two benzoate transport proteins, were detected. Similar to the findings of previous studies, catA is chromosomally linked with the ben genes in strain AQ5-05 [56–58]. Two steps are required to convert benzoate to catechol. BenABC encodes for benzoate dioxygenase, which is thought to catalyse the oxidation of benzoate to benzoate diol, followed by decarboxylation to catechol by 1,6-dihydroxycyclohexa-2,4-

diene-1-carboxylate dehydrogenase encoded by benD [59]. Genes encoding pimeloyl-ACP methyl ester carboxylesterase and flavin reductase were also detected within the gene cluster.

According to Figure 2A, both genes encoding putative LuxR family transcriptional regulators were detected in the gene clusters of phenol 2-monooxygenase and C12O. Previous studies have shown that LuxR family transcriptional regulators are activators of the expression of genes encoding enzymes associated with ortho-cleavage through the β-ketoadipate pathway in an aromatic-degrading *Corynebacterium glutamicum* [60,61]. Many transcriptional regulators, including GntR, LuxR, CRP, FNR, TetR, and the Fis family transcriptional regulator, have been identified to be potentially involved in phenol biodegradation in the phenol-degrading *Acinetobacter* sp. DW-1 [62].

Based on the gene annotations, genes encoding the phenol 2-monooxygenase and ortho cleavage enzyme for catechol were also detected in the genome of strain AQ5-06. Figure 2B displays the physical map of the gene cluster containing genes coding for phenol 2-monooxygenase, catechol 1,2 dioxygenase, muconolactone isomerase, and muconate cycloisomerase, and the putative functions of the genes are listed in Table 3.

Table 3. The putative functions of the genes in phenol and catechol catabolic gene clusters from *Arthrobacter* sp. strain AQ5-06.

ORF	Predicted Functions
1	GntR family transcriptional regulator
2	Hypothetical protein
3	Phenol 2-monooxygenase, PheA/tdfB family
4	Hypothetical protein
5	Benzoate transport protein, benE
6	Muconolactone isomerase, catC
7	Muconate cycloisomerase, catB
8	Catechol 1,2- dioxygenase, catA
9	Benzoate 1,2-dioxygenase alpha subunit, benA
10	Benzoate 1,2-dioxygenase beta subunit, benB
11	Benzoate 1,2-dioxygenase reductase subunit, benC
12	1,6-dihydroxycyclohexa-2,4-diene-1-carboxylate dehydrogenase, benD
13	LuxR family transcriptional regulator

Genetic annotation predicted that strain AQ5-06 is capable of degrading phenol only through the ortho pathway as genes encoding for meta-cleavage were absent in the genome. According to Figure 2B, a transcriptional regulator GntR family was detected within the gene cluster, shown as ORF1. Previous studies have reported the involvement of the GntR transcriptional regulator family in the degradation pathways of aromatic compounds [62–64]. For instance, Gu et al. [62] claimed that a GntR-type regulator was potentially involved in phenol degradation in *Acinetobacter* sp. DW-1. Moreover, Rucká et al. [65] reported that the GntR regulator is required besides the XylR regulator for efficient regulation of phenol degradation operons. Teramoto et al. [66] and Arai et al. [67] reported the presence of the GntR family of transcriptional regulators in the expression of the multicomponent phenol hydroxylase (mPH) of *Comamonas testosteroni* spp.

The gene encoding phenol 2-monooxygenase is shown as ORF3, located between two hypothetical proteins with unknown functions. According to the result from BLASTP searches, the amino acid sequence of phenol 2-monooxygenase from strain AQ5-06 shared the highest identity (88.1%) with phenol 2-monooxygenase from *Arthrobacter* sp. B6 (accession no. WP_066285143). Phenol 2-monooxygenase serves as the first enzyme in the ring cleavage of phenol by catalysing the ortho-hydroxylation of phenol to catechol [68,69].

The phenol 2-monooxygenase gene cluster of strain AQ5-06 also harbours the genes coding for both the catechol and benzoate pathways. Catechol catabolic genes, catABC, were detected within the gene cluster and are shown as ORF6-8. Previous studies have demonstrated that several bacteria have catABC genes in one operon responsible for the first three steps of β-ketoadipade, similar to the gene organisation in strain AQ5-06 [70–72]. The amino acid sequence of catA for strain AQ5-06 showed the highest (94.9%) identity with the catA amino acid sequence of *Arthrobacter* sp. AGTC131 (accession no. WP_104138975.1). Contiguous to the strain AQ5-06 cat operon, benzoate catabolic genes, benABCD, and the benzoate transport protein, benE, were detected within the gene cluster, which is responsible for metabolising benzoate to catechol. Several studies have noted that cat genes are chromosomally linked with ben genes forming a cluster in bacteria [56,73,74]. Adjacent to the benABCD operon, a LuxR family transcriptional regulator was detected. Brinkrolf et al. [60] noted that LuxR family transcriptional regulators are the activators of the expression of genes encoding enzymes associated with ortho-cleavage through the β-ketoadipate pathway in an aromatic-degrading *Corynebacterium glutamicum*.

Bioinformatics analysis revealed the presence of genes related to the phenol and catechol degradation pathways in strains AQ5-06 and AQ5-05 (Figure 3). Based on gene annotation (Figure 3A), the phenol degradation pathway of strain AQ5-05 is most likely to be the ortho-pathway. Detailed information, including the gene name, gene products, locus tags, length of amino acid sequence, KEGG, COG, and Enzyme Commission number (EC No.), is given in Table 4 for strain AQ5-05 and in Table 5 for strain AQ5-06. The PheA/tfdB family gene in strains AQ5-05 and AQ5-06 were predicted to encode phenol 2-monooxygenase, containing 634 amino acid residues that catalysed the oxidation of phenol to catechol.

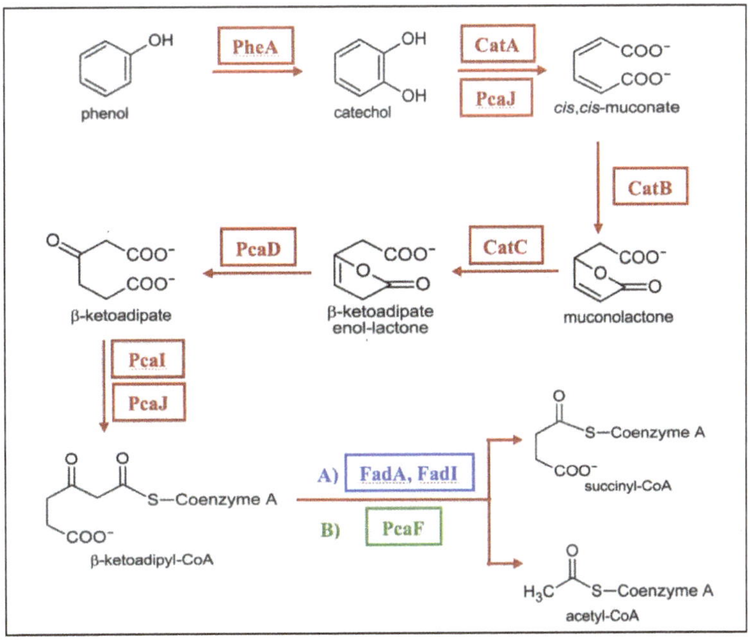

Figure 3. Predicted pathways for the degradation of phenol in *Arthrobacter* sp. strains AQ5-05 (**A**) and AQ5-06 (**B**) based on the presence of gene sequences shown in red fonts.

Table 4. Characteristics of phenol catabolic genes detected in the genome of *Arthrobacter* sp. strain AQ5-05.

Gene Name (Locus Tag)	Gene Product	Amino Acid Residues (aa)	COG No.	KEGG No.	EC No.
PheA (STRA2_03637)	Phenol 2-monooxygenase	634	COG2871	K03380	EC:1.14.13.7
CatA (STRA2_03962)	Catechol 1,2-dioxygenase	284	COG3485	K03381	EC:1.13.11.1
CatB (STRA2_03993)	Muconate cycloisomerase	366	COG4948	K01856	EC:5.5.1.1
CatC (STRA2_03992)	Muconolactone isomerase	92	COG4829	K03464	EC:5.3.3.4
PcaD (STRA2_02926)	3-oxoadipate enol-lactonase	271	COG0596	K01055	EC:3.1.1.24
PcaI (STRA2_02962)	3-oxoadipate coa-transferase, alpha subunit	224	COG1788	K01031	EC:2.8.3.6
PcaJ (STRA2_02963)	3-oxoadipate coa-transferase, beta subunit	223	COG2057	K01032	EC:2.8.3.6
FadA, FadI(STRA2_00274)	Acetyl-coa acyltransferase	412	COG0183	K00632	EC:2.3.1.16

Table 5. Characteristics of phenol catabolic genes found in the genome of *Arthrobacter* sp. strain AQ5-06.

Gene Name (Locus Tag)	Gene Products	Amino Acid Residues (aa)	COG No.	KEGG No.	EC No.
PheA (ASC_01925)	Phenol 2-monooxygenase	640	COG0654	K03380	EC:1.14.13.7
CatA (ASC_01930)	Catechol 1,2-dioxygenase	294	COG3485	K03381	EC:1.13.11.1
CatB (ASC_01929)	Muconate cycloisomerase	359	COG4948	K01856	EC:5.5.1.1
CatC (ASC_01928)	Muconolactone isomerase	92	COG4829	K03464	EC: 5.3.3.4
PcaD (ASC_02352)	3-oxoadipate enol-lactonase	272	COG0596	K01055	EC:3.1.1.24
PcaI (ASC_02349)	3-oxoadipate coa-transferase, alpha subunit	219	COG1788	K01031	EC:2.8.3.6
PcaJ (ASC_02348)	3-oxoadipate coa-transferase, beta subunit	226	COG2057	K01032	EC:2.8.3.6
PcaF (ASC_03031)	3-oxoadipyl-CoA thiolase	413	COG0183	K07823	EC:2.3.1.174

Phenol 2-monooxygenase, also known as phenol hydroxylase, is the first enzyme in phenol degradation that catalyses the hydroxylation of phenol to catechol [50]. The presence of genes coding for phenol 2-monooxygenase or phenol hydroxylase in several *Arthrobacter* spp. has been reported in previous studies [32,75,76]. For example, Nordin et al. [75] revealed that the gene-encoding single-component phenol 2-monooxygenase of 4-chlorophenol-degrading *Arthrobacter chlorophenolicus* A6 shared the highest identity with the PheA gene from the *Pseudomonas* sp. strain EST1001. Qu et al. [76] described a phenol hydroxylase gene cluster (4606 bp), containing six components in the order of KLMNOP from *Arthrobacter* sp. W1. The phenol 2-monooxygenase from strains AQ5-05 and AQ5-06 are highly similar to each other, with 490 identical residues and 53 similar residues within 639 aligned residues (83.8% similarity). A domain for FAD-binding was detected within both proteins, between residue 32 and residue 405. This domain is known for binding to flavin-adenine dinucleotide, which is the cofactor for various flavoprotein oxidoreductase enzymes, including aromatic hydrolases such as salicylate hydrolase, *p*-hydroxybenzoate, and 2,4-dichlorophenol hydroxylase [77]. A conserved domain for the phenol hydroxylase

C-terminal dimerization (thioredoxin-like fold) domain and a highly conserved sequence (DGxxSxxR) for identifying the FAD/NAD(P)H-binding flavoprotein [78] were also identified using Uniprot peptide search utility. These findings suggested that both strains employ class A single-component flavoprotein monooxygenase for phenol hydroxylation. Class A single-component flavoprotein monooxygenases contain a FAD cofactor and depend on NAD(P)H as the external electron donor. Unlike the two-component flavoprotein monooxygenase, single-component flavoprotein monooxygenases can utilize NAD(P)H directly as flavin cofactors.

Catechol is often produced as the intermediate from a wide range of aromatic catabolic pathways [50]. Subsequent to phenol hydroxylation, the catechol intermediate formed is cleaved by C12O encoded by the catA gene in the ortho-pathway. C12O (EC 1.13.11.1) is an intradiol-cleaving enzyme with an Fe^{3+} prosthetic group, whereas C23O (EC 1.13.11.2) has a catalytic iron ion (Fe^{2+}) in each of its four identical subunits and performs extradiol cleavage [79]. As shown in Figure 3, C12O is the key enzyme in the catechol ortho-pathway, initiating an intradiol cleavage reaction of catechol to form cis,cis-muconate (CCMA), with further processing as described earlier. Similar to the catechol pathway of strains AQ5-05 and AQ5-06, Hong et al. [80] suggested the ortho-pathway for the degradation of catechol in hydrocarbon-degrading *Achromabacter* sp. HZ01 isolated from oil-polluted seawater in the South China Sea, which was predicted using genome sequencing. Moreover, Li et al. [81] confirmed the predicted ortho-pathway of phenol degradation for *Arthrobacter sulfureus* strain XSH(1) based on proteomic analysis.

To validate the pathway of phenol degradation in strain AQ5-06 experimentally, the two routes for the degradation of phenol via the ortho- and meta-pathways were tested by quantifying the resulting products. Prior to the enzyme assay, the bacterium was cultured in the presence of phenol to induce the production of catechol dioxygenases. The catechol degradation pathways of strains AQ5-05 and AQ5-06 were experimentally confirmed through enzyme assays of C12O and C23O by measuring the amount of CCMA and 2-HMSA produced, using colorimetric methods [82–84]. The amounts of CCMA and 2-HMSA formed were determined spectrophotometrically at 260 nm and 370 nm, respectively (Figure 4). The data obtained revealed that only CCMA was produced in the presence of catechol, consistent with the activity of only the C12O in both strains. The average rates of CCMA production over 1 h were 0.52 and 0.67 µmol/min, with a maximum specific activity of the C120 of 24 and 20 U/mg for strains AQ5-05 and AQ5-06, respectively.

The results from both the bioinformatics analysis and enzyme assays confirmed that both studied strains can degrade phenol via the ortho-pathway. A recent study concluded that C12O was detected in all of the fifteen cold-adapted phenol-degrading strains isolated and examined from cold regions, while only eight strains produced C23O [85]. Previous studies have reported several C12O enzymes produced by members of the genera *Arthrobacter*, *Acinetobacter*, *Pseudomonas*, and *Rhodococcus* [83,86–88].

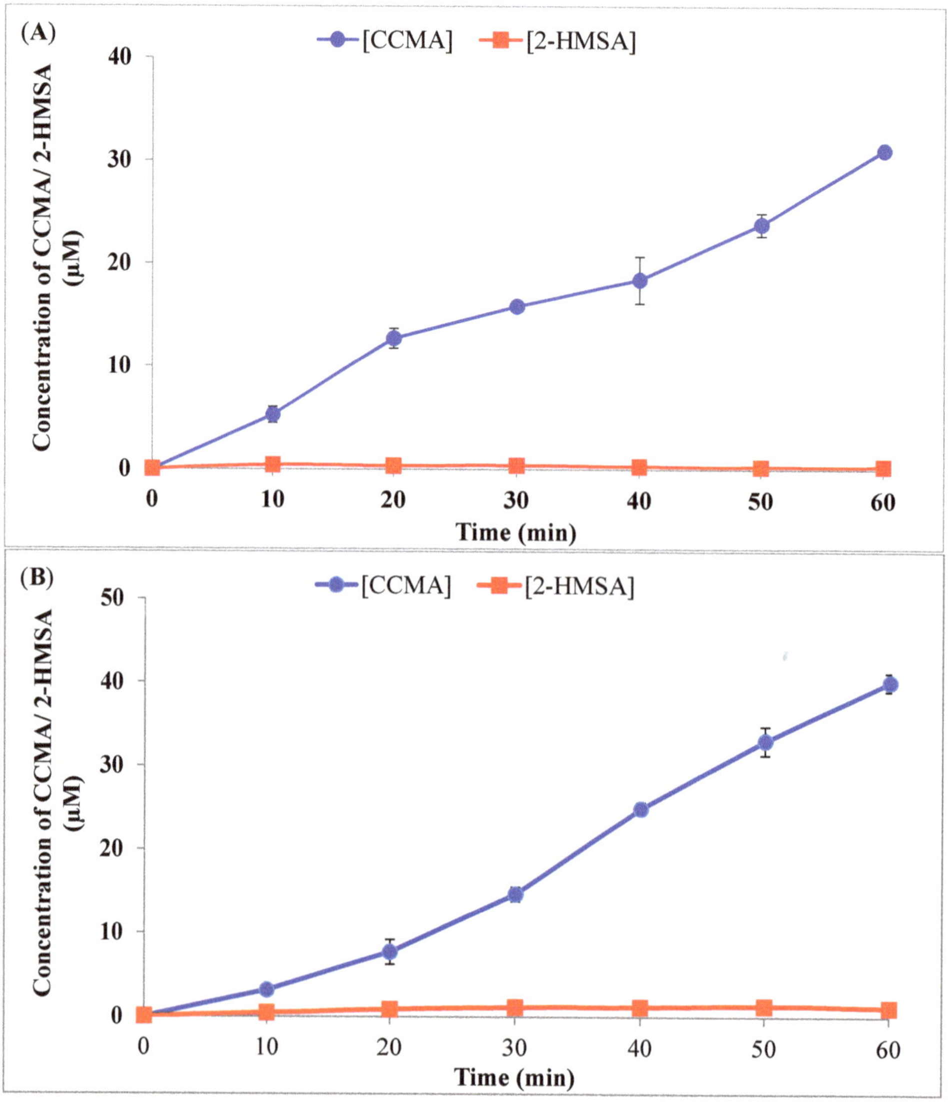

Figure 4. Formation of cis,cis-muconate and 2-hydroxymuconic semialdehyde by catechol dioxygenases of *Arthrobacter* sp. strains AQ5-05 (**A**) and AQ5-06 (**B**). Error bars represent mean ± standard deviation for three replicates.

3. Materials and Methods

3.1. Bacterial Culture and Medium

Phenol-degrading *Arthrobacter* sp. strains AQ5-05 and AQ5-06 were originally isolated from soil obtained on King George Island, South Shetland Islands (maritime Antarctic) [89]. A phenol medium (0.5 g/L) was prepared in 1 L of volume by adding 0.4 g KH_2PO_4, 0.2 g K_2HPO_4, 0.1 g $MgSO_4$, 0.1 g NaCl, 0.01 g $MnSO_4.H_2O$, 0.01 g $Fe_2(SO_4)_3.H_2O$, 0.01 g $Na_2MoO_4.H_2O$, and 0.4 g $(NH_4)_2SO_4$ to distilled water at pH 7.2. The medium was autoclaved for 15 min at 121 °C. The sterilised medium was then augmented with 0.5 g of crystalline phenol.

3.2. Whole Genome Sequencing

3.2.1. Extraction of Genomic DNA

Strains AQ5-05 and AQ5-06 were each cultured in 50 mL of NB in a shaking incubator at 15 °C for 48 h. The cultured broth (1.5 mL) was transferred to an Eppendorf tube and centrifuged at 15,000× *g* for 2 min in order to remove the supernatant. Genomic DNA was extracted using a GeneJET Genomic DNA Extraction Kit (Thermo Scientific, Waltham, MA, USA) and following the manufacturer's protocol. The extracted DNA was examined using 1.0% (*w*/*v*) agarose gel electrophoresis stained with 0.5 μg/mL ethidium bromide (Vivantis, Oceanside, CA, USA) at 100 V for 40 min. Gel electrophoresis was performed using a Lambda/Hind III marker (Vivantis, Shah Alam, Selangor, Malaysia) as the ladder to examine the size of the extracted genomic DNA prior to observation of the gel under UV light. Subsequently, the purity and concentration of extracted genomic DNA were assessed using a Nanodrop spectrophotometer (Bio-Rad, Hercules, CA, USA). The extracted genomic DNA of both strains was sent to Japan's Shizuoka University for whole genome sequencing.

3.2.2. Genome Sequencing and Assembly

The whole genomes of both strains were sequenced using an Illumina MiSeq platform (301 base paired-end reads) at the Instrumental Research Support Office, Research Institute of Green Science and Technology, Shizuoka University, Japan, and the high-quality reads were then assembled using SPAdes version 3.6.2 [90].

3.2.3. Gene Prediction and Annotation

Gene predictions of the genomes of both strains were performed using Prokka 1.12-beta software [91] and the Expert Review version of the Integrated Microbial Genomes (IMG/ER) [92]. Interspersed repetitive sequences were predicted using RepeatMasker [93]. Tandem repeats were analysed by a TRF (Tandem repeats finder) [94]. Transfer RNA (tRNA) genes were predicted by tRNAscan-SE [95]. Ribosomal RNA (rRNA) genes were analysed by rRNAmmer [96]. Small nuclear RNAs (snRNA) were predicted by BLAST against the Rfam database [97]. Annotations of the whole genome sequences were performed using the automated web-based tool, Rapid Annotations using subsystems Technology (RAST) server, with the SEED database [98]. To screen for genes involved in phenol degradation, the amino acid sequences were searched against the protein sequence database from the Basic Local Alignment Search Tool (BLAST) [99], Uniprot [100], the Kyoto Encyclopedia of Genes and Genomes (KEGG) [101], and Clusters of Orthologous Groups (COG) databases [102].

3.3. Preparation of Cell Extracts

Phenol-degrading bacterial strains were cultured in 10 mL of MSM containing 0.7 g/L phenol on a shaking incubator at 150 rpm at 15 °C for 5 d. Cells (2 mL) were then harvested by centrifugation at 4500× *g* for 15 min. This was followed by washing with a 50 mM phosphate buffer at pH 7.5 and resuspension in 2 mL of the same buffer. Eventually, cells were disrupted by sonication treatment with 30-s intervals of sonication and 30-s intervals of interruption for a total of 6 min in an ice-cooled bath [83]. Pellets were removed by centrifugation at 9000× *g* for 30 min at 4 °C. The collected supernatants were used in the subsequent enzyme assays [82,86].

3.4. Enzyme Assay of Catechol 1,2 Dioxygenase (C12O) and Catechol 2,3 Dioxygenase (C23O)

Enzyme activity of C12O was determined by adding the cell-free extract (20 μL) to a 50 mM phosphate buffer (pH 7.5) containing 20 mM Na_2EDTA and 50 mM catechol to give a final volume of 1 mL, according to the protocols previously reported [82–84]. The enzyme activity of C23O was determined spectrophotometrically at 370 nm, based on the formation of 2-hydroxymuconic semialdehyde (2-HMS). Product formation was measured spectrophotometrically for 1 h at 10-min intervals in an ice bath.

4. Conclusions

Bioinformatics analysis of WGS data was used to predict the pathways of phenol degradation based on the gene-associated functions in the phenol degradation of two Antarctic strains of *Arthrobacter* sp. The studied strains originated from Antarctic soils and showed phenol degradation activity in cultures at low temperatures, supporting the potential use of such bacterial strains in future bioremediation protocols for phenol-contaminated sites in cold regions. Based on genome annotation, a number of genes involved in the metabolism of aromatic hydrocarbons were predicted in both strains, consistent with the results of enzymatic analyses which confirmed the presence only of enzymes in the ortho-pathway of the phenol degradation. Genes encoding cold-shock proteins and antifreeze proteins were also detected in the genome annotation.

Author Contributions: Conceptualization, S.A.A. and H.F.; methodology, S.A.A., G.L.Y.L. and H.F.; Software, G.L.Y.L., H.F. and K.S.; validation, H.F., P.C., A.Z., N.A.S. and S.A.A.; formal analysis, G.L.Y.L., H.F., K.S., N.N.Z. and K.N.M.Z.; investigation, G.L.Y.L.; resources, S.A.A.; data curation, G.L.Y.L., H.F. and A.Z.; original draft preparation, G.L.Y.L.; writing-review and editing, S.A.A., N.N.Z., H.F., P.C. and N.A.S.; visualization, S.A.A.; supervision, S.A.A., P.C., A.Z. and H.F.; project administration, S.A.A.; funding acquisition, S.A.A. and H.F. All authors have read and agreed to the published version of the manuscript.

Funding: This work was supported by the Matching Grants PUTRA (UPM-YPASM 9300430), YPASM Berth Support, PUTRA-IPS (9631800), and PUTRA-Berimpak (9660000). P. Convey is supported by NERC core funding for the British Antarctic Survey's 'Biodiversity, Evolution and Adaptation' Team.

Data Availability Statement: Not applicable.

Acknowledgments: The authors would like to thank the Malaysia Ministry of High Education for providing a My Ph.D. scholarship to Gillian Li Yin Lee. This paper also contributes to the international SCAR research programme, 'State of the Antarctic Ecosystem'. We acknowledge the Eco-Remediation Technology Laboratory, Universiti Putra Malaysia and Instrumental Research Support Office, the Research Institute of Green Science and Technology, Shizuoka University, Japan for providing an enabling environment to conduct the research. We also thank S.A. Alias (Universiti Malaya) and G. Gonzalez-Rocha (Universidad de Concepcion) for their previous help.

Conflicts of Interest: The authors declare no conflict of interest.

References

1. Mazuki, T.A.T.; Shukor, M.Y.; Ahmad, S.A. Bioremediation of phenol in Antarctic: A mini review. *Malays. J. Biochem. Mol. Biol.* **2019**, *22*, 1–6.
2. Chang, Y.-C.; Fuzisawa, S.; Reddy, M.V.; Kobayashi, H.; Yoshida, E.; Yajima, Y.; Hoshino, T.; Choi, D. Degradation of Toxic Compounds at Low and Medium Temperature Conditions Using Isolated Fungus. *CLEAN Soil Air Water* **2016**, *44*, 992–1000. [CrossRef]
3. De Jesus, H.E.; Peixoto, R.S.; Cury, J.; Van Elsas, J.D.; Rosado, A.S. Evaluation of soil bioremediation techniques in an aged diesel spill at the Antarctic Peninsula. *Appl. Microbiol. Biotechnol.* **2015**, *99*, 10815–10827. [CrossRef] [PubMed]
4. Roslee, A.F.A.; Zakaria, N.N.; Convey, P.; Zulkharnain, A.; Lee, G.L.Y.; Gomez-Fuentes, C.; Ahmad, S.A. Statistical optimisation of growth conditions and diesel degradation by the Antarctic bacterium, *Rhodococcus* sp. strain AQ5-07. *Extremophiles* **2019**, *24*, 277–291. [CrossRef]
5. Thompson, B.A.W.; Goldsworthy, P.M.; Riddle, M.J.; Snape, I.; Stark, J.S. Contamination effects by a 'conventional' and a 'biodegradable' lubricant oil on infaunal recruitment to Antarctic sediments: A field experiment. *J. Exp. Mar. Biol. Ecol.* **2007**, *340*, 213–226. [CrossRef]
6. Zakaria, N.N.; Man, Z.; Zulkharnain, A.; Ahmad, S.A. Psychrotolerant biosurfactant-producing bacteria for hydrocarbon degradation: A mini review. *Malays. J. Biochem. Mol. Biol.* **2019**, *21*, 52–59.
7. Powell, S.M.; Harvey, P.M.; Stark, J.S.; Snape, I.; Riddle, M.J. Biodegradation of petroleum products in experimental plots in Antarctic marine sediments is location dependent. *Mar. Pollut. Bull.* **2007**, *54*, 434–440. [CrossRef]
8. Abdulrasheed, M.; Zakaria, N.N.; Roslee, A.F.A.; Shukor, M.Y.; Zulkharnain, A.; Napis, S.; Convey, P.; Alias, S.A.; Gonzalez-Rocha, G.; Ahmad, S.A. Biodegradation of diesel oil by cold-adapted bacterial strains of *Arthrobacter* spp. from Antarctica. *Antarct. Sci.* **2020**, *32*, 341–353. [CrossRef]
9. Lana, N.B.; Berton, P.; Covaci, A.; Ciocco, N.F.; Barrera-Oro, E.; Atencio, A.; Altamirano, J.C. Fingerprint of persistent organic pollutants in tissues of Antarctic notothenioid fish. *Sci. Total Environ.* **2014**, *499*, 89–98. [CrossRef]

10. Subramaniam, K.; Ahmad, S.A.; Shaharuddin, N.A. Mini review on phenol biodegradation in Antarctica using native microorganisms. *Asia Pac. J. Mol. Biol. Biotechnol.* **2020**, *28*, 77–89. [CrossRef]
11. Vecchiato, M.; Zambon, S.; Argiriadis, E.; Barbante, C.; Gambaro, A.; Piazza, R. Polychlorinated biphenyls (PCBs) and polybrominated diphenyl ethers (PBDEs) in Antarctic ice-free areas: Influence of local sources on lakes and soils. *Microchem. J.* **2015**, *120*, 26–33. [CrossRef]
12. Hill, J.; Nelson, E.; Tilman, D.; Polasky, S.; Tiffany, D. Environmental, economic, and energetic costs and benefits of biodiesel and ethanol biofuels. *Proc. Natl. Acad. Sci. USA* **2006**, *103*, 11206–11210. [CrossRef] [PubMed]
13. Koshy, J.; Chandran, N.; Nambisan, P. Biodegradation of phenol using spent substrate of *Pleurotus* sp. *World J. Pharm. Pharm. Sci.* **2012**, *1*, 656–661.
14. Jame, S.A.; Alam, A.R.; Alam, M.K.; Fakhruddin, A. Isolation and Identification of Phenol and Monochlorophenols-Degrading Bacteria: Pseudomonas and Aeromonas Species. *Bangladesh J. Microbiol.* **1970**, *25*, 41–44. [CrossRef]
15. Ahmad, S.A.; Syed, M.A.; Arif, N.M.; Shukor, M.Y.A.; Shamaan, N.A. Isolation, identification and characterization of elevated phenol degrading *Acinetobacter* sp. strain AQ5NOL 1. *Aust. J. Basic Appl. Sci.* **2011**, *5*, 1035–1045.
16. Villegas, L.G.C.; Mashhadi, N.; Chen, M.; Mukherjee, D.; Taylor, K.E.; Biswas, N. A Short Review of Techniques for Phenol Removal from Wastewater. *Curr. Pollut. Rep.* **2016**, *2*, 157–167. [CrossRef]
17. Subramaniam, K.; Shaharuddin, N.A.; Tengku-Mazuki, T.A.; Zulkharnain, A.; Khalil, K.A.; Convey, P.; Ahmad, S.A.; Survey, N.B.A. Statistical optimisation for enhancement of phenol biodegradation by the Antarctic soil bacterium *Arthrobacter* sp. strain AQ5-15 using response surface methodology. *J. Environ. Biol.* **2020**, *41*, 1560–1569. [CrossRef]
18. O'Neill, T.A.; Aislabie, J.; Balks, M.R. Human Impacts on Soils. In *The Soils of Antarctica*; Bockheim, J.G., Ed.; Springer International Publishing: Cham, Switzerland, 2015; pp. 281–303. [CrossRef]
19. Basha, K.M.; Rajendran, A.; Thangavelu, V. Recent advances in the biodegradation of phenol: A review. *Asian J. Exp. Biol. Sci.* **2010**, *1*, 219–234.
20. Zakaria, N.N.; Ahmad, S.A.; Yin, G.L.L.; Yasid, N.A.; Manogaran, M.; Nawawi, N.M.; Shukor, M.Y. Biodegradation of phenol by Antarctic bacterium *Rhodococcus baikonurensis* strain AQ5-001 in the presence of heavy metals. *Malays. J. Biochem. Mol. Biol.* **2018**, *21*, 29–36.
21. Liu, Z.; Xie, W.; Li, D.; Peng, Y.; Li, Z.; Liu, S. Biodegradation of Phenol by Bacteria Strain *Acinetobacter calcoaceticus* PA Isolated from Phenolic Wastewater. *Int. J. Environ. Res. Public Health* **2016**, *13*, 300. [CrossRef]
22. Ahmad, S.A.; Asokan, G.; Yasid, N.A.; Nawawi, N.M.; Subramaniam, K.; Zakaria, N.N.; Shukor, M.Y. Effect of heavy metals on biodegradation of phenol by Antarctic bacterium: *Arthrobacter bambusae* strain AQ5-003. *Malays. J. Biochem. Mol. Biol.* **2018**, *21*, 47–51.
23. Fernández, P.M.; Martorell, M.M.; Blaser, M.G.; Ruberto, L.A.M.; de Figueroa, L.I.C.; Mac Cormack, W.P. Phenol degradation and heavy metal tolerance of Antarctic yeasts. *Extremophiles* **2017**, *21*, 445–457. [CrossRef] [PubMed]
24. Abdulrasheed, M.; Zulkharnain, A.; Zakaria, N.N.; Roslee, A.; Khalil, K.A.; Napis, S.; Convey, P.; Gomez-Fuentes, C.; Ahmad, S. Response Surface Methodology Optimization and Kinetics of Diesel Degradation by a Cold-Adapted Antarctic Bacterium, *Arthrobacter* sp. Strain AQ5-05. *Sustainability* **2020**, *12*, 6966. [CrossRef]
25. Tengku-Mazuki, T.A.; Subramaniam, K.; Zakaria, N.N.; Convey, P.; Khalil, K.A.; Lee, G.L.Y.; Zulkharnain, A.; Shaharuddin, N.A.; Ahmad, S.A. Optimization of phenol degradation by Antarctic bacterium *Rhodococcus* sp. *Antarct. Sci.* **2020**, *32*, 486–495. [CrossRef]
26. Shibata, A.; Inoue, Y.; Katayama, A. Aerobic and anaerobic biodegradation of phenol derivatives in various paddy soils. *Sci. Total Environ.* **2006**, *367*, 979–987. [CrossRef]
27. Arora, P.K.; Bae, H. Integration of bioinformatics to biodegradation. *Biol. Proced. Online* **2014**, *16*, 8. [CrossRef]
28. Darham, S.; Gomez-Fuentes, C.; Zulkharnain, A.; Sabri, S.; Calisto-Ulloa, N.; Ramírez-Moreno, N.; Ahmad, S.A. Isolation and identification of molybdenum-reducing cold-adapted marine bacteria isolated from Bernardo O'Higgins Riquelme Base Station, Antarctica. *Malay. J. Biochem. Mol. Biol.* **2019**, *22*, 16–19.
29. Ibrahim, S.; Zahri, K.N.M.; Convey, P.; Khalil, K.A.; Gomez-Fuentes, C.; Zulkarnain, A.; Alias, S.A.; González-Rocha, G.; Ahmad, S.A. Optimisation of biodegradation conditions for waste canola oil by cold-adapted *Rhodococcus* sp. AQ5-07 from Antarctica. *Electron. J. Biotechnol.* **2020**, *48*, 1–12. [CrossRef]
30. Kumar, R.; Singh, D.; Swarnkar, M.K.; Singh, A.K.; Kumar, S. Complete genome sequence of *Arthrobacter alpinus* ERGS4:06, a yellow pigmented bacterium tolerant to cold and radiations isolated from Sikkim Himalaya. *J. Biotechnol.* **2016**, *220*, 86–87. [CrossRef]
31. Nakatsu, C.H.; Barabote, R.; Thompson, S.; Bruce, D.; Detter, C.; Brettin, T.; Han, C.; Beasley, F.; Chen, W.; Konopka, A.; et al. Complete genome sequence of *Arthrobacter* sp. strain FB24. *Stand. Genom. Sci.* **2013**, *9*, 106–116. [CrossRef]
32. Couger, M.B.; Hanafy, R.A.; Edens, C.; Budd, C.; French, D.P.; Hoff, W.D.; Elshahed, M.S.; Youssef, N.H. Draft Genome of the *Arthrobacter* sp. Strain Edens01. *Genome Announc.* **2015**, *3*, e01475-15. [CrossRef] [PubMed]
33. Kincheloe, G.N.; Eisen, J.A.; Coil, D.A. Draft Genome Sequence of *Arthrobacter* sp. Strain UCD-GKA (Phylum *Actinobacteria*). *Genome Announc.* **2017**, *5*, e01599-16. [CrossRef] [PubMed]
34. Manzanera, M.; Santa-Cruz-Calvo, L.; Vílchez, J.I.; García-Fontana, C.; Silva-Castro, G.A.; Calvo, C.; González-López, J. Genome Sequence of *Arthrobacter siccitolerans* 4J27, a Xeroprotectant-Producing Desiccation-Tolerant Microorganism. *Genome Announc.* **2014**, *2*, e00526-14. [CrossRef] [PubMed]

35. Sastre, D.E.; Santos, L.P.; Kagohara, E.; Andrade, L.H. Draft Whole-Genome Sequence of Psychrotrophic *Arthrobacter* sp. Strain 7749, Isolated from Antarctic Marine Sediments with Applications in Enantioselective Alcohol Oxidation. *Genome Announc.* **2017**, *5*, e01197-17. [CrossRef]
36. Kim, J.; Kim, S.J.; Kim, S.H.; Moon, Y.-J.; Park, S.-J.; Kim, S.I.; Kahng, H.-Y.; Chung, Y.-H. Genome Sequence of *Arthrobacter* sp. MWB30, Isolated from a Crude Oil-Contaminated Seashore. *Genome Announc.* **2015**, *3*, e00013-15. [CrossRef]
37. Lau, Y.Y.; Yin, W.-F.; Chan, K.-G. *Enterobacter asburiae* Strain L1: Complete Genome and Whole Genome Optical Mapping Analysis of a Quorum Sensing Bacterium. *Sensors* **2014**, *14*, 13913–13924. [CrossRef]
38. Keto-Timonen, R.; Hietala, N.; Palonen, E.; Hakakorpi, A.; Lindström, M.; Korkeala, H. Cold Shock Proteins: A Minireview with Special Emphasis on Csp-family of Enteropathogenic Yersinia. *Front. Microbiol.* **2016**, *7*, 1151. [CrossRef]
39. Kim, M.-J.; Lee, Y.K.; Lee, H.K.; Im, H. Characterization of Cold-Shock Protein A of Antarctic *Streptomyces* sp. AA8321. *J. Protein Chem.* **2007**, *26*, 51–59. [CrossRef]
40. Uh, J.-H.; Jung, Y.H.; Lee, Y.K.; Lee, H.K.; Im, H. Rescue of a cold-sensitive mutant at low temperatures by cold shock proteins from *Polaribacter irgensii* KOPRI 22228. *J. Microbiol.* **2010**, *48*, 798–802. [CrossRef]
41. Dsouza, M.; Taylor, M.W.; Turner, S.J.; Aislabie, J. Genomic and phenotypic insights into the ecology of Arthrobacter from Antarctic soils. *BMC Genom.* **2015**, *16*, 36. [CrossRef]
42. Uppal, S.; Akkipeddi, V.S.R.; Jawali, N. Posttranscriptional regulation of *cspE* in *Escherichia coli*: Involvement of the short 5′-untranslated region. *FEMS Microbiol. Lett.* **2008**, *279*, 83–91. [CrossRef] [PubMed]
43. Jung, Y.H.; Yi, J.-Y.; Jung, H.J.; Lee, Y.K.; Lee, H.K.; Naicker, M.C.; Uh, J.-H.; Jo, I.S.; Jung, E.J.; Im, H. Overexpression of Cold Shock Protein A of *Psychromonas arctica* KOPRI 22215 Confers Cold-Resistance. *J. Protein Chem.* **2010**, *29*, 136–142. [CrossRef]
44. Cohen-Or, I.; Shenhar, Y.; Biran, D.; Ron, E.Z. CspC regulates rpoS transcript levels and complements hfq deletions. *Res. Microbiol.* **2010**, *161*, 694–700. [CrossRef] [PubMed]
45. Phadtare, S.; Inouye, M. Role of CspC and CspE in Regulation of Expression of RpoS and UspA, the Stress Response Proteins in *Escherichia coli*. *J. Bacteriol.* **2001**, *183*, 1205–1214. [CrossRef] [PubMed]
46. Shenhar, Y.; Biran, D.; Ron, E.Z. Resistance to environmental stress requires the RNA chaperones CspC and CspE. *Environ. Microbiol. Rep.* **2012**, *4*, 532–539. [CrossRef] [PubMed]
47. Jiang, Y.; Qu, Y.; Xu, P.; Tang, H. Genome Sequence of a Versatile Aromatic Hydrocarbon-Degrading Bacterium, *Arthrobacter* sp. W1. *Genome Announc.* **2015**, *3*, e00387-15. [CrossRef] [PubMed]
48. Herschend, J.; Raghupathi, P.K.; Røder, H.L.; Sørensen, S.J.; Burmølle, M. Genome Sequence of *Arthrobacter antarcticus* Strain W2, Isolated from a Slaughterhouse. *Genome Announc.* **2016**, *4*, e00073-16. [CrossRef]
49. See-Too, W.-S.; Ee, R.; Lim, Y.-L.; Convey, P.; Pearce, D.A.; Mohidin, T.B.M.; Yin, W.-F.; Chan, K.G. Complete genome of *Arthrobacter alpinus* strain R3.8, bioremediation potential unraveled with genomic analysis. *Stand. Genom. Sci.* **2017**, *12*, 52. [CrossRef]
50. Nair, C.I.; Jayachandran, K.; Shashidhar, S. Biodegradation of phenol. *Afr. J. Biotechnol.* **2008**, *7*, 4951–4958.
51. Batra, R.; Christendat, D.; Edwards, A.; Arrowsmith, C.; Tong, L. Crystal structure of MTH169, a crucial component of phosphoribosylformylglycinamidine synthetase. *Proteins Struct. Funct. Bioinform.* **2002**, *49*, 285–288. [CrossRef]
52. Lai, C.-Y.; Cronan, J.E. Isolation and Characterization of β-Ketoacyl-Acyl Carrier Protein Reductase (*fabG*) Mutants of *Escherichia coli* and *Salmonella enterica* Serovar Typhimurium. *J. Bacteriol.* **2004**, *186*, 1869–1878. [CrossRef] [PubMed]
53. Machovina, M.M.; Usselman, R.J.; DuBois, J.L. Monooxygenase Substrates Mimic Flavin to Catalyze Cofactorless Oxygenations. *J. Biol. Chem.* **2016**, *291*, 17816–17828. [CrossRef] [PubMed]
54. Joshi, M.N.; Pandit, A.S.; Sharma, A.; Pandya, R.V.; Desai, S.M.; Saxena, A.K.; Bagatharia, S.B. Draft Genome Sequence of *Arthrobacter crystallopoietes* Strain BAB-32, Revealing Genes for Bioremediation. *Genome Announc.* **2013**, *1*, e00452-13. [CrossRef]
55. Wietz, M.; Månsson, M.; Bowman, J.S.; Blom, N.S.; Ng, Y.; Gram, L. Wide Distribution of Closely Related, Antibiotic-Producing Arthrobacter Strains throughout the Arctic Ocean. *Appl. Environ. Microbiol.* **2012**, *78*, 2039–2042. [CrossRef] [PubMed]
56. Li, D.; Yan, Y.; Ping, S.; Chen, M.; Zhang, W.; Li, L.; Lin, W.; Geng, L.; Liu, W.; Lin, M. Genome-wide investigation and functional characterization of the β-ketoadipate pathway in the nitrogen-fixing and root-associated bacterium *Pseudomonas stutzeri*A1501. *BMC Microbiol.* **2010**, *10*, 36. [CrossRef]
57. Moreno, M.D.L.; Sânchez-Porro, C.; Piubeli, F.; Frias, L.; García, M.T.; Mellado, E. Cloning, Characterization and Analysis of cat and ben Genes from the Phenol Degrading Halophilic Bacterium *Halomonas organivorans*. *PLoS ONE* **2011**, *6*, e21049. [CrossRef]
58. Suzuki, K.; Ichimura, A.; Ogawa, N.; Hasebe, A.; Miyashita, K. Differential Expression of Two Catechol 1,2-Dioxygenases in *Burkholderia* sp. Strain TH2. *J. Bacteriol.* **2002**, *184*, 5714–5722. [CrossRef]
59. Haußmann, U.; Qi, S.-W.; Wolters, D.; Rögner, M.; Liu, S.-J.; Poetsch, A. Physiological adaptation of *Corynebacterium glutamicum* to benzoate as alternative carbon source—A membrane proteome-centric view. *Proteomics* **2009**, *9*, 3635–3651. [CrossRef]
60. Brinkrolf, K.; Brune, I.; Tauch, A. Transcriptional regulation of catabolic pathways for aromatic compounds in *Corynebacterium glutamicum*. *Genet. Mol. Res.* **2006**, *5*, 773–789.
61. Cramer, A.; Gerstmeir, R.; Schaffer, S.; Bott, M.; Eikmanns, B.J. Identification of RamA, a Novel LuxR-Type Transcriptional Regulator of Genes Involved in Acetate Metabolism of *Corynebacterium glutamicum*. *J. Bacteriol.* **2006**, *188*, 2554–2567. [CrossRef]
62. Gu, Q.; Wu, Q.; Zhang, J.; Guo, W.; Wu, H.; Sun, M. *Acinetobacter* sp. DW-1 immobilized on polyhedron hollow polypropylene balls and analysis of transcriptome and proteome of the bacterium during phenol biodegradation process. *Sci. Rep.* **2017**, *7*, 4863. [CrossRef] [PubMed]

63. Ferrández, A.; Miñambres, B.; García, B.; Olivera, E.R.; Luengo, J.M.; García, J.L.; Diaz, E. Catabolism of Phenylacetic Acid in *Escherichia coli* characterization of a new aerobic hybrid pathway. *J. Biol. Chem.* **1998**, *273*, 25974–25986. [CrossRef] [PubMed]
64. Tropel, D.; van der Meer, J.R. Bacterial Transcriptional Regulators for Degradation Pathways of Aromatic Compounds. *Microbiol. Mol. Biol. Rev.* **2004**, *68*, 474–500. [CrossRef]
65. Rucká, L.; Nešvera, J.; Pátek, M. Biodegradation of phenol and its derivatives by engineered bacteria: Current knowledge and perspectives. *World J. Microbiol. Biotechnol.* **2017**, *33*, 174. [CrossRef] [PubMed]
66. Teramoto, M.; Harayama, S.; Watanabe, K. PhcS Represses Gratuitous Expression of Phenol-Metabolizing Enzymes in *Comamonas testosteroni* R5. *J. Bacteriol.* **2001**, *183*, 4227–4234. [CrossRef]
67. Arai, H.; Akahira, S.; Ohishi, T.; Kudo, T. Adaptation of *Comamonas testosteroni* TA441 to utilization of phenol by spontaneous mutation of the gene for a trans-acting factor. *Mol. Microbiol.* **2002**, *33*, 1132–1140. [CrossRef]
68. Al-Khalid, T.; El-Naas, M. Aerobic Biodegradation of Phenols: A Comprehensive Review. *Crit. Rev. Environ. Sci. Technol.* **2012**, *42*, 1631–1690. [CrossRef]
69. Kirchner, U.; Westphal, A.H.; Müller, R.; van Berkel, W.J.H. Phenol Hydroxylase from *Bacillus thermoglucosidasius* A7, a Two-protein Component Monooxygenase with a Dual Role for FAD. *J. Biol. Chem.* **2003**, *278*, 47545–47553. [CrossRef]
70. Eulberg, D.; Schlömann, M. The putative regulator of catechol catabolism in *Rhodococcus opacus* 1CP–an IclR-type, not a LysR-type transcriptional regulator. *Antonie Leeuwenhoek* **1998**, *74*, 71–82. [CrossRef]
71. Kim, S.I.; Ha, K.-S.; Leem, S.-H. Differential organization and transcription of the cat2 gene cluster in aniline-assimilating *Acinetobacter lwoffii* K24. *J. Biosci. Bioeng.* **1999**, *88*, 250–257. [CrossRef]
72. Matsumura, E.; Sakai, M.; Hayashi, K.; Murakami, S.; Takenaka, S.; Aoki, K. Constitutive expression of *catABC* genes in the aniline-assimilating bacterium *Rhodococcus* species AN-22: Production, purification, characterization and gene analysis of CatA, CatB and CatC. *Biochem. J.* **2005**, *393*, 219–226. [CrossRef] [PubMed]
73. Kim, D.; Kim, S.W.; Choi, K.Y.; Lee, J.S.; Kim, E. Molecular cloning and functional characterization of â€ƒthe genes encoding benzoate and p-hydroxybenzoate degradation by the halophilic *Chromohalobacter* sp. strain HS-2. *FEMS Microbiol. Lett.* **2008**, *280*, 235–241. [CrossRef] [PubMed]
74. Patrauchan, M.A.; Florizone, C.; Dosanjh, M.; Mohn, W.W.; Davies, J.; Eltis, L.D. Catabolism of Benzoate and Phthalate in *Rhodococcus* sp. Strain RHA1: Redundancies and Convergence. *J. Bacteriol.* **2005**, *187*, 4050–4063. [CrossRef] [PubMed]
75. Nordin, K.; Unell, M.; Jansson, J.K. Novel 4-Chlorophenol Degradation Gene Cluster and Degradation Route via Hydroxyquinol in *Arthrobacter chlorophenolicus* A6. *Appl. Environ. Microbiol.* **2005**, *71*, 6538–6544. [CrossRef]
76. Qu, Y.; Shi, S.; Zhou, H.; Ma, Q.; Li, X.; Zhang, X.; Zhou, J. Characterization of a Novel Phenol Hydroxylase in Indoles Biotranformation from a Strain *Arthrobacter* sp. W1. *PLoS ONE* **2012**, *7*, e44313. [CrossRef]
77. Kälin, M.; Neujahr, H.Y.; Weissmahr, R.N.; Sejlitz, T.; Jöhl, R.; Fiechter, A.; Reiser, J. Phenol hydroxylase from *Trichosporon cutaneum*: Gene cloning, sequence analysis, and functional expression in *Escherichia coli*. *J. Bacteriol.* **1992**, *174*, 7112–7120. [CrossRef]
78. Eppink, M.H.M.; Van Berkel, W.J.H.; Schreuder, H.A. Identification of a novel conserved sequence motif in flavoprotein hydroxylases with a putative dual function in FAD/NAD(P)H binding. *Protein Sci.* **2008**, *6*, 2454–2458. [CrossRef]
79. Silva, A.S.; Camargo, F.; Andreazza, R.; Jacques, R.J.S.; Baldoni, D.B.; Bento, F.M. Enzymatic activity of catechol 1,2-dioxygenase and catechol 2,3-dioxygenase produced by *Gordonia polyisoprenivorans*. *Química Nova* **2012**, *35*, 1587–1592. [CrossRef]
80. Hong, Y.-H.; Ye, C.-C.; Zhou, Q.-Z.; Wu, X.-Y.; Yuan, J.-P.; Peng, J.; Deng, H.; Wang, J.-H. Genome Sequencing Reveals the Potential of *Achromobacter* sp. HZ01 for Bioremediation. *Front. Microbiol.* **2017**, *8*, 1507. [CrossRef]
81. Li, F.; Song, W.; Wei, J.; Liu, C.; Yu, C. Comparative proteomic analysis of phenol degradation process by Arthrobacter. *Int. Biodeterior. Biodegrad.* **2016**, *110*, 189–198. [CrossRef]
82. Ahmad, S.A.; Shamaan, N.A.; Syed, M.A.; Khalid, A.; Ab Rahman, N.A.; Khalil, K.A.; Dahalan, F.A.; Shukor, M.Y. Meta-cleavage pathway of phenol degradation by *Acinetobacter* sp. strain AQ5NOL 1. *Rendiconti Lincei* **2016**, *28*, 1–9. [CrossRef]
83. Margesin, R.; Bergauer, P.; Gander, S. Degradation of phenol and toxicity of phenolic compounds: A comparison of cold-tolerant *Arthrobacter* sp. and mesophilic Pseudomonas putida. *Extremophiles* **2004**, *8*, 201–207. [CrossRef]
84. Tavakoli, A.; Hamzah, A. Characterization and evaluation of catechol oxygenases by twelve bacteria, isolated from oil contaminated soils in Malaysia. *Biol. J. Microorg.* **2017**, *5*, 71–80.
85. Margesin, R.; Gander, S.; Zacke, G.; Gounot, A.M.; Schinner, F. Hydrocarbon degradation and enzyme activities of cold-adapted bacteria and yeasts. *Extremophiles* **2003**, *7*, 451–458. [CrossRef] [PubMed]
86. Guzik, U.; Hupert-Kocurek, K.; Sitnik, M.; Wojcieszyńska, D. High activity catechol 1,2-dioxygenase from *Stenotrophomonas maltophilia* strain KB2 as a useful tool in cis,cis-muconic acid production. *Antonie Leeuwenhoek* **2013**, *103*, 1297–1307. [CrossRef]
87. Briganti, F.; Pessione, E.; Giunta, C.; Mazzoli, R.; Scozzafava, A. Purification and catalytic properties of two catechol 1,2-dioxygenase isozymes from benzoate-grown cells of *Acinetobacter radioresistens*. *J. Protein Chem.* **2000**, *19*, 709–716. [CrossRef]
88. Wang, Z.; Sun, Y.; Shi, Y.; Song, W.; Zhang, C. Cloning, Expression and Characterization of a Mesophilic Catechol 1,2-dioxygenase from *Rhodococcus ruber* OA1. *Biotechnol.* **2016**, *16*, 10–18. [CrossRef]
89. Lee, G.L.Y.; Ahmad, S.A.; Yasid, N.A.; Zulkharnain, A.; Convey, P.; Johari, W.L.W.; Alias, S.A.; Gonzalez-Rocha, G.; Shukor, M.Y. Biodegradation of phenol by cold-adapted bacteria from Antarctic soils. *Polar Biol.* **2017**, *41*, 553–562. [CrossRef]
90. Bankevich, A.; Nurk, S.; Antipov, D.; Gurevich, A.A.; Dvorkin, M.; Kulikov, A.S.; Lesin, V.M.; Nikolenko, S.I.; Pham, S.; Prjibelski, A.D.; et al. SPAdes: A new genome assembly algorithm and its applications to single-cell sequencing. *J. Comput. Biol.* **2012**, *19*, 455–477. [CrossRef]

91. Seemann, T. Prokka: Rapid Prokaryotic Genome Annotation. *Bioinformatics* **2014**, *30*, 2068–2069. [CrossRef]
92. Markowitz, V.M.; Mavrommatis, K.; Ivanova, N.; Chen, I.-M.A.; Chu, K.; Kyrpides, N. IMG ER: A system for microbial genome annotation expert review and curation. *Bioinformatics* **2009**, *25*, 2271–2278. [CrossRef]
93. Bedell, J.A.; Korf, I.; Gish, W. MaskerAid: A performance enhancement to RepeatMasker. *Bioinformatics* **2000**, *16*, 1040–1041. [CrossRef] [PubMed]
94. Benson, G. Tandem repeats finder: A program to analyze DNA sequences. *Nucleic Acids Res.* **1999**, *27*, 573–580. [CrossRef] [PubMed]
95. Lowe, T.M.; Eddy, S. tRNAscan-SE: A Program for Improved Detection of Transfer RNA Genes in Genomic Sequence. *Nucleic Acids Res.* **1997**, *25*, 955–964. [CrossRef] [PubMed]
96. Lagesen, K.; Hallin, P.; Rødland, E.A.; Staerfeldt, H.-H.; Rognes, T.; Ussery, D.W. RNAmmer: Consistent and rapid annotation of ribosomal RNA genes. *Nucleic Acids Res.* **2007**, *35*, 3100–3108. [CrossRef]
97. Griffiths-Jones, S.; Moxon, S.; Marshall, M.; Khanna, A.; Eddy, S.R.; Bateman, A. Rfam: Annotating non-coding RNAs in complete genomes. *Nucleic Acids Res.* **2004**, *33*, D121–D124. [CrossRef]
98. Overbeek, R.; Olson, R.; Pusch, G.D.; Olsen, G.J.; Davis, J.J.; Disz, T.; Edwards, R.A.; Gerdes, S.; Parrello, B.; Shukla, M.; et al. The SEED and the Rapid Annotation of microbial genomes using Subsystems Technology (RAST). *Nucleic Acids Res.* **2014**, *42*, D206–D214. [CrossRef]
99. Altschul, S.F.; Gish, W.; Miller, W.; Myers, E.W.; Lipman, D.J. Basic local alignment search tool. *J. Mol. Biol.* **1990**, *215*, 403–410. [CrossRef]
100. Apweiler, R.; Bairoch, A.; Wu, C.H.; Barker, W.C.; Boeckmann, B.; Ferro, S. UniProt: The universal protein knowledgebase. *Nucleic Acids Res.* **2004**, *32*, D115–D119. [CrossRef]
101. Kanehisa, M.; Goto, S. KEGG: Kyoto Encyclopedia of Genes and Genomes. *Nucleic Acids Res.* **2000**, *28*, 27–30. [CrossRef]
102. Tatusov, R.L.; Galperin, M.Y.; Natale, D.A.; Koonin, E.V. The COG database: A tool for genome-scale analysis of protein functions and evolution. *Nucleic Acids Res.* **2000**, *28*, 33–36. [CrossRef] [PubMed]

catalysts

MDPI

Article

Production of *Trans*-Cinnamic and *p*-Coumaric Acids in Engineered *E. coli*

Yuqi Liu [1], Weizhuo Xu [2,*] and Wei Xu [2,*]

[1] School of Life Sciences and Biopharmaceuticals, Shenyang Pharmaceutical University, 103 Wenhua Road, Shenhe District, Shenyang 110016, China
[2] School of Functional Food and Wine, Shenyang Pharmaceutical University, 103 Wenhua Road, Shenhe District, Shenyang 110016, China
* Correspondence: weizhuo.xu@syphu.edu.cn (W.X.); shxuwei8720@163.com (W.X.)

Abstract: *Trans*-cinnamic acid and *p*-coumaric acid are valuable intermediates in the synthesis of flavonoids and are widely employed in food, flavor and pharmaceutical industries. These products can be produced by the deamination of L-phenylalanine and L-tyrosine catalyzed by phenylalanine ammonia lyase or tyrosine ammonia lyase. Phenylalanine ammonia-lyase (PAL, EC 4.3.1.5) from *Rhodotorula glutinis* do not exhibit strong substrate specificity and can convert both L-phenylalanine and L-tyrosine. In this study, the PAL was utilized as the whole-cell biocatalyst, and the reaction conditions were optimized, and the production of *trans*-cinnamic acid and *p*-coumaric acid of 597 mg/L and 525 mg/L were achieved with high purity (>98%).

Keywords: *trans*-cinnamic acid; *p*-coumaric acid; phenylalanine ammonia-lyase; whole-cell biotransformation

Citation: Liu, Y.; Xu, W.; Xu, W. Production of *Trans*-Cinnamic and *p*-Coumaric Acids in Engineered *E. coli*. *Catalysts* **2022**, *12*, 1144. https://doi.org/10.3390/catal12101144

Academic Editors: Zhilong Wang and Tao Pan

Received: 3 September 2022
Accepted: 27 September 2022
Published: 30 September 2022

1. Introduction

Phenylalanine ammonia-lyase (PAL, EC 4.3.1.5) is a key enzyme in the phenylpropane pathway of higher plants and functioned in the biosynthesis of various secondary metabolites, such as lignans, flavonoids and coumarins [1]. The enzyme is widely found in plants [2–7], fungi [8–12] and prokaryotes [13]. In plants, PAL is widely present in monocots [4], dicots [5], ferns [6] and algae [7]; in fungi, it mainly exits in *Saccharomyces* [9,10], *Ascomycetes* [11] and *Basidiomycetes* [12]; and in prokaryotes, such as *Streptomycetes* [13]. Some species of enzymes, such as *Rhodotorula glutinis*, can not only catalyze the non-oxidative deamination of L-phenylalanine (L-Phe) to *trans*-cinnamic acid and ammonia but also convert L-tyrosine (L-Tyr) to *p*-coumaric acid (Figure 1). Therefore, PAL has become an important therapeutic enzyme being used for the treatment of phenylalanine- and tyrosine-related complications in recent years [14].

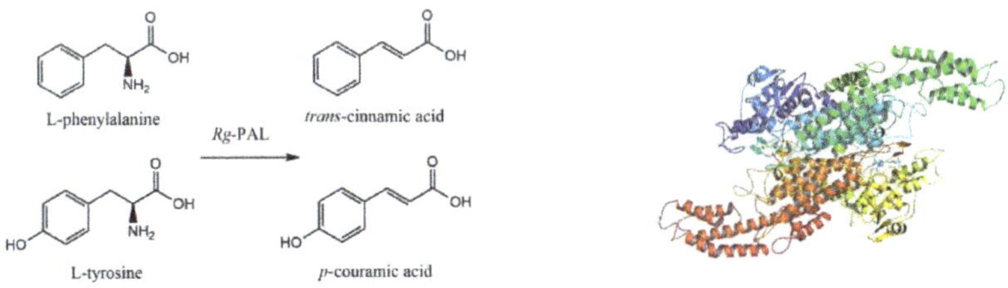

Figure 1. Reactions catalyzed by *Rg*-PAL.

PAL has a cofactor named 4-methylideneimidazole-5-one (MIO), which is a prosthetic group formed by the cyclization and elimination of the translated Ala-Ser-Gly tripeptide [14]. Two reaction mechanisms have been proposed for PAL to catalyze amino acid deamination. The first is the E1cb mechanism based on the electrophilic attack of the MIO moiety on the amino group of the substrate [15]. The second is the Friedel-Crafts's mechanism with electrophilic attack of MIO on the aromatic ring to eliminate α-NH_3 and β-H from the substrate (Figure 2) [16].

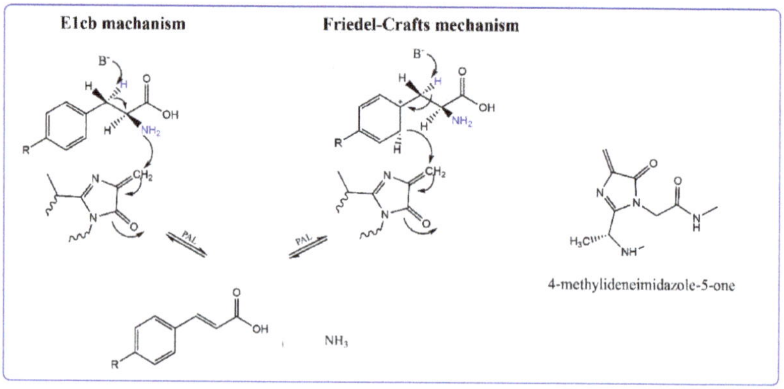

Figure 2. Reaction mechanism of PAL and the chemical structure of MIO cofactor.

Trans-cinnamic acid, the first product in the phenylpropane pathway, plays an important role in plant growth and development. *Trans*-cinnamic acid and its derivatives have been shown not only to be antioxidants but also to treat diseases, such as atherosclerosis and inflammatory damage [17]. Additionally, *trans*-cinnamic acid has been found to inhibit the formation of late glycosylation and products [18–20]. As an essential intermediate in the flavonoids synthesis pathway, *p*-coumaric acid exhibited antibacterial, anti-inflammatory effects and may contribute to the cardiovascular disease prevention [21,22]. Both *trans*-cinnamic acid and *p*-coumaric acid can be obtained via plant extraction, chemical synthesis and biosynthesis. The biosynthetic method has the advantages of low cost, high yield, sustainability and environmental friendliness. The promising microbial synthesis method was chosen to produce *trans*-cinnamic acid and *p*-coumaric acid in this work.

In previous work, Liang and co-workers also utilized *E. coli* BL21(DE3)/pMD18-*RgPAL* as a whole-cell catalyst to convert L-Phe and L-Tyr and investigated condition optimization obtaining 78.81 mg/L of *trans*-cinnamic acid and 34.67 mg/L of *p*-coumaric acid [23]. The recombinant *Zea mays* phenylalanine ammonia-lyase harboring *E. coli* BL21(DE3) was employed as whole-cell biocatalyst to transform L-Phe to *trans*-cinnamic acid in Zang' study, 5 g/L *trans*-cinnamic acid could be obtained from 10 g/L L-Phe under optimized conditions [24]. In Xue's experiment, Strain *DPD5124* expressing PAL/TAL from *Phanerochaete chrysosporium* and *DPD5154* expressing PAL/TAL from *Rhodotorula glutinis* were used as a whole-cell catalyst for the bioconversion of L-Tyr to *p*-coumaric acid, and the product yields of *p*-coumaric acid were achieved to 0.44 g/g dcw/L and 1.14 g/g dcw/L from 50 g/L L-tyrosine after optimizing reaction conditions [25]. The low solubility of tyrosine is an important reason for the low yield of *p*-coumaric acid. Based on our previous works, the catalytic activities of three phenylalanine ammonia-lyase from *Zea mays*, *Rhodotorula glutinis* and *Petroselinum crispum* were compared to transform L-Phe and L-Tyr. The results showed that the activity of converting L-Phe was *Pc*-PAL > *Rg*-PAL > *Zm*-PAL. Among these three enzymes, an excellent conversion of L-Tyr to *p*-coumaric acid was achieved by *Rg*-PAL, while the conversion of L-Tyr was extremely low or even absent with *Zm*-PAL and *Pc*-PAL as biocatalysts. Herein, Glycine-NaOH was chosen as reaction medium and *Rg*-PAL harboring *E. coli* BL21(DE3) as whole-cell biocatalyst for the conversions of L-Phe and L-Tyr, while other reaction conditions were optimized.

2. Results

2.1. Optimization of Temperature

The reaction temperature is an important parameter affecting the catalytic reaction; either higher or lower temperature will affect the enzymatic efficiency, and appropriately increasing the temperature can reduce the viscosity of the mixture and enhance opportunities for the enzyme to collide with the substrate [26,27]. *Rg*-PAL was reacted with L-Phe and L-Tyr at different temperatures, ranging from 20 to 50 °C. The results are shown in the Figure 3. Although the enzyme was still highly active at 50 °C, the reaction temperature of 42 °C was finally chosen considering the stability and reusability of whole cells.

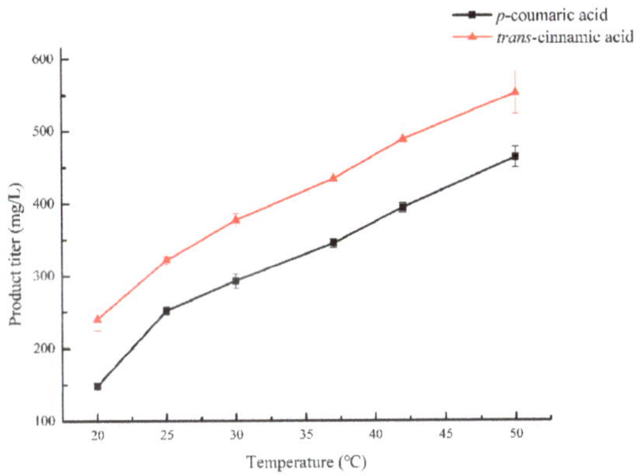

Figure 3. Effect of temperature on conversion of L-Phe and L-Tyr to *trans*-cinnamic acid and *p*-coumaric acid by *E. coli* BL21(DE3)/pETDuet-1-*RgPAL*. Reaction conditions: Time: 24 h; Medium: pH 9.0 10 mM Glycine-NaOH; L-Phe and L-Tyr 1 mg/mL; Cell amounts: 6.67 g/L.

2.2. Optimization of Time

The duration of whole-cell catalysis has an essential effect on the reaction. If the reaction time is too short, the reaction will be incomplete. With too long a reaction time, product inhibition may occur and affect products' yields. The experiment investigated the effect of conversion time on the synthesis of products and the results were shown in Figure 4. When L-Phe was the substrate, the peak yield of *trans*-cinnamic acid was 553 mg/L for 20 h; when L-Tyr was the substrate, the yield of *p*-coumaric acid was up to 393 mg/L for 24 h.

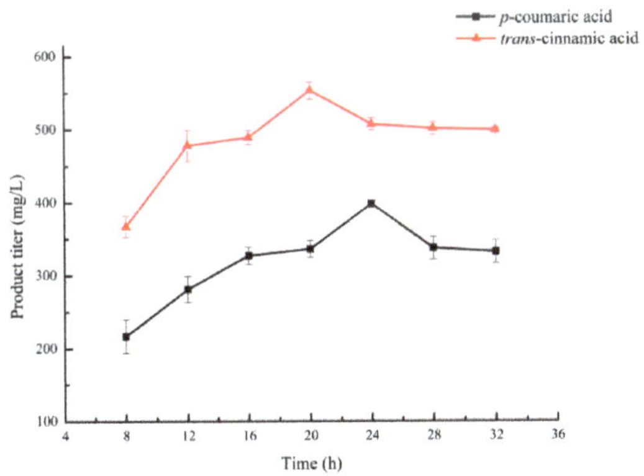

Figure 4. Effect of time on conversion of L-Phe and L-Tyr to *trans*-cinnamic acid and *p*-coumaric acid by *E. coli* BL21(DE3)/pETDuet-1-*RgPAL*. Reaction conditions: Temperature: 42 °C; Medium: pH 9.0 10 mM Glycine-NaOH; L-Phe and L-Tyr 1 mg/mL; Cell amounts: 6.67 g/L.

2.3. Optimization of Buffer Concentration

The amino acid deamination reaction occurs under alkaline conditions, and Glycine-NaOH was chosen as the reaction media in this study. Glycine, a non-polar amino acid, has both acidic and basic groups, is ionizable in water, and is highly hydrophilic. The polarity of the solution was changed by adjusting the addition of glycine in order to provide the ideal reaction environment for the substrate and enzyme. In this study, the impact of different glycine additions, ranging from 0 to 50 mM on the whole-cell catalytic reaction was evaluated at pH 9.0. Results were presented in Figure 5, the peak productions of *trans*-cinnamic acid and *p*-coumaric acid were 538 mg/L and 419 mg/L, respectively, at pH 9.0 20 mM Glycine-NaOH, respectively.

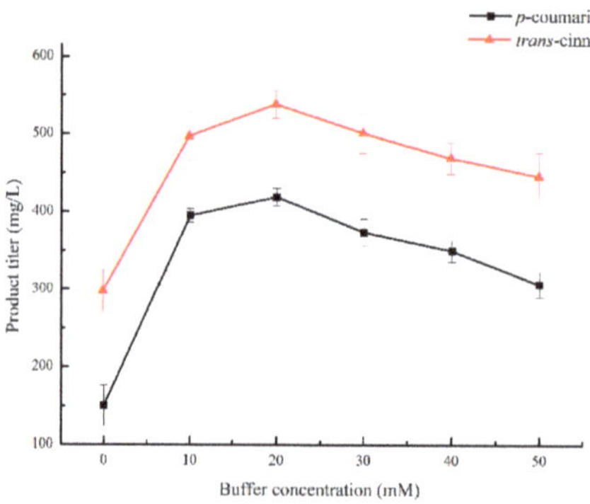

Figure 5. Effect of buffer concentration on conversion of L-Phe and L-Tyr to *trans*-cinnamic acid and *p*-coumaric acid by *E. coli* BL21(DE3)/pETDuet-1-*RgPAL*. Reaction conditions: Temperature: 42 °C; Medium: pH 9.0 Glycine-NaOH; L-Phe and L-Tyr 1 mg/mL; L-Phe conversion time: 24 h and L-Tyr conversion time: 20 h; Cell amounts: 6.67 g/L.

2.4. Optimization of pH

Changes in pH of the reaction medium can affect the charge density and molecular structure of the cell surface of enzyme molecules, resulting in changes in the rate and exit of substances into and out of the cell and in the catalytic efficiency of the enzyme. Therefore, it is important to regulate the pH of reaction medium. In this study, the effect of different pH (8.0–12.0) on the catalytic reaction of whole-cell was investigated. The results were shown in Figure 6. The maximum yield of *trans*-cinnamic acid was 552 mg/L at pH 10.0 and that of *p*-coumaric acid reached 455 mg/L at pH 11.0.

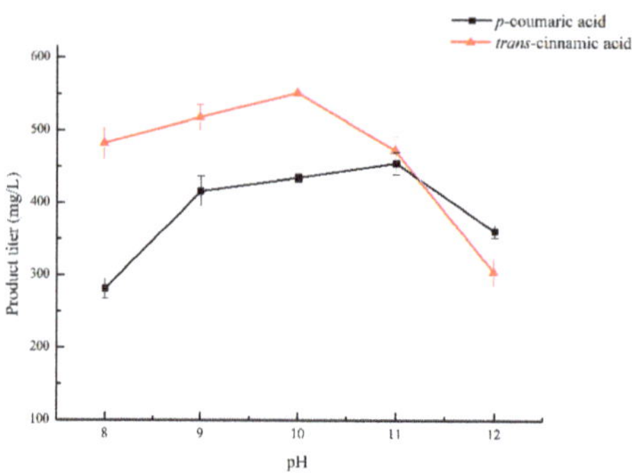

Figure 6. Effect of pH on conversion of L-Phe and L-Tyr to *trans*-cinnamic acid and *p*-coumaric acid by *E. coli* BL21(DE3)/pETDuet-1-*RgPAL*. Reaction conditions: Temperature: 42 °C; Medium: 20 mM Glycine-NaOH; L-Phe and L-Tyr 1 mg/mL; L-Phe conversion time: 24 h and L-Tyr conversion time: 20 h; Cell amounts: 6.67 g/L.

2.5. Optimization of Cell Amount

As a biocatalyst, the rate of reaction catalyzed by free cells directly depends on the concentration of cells, and, as shown from Figure 7, the rate of product generation catalyzed by cells increases linearly when the concentration of cells is relatively low. The production of *trans*-cinnamic acid and *p*-coumaric acid reached the maximum value of 595 mg/L and 525 mg/L when the amount of the cell increased to 10 g/L. The reaction rate decreased when the cell amount was higher than 10 g/L. The reason was that the cell concentration was too high and the cells clustered with each other, which reduced the opportunity for the active center of the enzyme in the cells to contact with the substrate, leading to a decrease in the catalytic reaction rate.

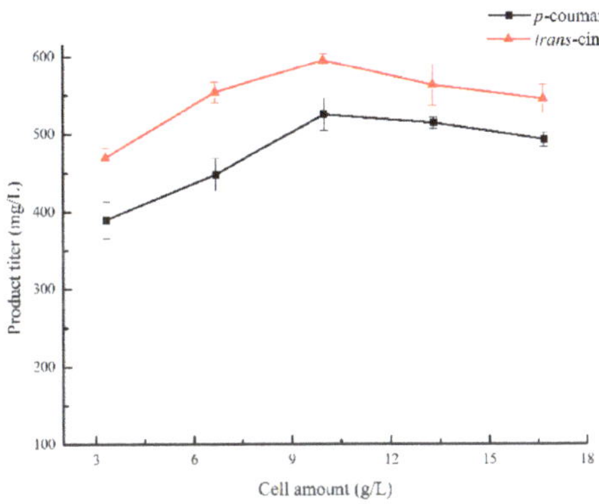

Figure 7. Effect of cell amounts on conversion of L-Phe and L-Tyr to *trans*-cinnamic acid and *p*-coumaric acid by *E. coli* BL21(DE3)/pETDuet-1-*RgPAL*. Reaction conditions: Temperature: 42 °C; L-Phe conversion time: 24 h and L-Tyr conversion time: 20 h Medium: L-Phe in pH 10 20 mM Glycine-NaOH and L-Tyr in pH 11 20 mM Glycine-NaOH; L-Phe and L-Tyr 1 mg/mL.

2.6. Reusability

In the batch reactions, the cells were collected by centrifugation, washed with pH 7.4 10 mM PBS buffer three times and reused for the next batch after the previous reaction ended. As shown in Figure 8, *E. coli* BL21(DE3)/pETDuet-1-*RgPAL* can be used for at least five cycles. With the increasing use of whole cells, the yield of trans-cinnamic acid decreased from 597 mg/L to 294 mg/L and the production of *p*-courmic acid reduced from 525 mg/L to 172 mg/L. Therefore, *E. coli* BL21(DE3)/pETDuet-1-*RgPAL* is a promising strain for production of *trans*-cinnamic acid and *p*-coumaric acid.

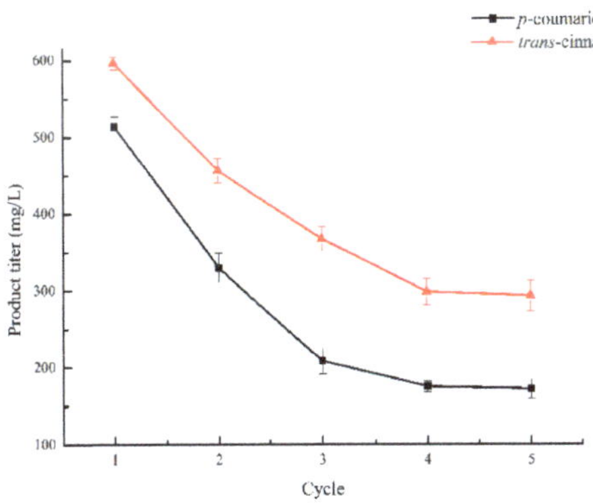

Figure 8. Batch reactions with repeatedly used *E. coli* BL21(DE3)/pETDuet-1-RgPAL cells at cell amounts of 10 g/L.

3. Discussion

Due to low solubility of L-tyrosine, Glycine-NaOH was chosen as the reaction medium to provide a more favorable reaction environment for the formation of *p*-coumaric acid by altering the glycine concentration. After optimizing reaction conditions, the product yields of *trans*-cinnamic acid and *p*-coumaric acid were 597 mg/L and 525 mg/L using free whole-cells as catalysts. Additionally, the catalytic activity of whole-cells remained well after five cycles under optimal reaction conditions. In further studies, the immobilized whole cells or enzymes as a catalyst will be utilized to produce *trans*-cinnamic acid and *p*-coumaric acid. Based on the optimized reaction conditions, we aim to further increase the enzyme activity through molecular modifications to achieve higher yields of products. Two reaction mechanisms of PAL have been reported so far. In the molecular docking results, some results were found to be more consistent with the Friedel–Crafts's mechanism, while others were more compliant with the E1cb mechanism (see in Supplementary Materials). According to this feature, targeted mutagenesis can be performed to improve the enzyme activity. In addition, the aim of improving their catalytic activity and reusability, as well as reducing the production cost, will be achieved.

4. Materials and Methods

4.1. Bacterial Strains and Plasmids

All bacterial strains and plasmids used in this study were listed in Table 1. *E. coli* DH5α was used to propagate plasmid. *E. coli* BL21(DE3) was used for functional expression of constructed plasmid. The *E. coli* BL21(DE3) harboring pETDuet-1-*RgPAL* was constructed by our laboratory.

Table 1. Strains and plasmids used in this study.

Names	Characteristics	Source
Plasmid		
pETDuet-1	pBR322ori with P_{T7}; Amp^R	Novagen
pETDuet-1-*Rg*-PAL	pETDuet-1 with *Rg*-PAL	This study
Strain		
E. coli DH5α	ΔLac U169 (Φ80 Lac Z ΔM15)	Invitrogen
E. coli BL21(DE3)	F-ompT hsdS (rB-mB-) gal dcm (DE3)	Invitrogen

4.2. Media and Cultivations

The engineered strains were precultured overnight at 37 °C in 3 mL LB liquid medium with ampicillin (100 mg/L), and then the preculture was inoculated into 30 mL LB liquid medium supplemented with ampicillin (100 mg/L) in 150 mL shake flasks at 1% of inoculum and grown at 37 °C again shaking at 180 rpm. When the culture reached an optical density of 0.6–0.8 at 600 nm, IPTG was added at a final concentration of 0.2 mM to induce gene expression and the cultivation was continued for additional 10 h at 37 °C and 180 rpm.

4.3. Whole-Cell Biotransformation of L-Phe to Trans-Cinnamic Acid and L-Tyr to p-Coumaric Acid by E. coli BL21(DE3)/pETDuet-1-RgPAL

E. coli BL21(DE3)/pETDuet-1-RgPAL was grown and induced with a final concentration of 0.2 mM IPTG. After induction at 37 °C for 10 h, the cells were collected by centrifugation and washed with 10 mM PBS buffer (8.0 g NaCl, 0.2 g KCl, 1.44 g Na_2HPO_4 and 0.24 g KH_2PO_4 dissolved in 1 L distilled water, pH 7.4) three times. Whole-cell biotransformation was performed in 150 mL Erlenmeyer flasks with 30 mL of the reaction mixtures containing Glycine-NaOH buffer, recovered cells and 1 mg/mL substrates under different conditions. In order to obtain higher yields of *trans*-cinnamic acid and *p*-coumaric acid, the reaction conditions were optimized. All experiments were carried out in triplicate and the mean values were calculated. The standard deviation for each test was calculated with excel and indicated as error bars.

4.4. Analytical Methods

Cell densities of cultures were determined by measuring their absorbance at 600 nm with a 2800 UV/visible spectrophotometer [UNICO (Shanghai) INSTRUMENT]. For analysis of *trans*-cinnamic acid and *p*-coumaric acid, the reaction mixtures were centrifuged at 3500× *g* for 10 min, and supernatants were collected to analyze the product concentration by HPLC using a Wondasil C18 Superb column (250 mm × 4.6 mm, 5 μm) with the column temperature of 30 °C. To detect the formation of *trans*-cinnamic acid, we used a solution (50% of 1% acetic acid and 50% of acetonitrile) as mobile phase at flow rate of 0.7 mL/min. The detection wavelength was set at 290 nm. To analyze the production of *p*-coumaric acid, the HPLC was eluted with the solution (60% of 0.1% formic acid and 40% of methyl alcohol) as mobile phase at flow rate of 0.6 mL/min. The detection wavelength was set at 310 nm. Meanwhile, the supernatants after centrifugation were adjusted to pH 4.2 by HCl and allowed to stand at room temperature for 30 min, then the acidified supernatants were centrifuged again to separate the *trans*-cinnamic acid precipitate. The supernatants after centrifugation were adjusted to pH 4.6 by H_3PO_4 to obtain *p*-coumaric acid and other steps were same as the isolation of *trans*-cinnamic acid.

5. Conclusions

Rg-PAL-harboring *E. coli* BL21(DE3) as the whole-cell biocatalyst, the productions of *trans*-cinnamic acid and *p*-coumaric acid reached to 597 mg/L and 525 mg/L, respectively, via optimizing reaction conditions. After the cells were repeatedly tested five times, the enzyme still had good catalytic activity. This engineered strain is a potential strain to produce *trans*-cinnamic acid and *p*-coumaric acid for industrial applications.

Supplementary Materials: The following are available online at https://www.mdpi.com/article/10.3390/catal12101144/s1, Figure S1: The molecular docking results of PAL and L-Phe, Figure S2: Solubilities of L-Phe and L-Tyr in NaOH, Gly-NaOH and Na_2CO_3.$NaHCO_3$ at pH 10, Figure S3: Results of two rounds transformation of trans-cinnamic acids and p-coumaric acid, Figure S4: The standard curves for quantification of the product concentration, Figure S5: The HPLC graph of trans-cinnamic acid and p-coumaric acid.

Author Contributions: Y.L. (Data Curation, Investigation, Methodology, Writing—Original Draft); W.X. (Weizhuo Xu) and W.X. (Wei Xu) (Resources, Supervision, Writing—Review and Editing). All authors have read and agreed to the published version of the manuscript.

Funding: This research received no external funding.

Data Availability Statement: Data are available upon reasonable request.

Conflicts of Interest: The authors declare no conflict of interest.

References

1. Hyun, M.W.; Yun, Y.H.; Kim, J.Y.; Kim, S.H. Fungal and Plant Phenylalanine Ammonia-Lyase. *Mycobiology* **2011**, *39*, 257–265. [CrossRef] [PubMed]
2. Whetten, R.W.; Sederoff, R.R. Phenylalanine Ammonia-Lyase from Loblolly Pine: Purification of the Enzyme and Isolation of Complementary DNA Clones. *Plant Physiol.* **1992**, *98*, 380–386. [CrossRef] [PubMed]
3. Koukol, J.; Conn, E.E. The Metabolism of Aromatic Compounds in Higher Plants. *J. Biol. Chem.* **1961**, *236*, 2692–2698. [CrossRef]
4. Wang, X.H.; Gong, M.; Tang, L.; Zheng, S.; Lou, J.D.; Ou, L.; Gomes-Laranjo, J.; Zhang, C. Cloning, Bioinformatics and the Enzyme Activity Analyses of a Phenylalanine Ammonia-Lyase Gene Involved in Dragon's Blood Biosynthesis in *Dracaena cambodiana*. *Mol. Biol. Rep.* **2013**, *40*, 97–107. [CrossRef] [PubMed]
5. Hou, X.; Shao, F.; Ma, Y.; Lu, S. The Phenylalanine Ammonia-Lyase Gene Family in *Salvia miltiorrhiza*: Genome-Wide Characterization, Molecular Cloning and Expression Analysis. *Mol. Biol. Rep.* **2013**, *40*, 4301–4310. [CrossRef] [PubMed]
6. Young, M.R.; Towers, G.H.N.; Neish, A.C. Taxonomic Distribution of Ammonia-Lyases for L-Phenylalanine and L-Tyrosine in Relation To Lignification. *Can. J. Bot.* **1966**, *44*, 341–349. [CrossRef]
7. Czichi, U.; Kindl, H. Formation of P-Coumaric Acid and o-Coumaric Acid from l-Phenylalanine by Microsomal Membrane Fractions from Potato: Evidence of Membrane-Bound Enzyme Complexes. *Planta* **1975**, *125*, 115–125. [CrossRef]
8. Sikora, L.A.; Marzluf, G.A. Regulation of L-Phenylalanine Ammonia-Lyase by L-Phenylalanine and Nitrogen in *Neurospora crassa*. *J. Bacteriol.* **1982**, *150*, 1287–1291. [CrossRef] [PubMed]

9. Marusich, W.C.; Jensen, R.A.; Zamir, L.O. Induction of L-Phenylalanine Ammonia-Lyase during Utilization of Phenylalanine as a Carbon or Nitrogen Source in *Rhodotorula glutinis*. *J. Bacteriol.* **1981**, *146*, 1013–1019. [CrossRef]
10. Orndorff, S.A.; Costantino, N.; Stewart, D.; Durham, D.R. Strain Improvement of *Rhodotorula graminis* for Production of a Novel l-Phenylalanine Ammonia-Lyase. *Appl. Environ. Microbiol.* **1988**, *54*, 996–1002. [CrossRef]
11. Vance, C.P.; Bandoni, R.; Towers, G.H.N. Further Observations on Phenylalanine in Fungi. *Phytochemistry* **1975**, *14*, 1513–1514. [CrossRef]
12. Bandoni, R.J.; Moore, K.; Subba Rao, P.V.; Towers, G.H.N. Phenylalanine and Tyrosine Ammonia-Lyase Activity in Some Basidiomycetes. *Phytochemistry* **1968**, *7*, 205–207. [CrossRef]
13. Xiang, L.; Moore, B.S. Inactivation, Complementation, and Heterologous Expression of EncP, a Novel Bacterial Phenylalanine Ammonia-Lyase Gene. *J. Biol. Chem.* **2002**, *277*, 32505–32509. [CrossRef]
14. Kawatra, A.; Dhankhar, R.; Mohanty, A.; Gulati, P. Biomedical Applications of Microbial Phenylalanine Ammonia Lyase: Current Status and Future Prospects. *Biochimie* **2020**, *177*, 142–152. [CrossRef]
15. Alunni, S.; Cipiciani, A.; Fioroni, G.; Ottavi, L. Mechanisms of Inhibition of Phenylalanine Ammonia-Lyase by Phenol Inhibitors and Phenol/Glycine Synergistic Inhibitors. *Arch. Biochem. Biophys.* **2003**, *412*, 170–175. [CrossRef]
16. Givot, I.L.; Smith, T.A.; Abeles, R.H. Studies on the Mechanism of Action and the Structure of the Electrophilic Center of Histidine Ammonia Lyase. *J. Biol. Chem.* **1969**, *244*, 6341–6353. [CrossRef]
17. Poppe, L.; Rétey, J. Friedel-Crafts-Type Mechanism for the Enzymatic Elimination of Ammonia from Histidine and Phenylalanine. *Angew. Chem.-Int. Ed.* **2005**, *44*, 3668–3688. [CrossRef]
18. Darvesh, A.S.; Carroll, R.T.; Bishayee, A.; Geldenhuys, W.J.; Van Der Schyf, C.J. Oxidative Stress and Alzheimer's Disease: Dietary Polyphenols as Potential Therapeutic Agents. *Expert Rev. Neurother.* **2010**, *10*, 729–745. [CrossRef]
19. Khazaei, M.R.; Bakhti, M.; Habibi-Rezaei, M. Nicotine Reduces the Cytotoxic Effect of Glycated Proteins on Microglial Cells. *Neurochem. Res.* **2010**, *35*, 548–558. [CrossRef]
20. Schmidt, A.M.; Stern, D. Atherosclerosis and Diabetes: The RAGE Connection. *Curr. Atheroscler. Rep.* **2000**, *2*, 430–436. [CrossRef]
21. Adisakwattana, S.; Sompong, W.; Meeprom, A.; Ngamukote, S.; Yibchok-Anun, S. Cinnamic Acid and Its Derivatives Inhibit Fructose-Mediated Protein Glycation. *Int. J. Mol. Sci.* **2012**, *13*, 1778–1789. [CrossRef] [PubMed]
22. Koopman, F.; Beekwilder, J.; Crimi, B.; van Houwelingen, A.; Hall, R.D.; Bosch, D.; van Maris, A.J.A.; Pronk, J.T.; Daran, J.M. De Novo Production of the Flavonoid Naringenin in Engineered *Saccharomyces cerevisiae*. *Microb. Cell Fact.* **2012**, *11*, 155. [CrossRef] [PubMed]
23. Abdel-Wahab, M.H.; El-Mahdy, M.A.; Abd-Ellah, M.F.; Helal, G.K.; Khalifa, F.; Hamada, F.M.A. Influence of P-Coumaric Acid on Doxorubicin-Induced Oxidative Stress in Rat's Heart. *Pharmacol. Res.* **2003**, *48*, 461–465. [CrossRef]
24. Liang, J.L.; Guo, L.; Sun, P.; Jiang, B.; Lin, J.; Guo, W.; Wan, H. A Novel Process for Obtaining Phenylpropanoic Acid Precursor Using *Escherichia coli* with a Constitutive Expression System. *Food Sci. Biotechnol.* **2016**, *25*, 795–801. [CrossRef] [PubMed]
25. Zang, Y.; Jiang, T.; Cong, Y.; Zheng, Z.; Ouyang, J. Molecular Characterization of a Recombinant *Zea mays* Phenylalanine Ammonia-Lyase (ZmPAL2) and Its Application in Trans-Cinnamic Acid Production from l-Phenylalanine. *Appl. Biochem. Biotechnol.* **2015**, *176*, 924–937. [CrossRef]
26. Xue, Z.; McCluskey, M.; Cantera, K.; Ben-Bassat, A.; Sariaslani, F.S.; Huang, L. Improved Production of P-Hydroxycinnamic Acid from Tyrosine Using a Novel Thermostable Phenylalanine/Tyrosine Ammonia Lyase Enzyme. *Enzyme Microb. Technol.* **2007**, *42*, 58–64. [CrossRef]
27. Abdullah, A.Z.; Sulaiman, N.S.; Kamaruddin, A.H. Biocatalytic Esterification of Citronellol with Lauric Acid by Immobilized Lipase on Aminopropyl-Grafted Mesoporous SBA-15. *Biochem. Eng. J.* **2009**, *44*, 263–270. [CrossRef]

catalysts

Article

Enhancing Acetophenone Tolerance of Anti-Prelog Short-Chain Dehydrogenase/Reductase EbSDR8 Using a Whole-Cell Catalyst by Directed Evolution

Hui Zhang [1,*,†], Bei Wang [2,†], Shengli Yang [2,*], Hongwei Yu [3,*] and Lidan Ye [3,*]

1 School of Biological and Chemical Engineering, Zhejiang University of Science and Technology, Hangzhou 310023, China
2 College of Pharmaceutical Science, Zhejiang University of Technology, Hangzhou 310014, China
3 Institute of Bioengineering, College of Chemical and Biological Engineering, Zhejiang University, Hangzhou 310027, China
* Correspondence: zhanghui0461@zust.edu.cn (H.Z.); yangshengli01@zjut.edu.cn (S.Y.); yuhongwei@zju.edu.cn (H.Y.); yelidan@zju.edu.cn (L.Y.); Tel.: +86-571-135-8876-4575 (H.Z.); +86-571-8832-0913 (S.Y.); +86-571-8827-3997 (H.Y.); +86-571-8795-1873 (L.Y.); Fax: +86-571-8832-0913 (S.Y.)
† These authors contributed equally to this work.

Citation: Zhang, H.; Wang, B.; Yang, S.; Yu, H.; Ye, L. Enhancing Acetophenone Tolerance of Anti-Prelog Short-Chain Dehydrogenase/Reductase EbSDR8 Using a Whole-Cell Catalyst by Directed Evolution. *Catalysts* **2022**, *12*, 1071. https://doi.org/10.3390/catal12091071

Academic Editors: Zhilong Wang and Tao Pan

Received: 17 August 2022
Accepted: 16 September 2022
Published: 19 September 2022

Abstract: The short-chain dehydrogenase/reductase (SDR) from *Empedobacter brevis* ZJUY-1401 (EbSDR8, GenBank: ALZ42979.1) is a promising biocatalyst for the reduction of acetophenone to (*R*)-1-phenylethanol, but its industrial application is restricted by its insufficient tolerance to acetophenone. In this paper, we developed a chromogenic reaction-based high-throughput screening method and employed directed evolution to enhance the acetophenone tolerance of EbSDR8. The resulting variant, M190V, showed 74.8% improvement over the wild-type in specific activity when catalyzing the reduction of 200 mM acetophenone. Kinetic analysis revealed a 70% enhancement in its catalytic efficiency (k_{cat}/K_m). Molecular docking was conducted to reveal the possible mechanism behind the improved acetophenone tolerance, and the result implied that the M190V mutation is conducive to the binding and release of coenzyme. Aside from the improved catalytic performance when dealing with a high concentration of acetophenone, other features of M190V, such as a broad pH range (6.0 to 10.5), low optimal cosubstrate concentration (1% isopropanol), and a temperature optimum close to that of *E. coli* cells (35 °C), also contribute to its practical application as a whole-cell catalyst. In this study, we first designed a directed evolution means to engineer the enzyme and obtained the positive variant which has a high activity under high concentrations of acetophenone. After that, we optimized the catalytic performance of the variant to adapt to industrial applications.

Keywords: acetophenone tolerance; whole-cell catalyst; directed evolution; (*R*)-1-phenylethanol; short-chain dehydrogenase/reductase

1. Introduction

The growing demand for chiral pharmaceuticals, flavors, agrochemicals, and functional materials in the past few decades have led to increasing interest in the production of optically pure enantiomers [1–3]. One of the most popular building blocks is the enantiomerically pure chiral alcohols, which are mainly prepared by chemical transformation or biotransformation at present [4]. Biological transformation is a sustainable and 'green' technique featured by remarkable stereoselectivity, mild reaction conditions, and environmental friendliness [5–7]. Furthermore, the theoretical yield for biocatalytic asymmetric reduction of prochiral ketones reaches 100% [8–10]. By increasing the size of the substrate-binding pocket and eliminating steric repulsion, the medium-chain dehydrogenase (MDR) from *Candida parapsilosis* was engineered to accept more space-demanding substrates with improved catalytic efficiency (Wang et al., 2014). The carbonyl reductase YueD derived from *Bacillus subtilis* was transformed by molecular docking and alanine screening technology, a mutant V181A was screened out, and the conversion rate of m-bromoacetophenone was increased from 61% of the wild-type to 97% (Naeem, M et al., 2018). By providing a suitable substrate binding pocket for the enzyme starting from the configuration of the target product, efficient asymmetric reduction mutants of o-halogenated acetophenones, phenylacetone, aromatic ketone ester, and diarylketone were designed with >99% conversion and >98% ee (Su et al., 2019).

Recently, short-chain dehydrogenases/reductases(SDRs) have attracted increasing attention ascribed to their broad substrate spectrum, considerable thermostability, and tolerance against organic solvents when applied in the asymmetric reduction of carbonyl compounds [11–14]. Among these enzymes, EbSDR8 has been reported as a powerful biocatalyst for the stereoselective production of anti-Prelog alcohols [1]. The EbSDR8$^{G94A/L153I/Y188A/Y202M}$ was applied in the asymmetric bioreduction of 3-chloro-1-phenyl-1-propanoneto generate the important chiral drug intermediate (*R*)-3-Chloro-1-phenyl-1-propanol, which is widely used in the synthesis of antidepressant drugs and premature ejaculation drugs [15]. However, it shows limited tolerance to a high concentration of acetophenone. Therefore, improving the substrate tolerance is of significant importance for the industrialization of EbSDR8. Directed evolution has been extensively used as a potent tool for the modification of complex properties such as thermostability [16], organic solvent (toluene) tolerance [17], product (ethanol) tolerance [18], and substrate (acetate) tolerance [19]. However, a suitable and efficient high-throughput screening (HTS) method is a necessity.

In the present study, to facilitate the directed evolution of EbSDR8 for improved acetophenone tolerance, we developed an HTS method by adopting the chromogenic reaction depicted in Figure 1a. Acetophenone is catalyzed by EbSDR8 to form (*R*)-1-phenylethanol, accompanied by oxidation of NADH to NAD^+, and the hydrogen of NADH is meanwhile transferred by phenazine methosulfate (PMS) to thiazolyl blue tetrazolium bromide (MTT), generating the blue-violet formazan. The UV absorption of formazan is recorded at 575 nm, which can then be easily converted to NADH concentration, and finally, the concentration of (*R*)-1-phenylethanol formed can be calculated based on the NADH consumed [20]. Based on this method, the error-prone PCR library of EbSDR8 was transformed into *E. coli* BL21 and screened for positive mutants with improved performance at high acetophenone concentration. The underlying mechanism responsible for the enhanced acetophenone tolerance was then analyzed by molecular docking. Finally, the impacts of cosubstrate, pH, and temperature on the whole cell-catalyzed bioreduction of acetophenone using the mutant were investigated.

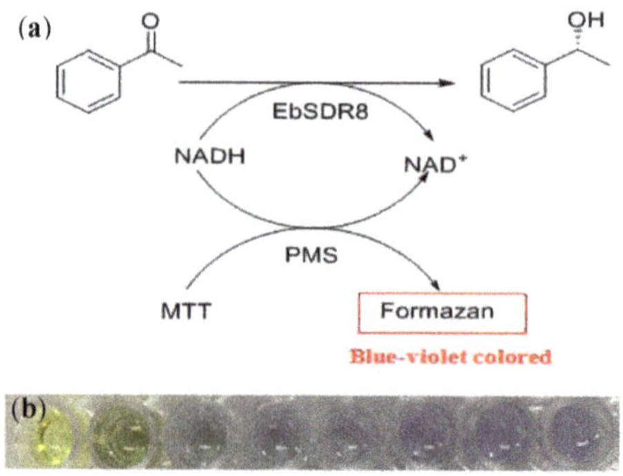

Figure 1. *Cont.*

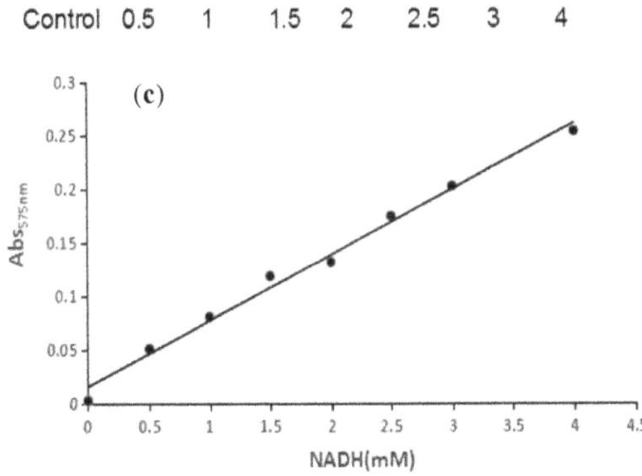

Figure 1. Development of the HTS method. (**a**). Principle of the HTS method. MTT, thiazolyl blue tetrazolium bromide; PMS, phenazine methosulfate. (**b**). The color variance of reaction mixture containing different concentrations of NADH. (**c**). The linear range of the measurement.

2. Results and Discussion

2.1. High-Throughput Screening Method and Activity Assay of the Mutants

To verify the linearity between the NADH concentration and the color variance, so as to validate the HTS method developed, NADH (0–4 mM) with different concentrations was added to the reaction system (Figure 1b). The results showed that the color turned blue-violet with increasing NADH amount, and the change of color was visible to the naked eye. Furthermore, the blue-violet formazan was quantified by recording the absorbance at 575 nm (Abs_{575}) [20] of the reaction systems in Figure 1b with a microplate reader. The linear range of the measurement outcome in accordance with NADH is shown in Figure 1c ($y = 0.0615x + 0.0148$, $R^2 = 0.99$). Therefore, if the value of Abs_{575} is within the linear range, we can compute the NADH amount using this standard curve. Since the bioreduction of acetophenone consumes NADH, a lower NADH amount indicates higher catalytic activity.

Based on the HTS method, directed evolution of EbSDR8 was conducted, and screening of about 10,000 mutants resulted in a positive mutant, M190V. Subsequent saturation mutagenesis with M190V as a template generated two mutants, M190F and M190W, with higher specific activity than the wild-type but not as high as that of M190V (Figure 2). Therefore, M190V was chosen for further study. As shown in Figure 2, the mutant M190V demonstrated 74.8% activity improvement in the reduction of 200 mM acetophenone as compared with the wild-type. This is consistent with the kinetic results in Table 1. The K_m value of the mutant M190V was 8% lower than that of the wild-type (4.57 vs. 4.95 mM), whereas the catalytic efficiency (k_{cat}/K_m) of M190V was 70% higher than that of the wild-type (870 vs. 511 s^{-1} M^{-1}). Therefore, the improved specific activity of the mutant can be commendably explained by these kinetics.

Table 1. Apparent kinetic parameters of the wild-type and the mutant M190V.

Mutants	*K*m (mM)	*K*cat (s^{-1})	*K*cat/*K*m ($S^{-1}M^{-1}$)
Wild-type	4.95	2.53	511
M190V	4.57	3.98	870

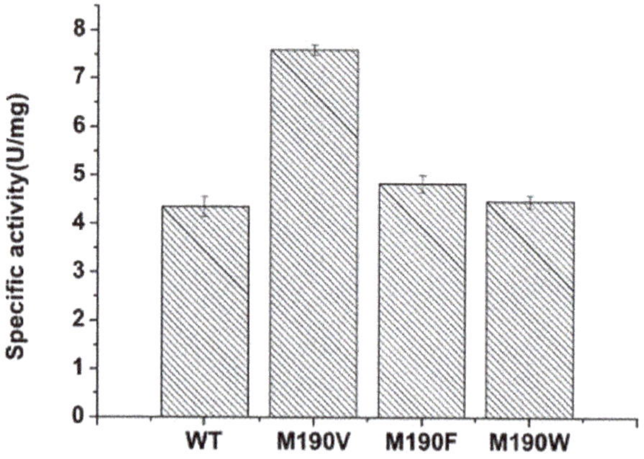

Figure 2. The specific activity of the wild-type and the mutants of M190X measured with purified proteins. Reaction conditions: 500 µL reaction media consisting of sodium phosphate buffer (100 mM, pH 7.5) and 50 µL pure protein, 5 mM NADH, 1% isopropanol, 200 mM of acetophenone, 30 °C, and 10 min. Error bars indicate the standard deviation.

2.2. Comparative Structure Analysis of the Mutant M190V and the Wild-Type

To gain insight into the molecular basis for the activity enhancement in acetophenone reduction, molecular dynamics simulations were conducted. The substrate-enzyme complexes of acetophenone and wild-type EbSDR8 or the variant M190V were employed as representative models (Figure 3a,b). On the basis of the suggested catalytic mechanism of SDR, the C4 atom in the nicotinamide ring of NADH and the hydroxy group of Tyr156 donate their hydrogen atoms to the carbon and oxygen atoms of the carbonyl group of the substrate acetophenone, respectively [21,22]. Moreover, the orientation of acetophenone ensures that NADH delivers its hydride to the Si face of acetophenone to produce (*R*)-1-phenylethanol, following antiPrelog's rule (Figure 3c). Compared with the wild-type, the M190V mutation in the variant resulted in altered protein flexibility, giving the coenzyme more space to select the appropriate conformation, which was conducive to the binding and release of the coenzyme, and thus improved the acetophenone tolerance. Compared with rational design, one of the greatest benefits of directed evolution is the ability to find important amino acids that are not predictable based on the known protein structure or reaction mechanism [23].

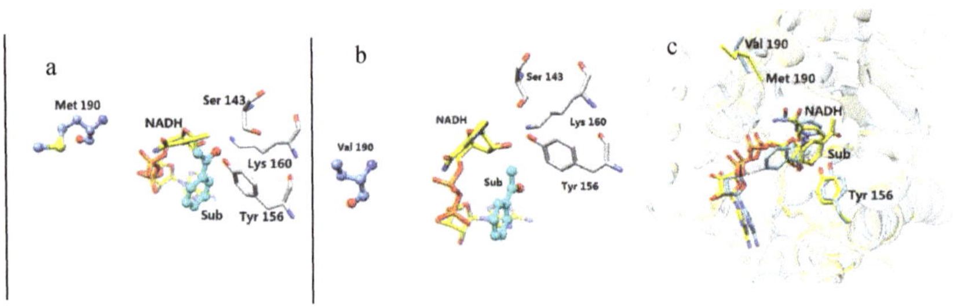

Figure 3. Wild-type EbSDR8-acetophenone complex models. (**a**). Complex models of acetophenone and wild-type EbSDR8. (**b**). Variant M190V-acetophenone complex models. (**c**). Binding pose of the complex models of acetophenone and the wild-type EbSDR8 or variant M190V. The wild-type EbSDR8 complex is shown in gold ribbons, while the variant M190V complex is shown in cyan ribbons.

2.3. *Process Optimization of the M190V Whole-Cell-Catalyzed Asymmetric Reduction of Acetophenone*

The whole-cell stereoselective biological reduction system consists of two parallel reactions, the asymmetric reduction of carbonyl compounds catalyzed by reductase accompanied by oxidation of reduced cofactors, and the regeneration of the reduced cofactor through the dissimilatory metabolism of a cosubstrate [24]. In such a cofactor regeneration recycling system, the cosubstrate is of significant importance and is essential for the sequential proceeding of biocatalytic reduction [25,26]. Isopropanol has been identified as a suitable cosubstrate in our previous work [1]. In order to find out the ideal isopropanol concentration in the M190V-catalyzed high-concentration acetophenone system, the influence of isopropanol concentration on the asymmetric reduction of acetophenone was subsequently tested. As illustrated in Figure 4a, when adding 1% isopropanol to the reaction system, the conversion increased from 9.73% to 23.66%, manifesting the effective utilization of isopropanol for cofactor recycling by the recombinant whole-cell system. However, further elevation of isopropanol concentration (up to 6%) negatively influenced the acetophenone conversion. Nevertheless, the conversion was still higher than the cosubstrate-free control when the isopropanol concentration was not higher than 5%.

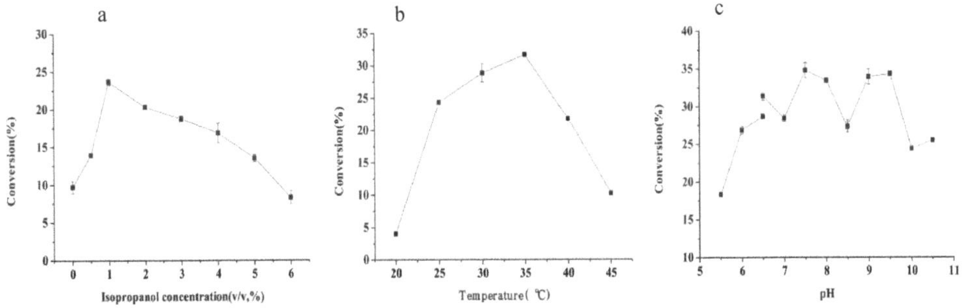

Figure 4. Characterization of EbSDR8 mutant M190V. Reaction condition: isopropanol, 0.025 g wet resting cells, 200 mM of acetophenone, and 500 μL reaction mixture. Error bars indicate the standard deviations. (**a**). The effect of isopropanol concentration (0–6%) on the M190V whole-cell-catalyzed asymmetric reduction of acetophenone. (**b**). The effect of temperature (20–45 °C) on the M190V whole-cell-catalyzed asymmetric reduction of acetophenone. (**c**). The effect of buffer pH (citrate buffer, 5.5–6.5; Sodium phosphate buffer, 6.5–8.5; glycine-NaOH buffer, 8.5–10.5) on the M190V whole-cell-catalyzed asymmetric reduction.

The reaction temperature is another key factor in biocatalysis [27]. As illustrated in Figure 4b, when the temperature rose from 20 °C to 35 °C, the conversion dramatically increased from 4.03% to 31.69%, but with further rising of the temperature to 45 °C, the conversion rate decreased. Nevertheless, the conversion of the acetophenone was maintained above 10.24% at a temperature as high as 45 °C, indicating the good thermostability of M190V.

In addition, variation of pH has been extensively shown to affect the activity of the biocatalyst [27]. To determine the optimal pH for the acetophenone asymmetric bioreduction system using M190V, a range of buffers was used, with pH varying from 5.5 to 10.5. As illustrated in Figure 4c, when the buffer pH increased from 5.5 to 6, the conversion of acetophenone was obviously improved, and relatively high conversions (24.38–34.79%) were obtained over a pH range of 6.5–10.5. The highest conversion was recorded in sodium phosphate buffer (pH 7.5). The considerable capability of M190V whole cells in catalyzing the reduction of acetophenone to (*R*)-1-phenylethanol over such a wide pH range (5.5–10.5) is of significance for its practical application. In fact, biocatalysts with such extraordinary pH adaptability are still few [28,29].

2.4. Time Course Study on the M190V Whole-Cell-Catalyzed Asymmetric Reduction of Acetophenone

To determine the maximum conversion that could be achieved by the asymmetric bioreduction of acetophenone using M190V whole cells, a time course study was conducted. As illustrated in Figure 5, under the optimal reaction condition, the conversion reached 87.52% by 135 min, whereas the conversion of the wild-type under the same condition was only 68.65% by the end (Figure 5). This result further demonstrated the improved catalytic performance of M190V at a high concentration of acetophenone.

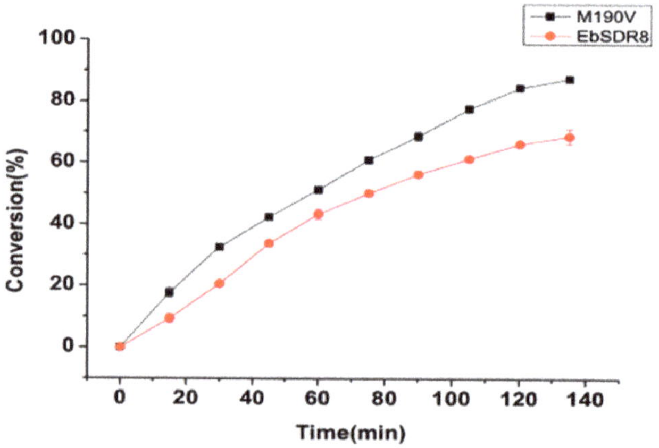

Figure 5. Time courses of M190V and wild-type EbSDR8 whole-cell-catalyzed asymmetric reduction of acetophenone. Reaction condition: 1% isopropanol, 0.05 g wet resting cells, and 200 mM of acetophenone in 500 μL reaction mixture, 35 °C. The conversion of wild-type EbSDR8 is shown in red circles, while that of the variant M190V is shown in black squares. Error bars indicate the standard deviations.

3. Materials and Methods

3.1. Materials

The short-chain dehydrogenase EbSDR8 was deposited in the native laboratory [1]. The plasmid pET30a (Novagen) was used as an expression vector, and *E. coli* strain BL21 (DE3) was employed as the expression host. *EasyTaq* DNA polymerase was purchased from TransGen Biotech Ltd. (Beijing, China). Restriction endonucleases (*BamHI* and *Xho*I) and T4 DNA ligase were purchased from TaKaRa (Dalian, China). Primers were synthesized in Sangon Biotech Ltd. Chemical reagents used in the experiments were all analytical grade and obtained from local companies.

3.2. Error-Prone PCR and Construction of Mutant Library

The error-prone PCR reaction mixture (50 μL) was made up of 10 × *EasyTaq* buffer (5 μL), dNTPs (2.5 mM, 4 μL), ddH_2O (30 μL), template plasmid (1 μL), forward primer (5′-GCTGAGGATCCATGTCAATATTAAAAGATAAGGTAGC-3′) (10 μM, 2 μL), reverse primer (5′-GCATCCTCGAGTTAAACTGCTGTATATCCTCCATC-3′) (10 μM, 2 μL), Mn^{2+} (1 mM, 5 μL), and *EasyTaq* DNA polymerase (5 U/μL, 1 μL). The PCR program was as follows: a denaturation step of 3 min at 94 °C, followed by 30 cycles of 30 s denaturation at 94 °C, 30 s annealing at 55 °C, and 1 min elongation at 72 °C. After double digestion with *BamHI* and *XhoI*, the error-prone PCR product was ligated with the expression vector pET30a and then transformed into *E. coli* BL21 (DE3) competent cells. Finally, the cells were cultured overnight at 37 °C on LB agar plates (containing 50 μg/mL kanamycin), creating the EbSDR8 mutant library.

3.3. High-Throughput Screening

The single colonies of EbSDR8 mutants were selected into conical deep 96-well plates containing 400 μL LB media supplemented with 50 μg/mL of kanamycin. Then, the 96-well plates were shaken at 200 rpm and 37 °C for about 16 h. Subsequently, 10 μL of each culture were transferred into a new 96 deep-well plate, with each well containing 500 μL LB media supplemented with 50 μg/mL of kanamycin. The daughter plates were incubated for 2 h in the same condition as above. In order to induce enzyme expression, isopropyl β-D-1-thiogalactopyranoside (IPTG) was added to obtain a final concentration of 0.1 mM. After 6 h incubation at 26 °C, cells were harvested by centrifugation at 4000 rpm for 15 min. Then, cells were lysed at 37 °C after being supplemented with 250 μL lysis buffer (100 mM Na_2HPO_4-NaH_2PO_4 buffer, 5 mM $MgCl_2$, 250 mg/mL lysozyme, pH 7.5). After 2 h, the cells were pelleted by centrifugation at 4000 rpm for 15 min, and 70 μL of the supernatant was transferred to 96-well microplates. Then, 10 μL of 20 mM NADH and 10 μL of 2 M acetophenone were added into each well of the microplate, followed by 10 min incubation at 30 °C, and the addition of 10 μL of color agent (4.2 mM MTT, 16.6 mM PMS, and the MTT:PMS = 1:2). Then, the UV absorption was determined at 575 nm using a microtiter plate reader [20]. Finally, the variants giving lower absorbance than the wild-type were selected for further verification.

3.4. Enzyme Activity Assay and Kinetic Characterization

The crude enzyme samples of the wild-type and positive mutants were prepared by cell lysis, as described above in Section 2.3., and then purified with a Ni^{2+}-charged His-tag affinity column (GE Healthcare) following the manufacturer's instructions. The concentrations of the protein were measured using the Bradford method [30] with BSA as the protein standard. The amount of enzyme catalyzing the oxidation of 0.3 mM NADH per minute under measurement conditions was defined by one unit of enzyme activity. Enzyme kinetic parameters were obtained using different acetophenone concentrations, and the constants were computed from the Lineweaver–Burk double-reciprocal plot [31].

3.5. Characterization of EbSDR8 Mutant

Using acetophenone as substrate, the catalytic properties of EbSDR8 mutant were analyzed. The biotransformations were carried out at 220 rpm, 30 °C. In total, 200 mM acetophenone, Na_2HPO_4-NaH_2PO_4 buffer (100 mM, pH7.5), and 0.025 mg of EbSDR8 mutant in a total volume of 500 μL constituted the standard reaction mixture. Samples were extracted with ethylacetate for about 2 min, then dried over the organic layer using anhydrous Na_2SO_4. Eventually, the conversion was measured by chiral gas chromatography(GC) analyses. In order to ensure the accuracy of the results, all experiments were conducted in triplicate.

The influence of isopropanol concentration on the asymmetric reduction of acetophenone was investigated with an isopropanol concentration varying from 0–6%. The effects of reaction temperature and pH were determined in a temperature range of 20 °C to 45 °C and a pH range of 5.5–10.5 (100 mM buffers: citrate buffer pH 5.5–6.5, sodium phosphate buffer pH 6.5–8.5, glycine-NaOH buffer pH 8.5–10.5).

3.6. Analytical Methods

Quantification of acetophenone and 1-phenylethanol was performed on a GC-9790 gas chromatography system (Fuli, Wenling, China) equipped with an FID detector using nitrogen as the carrier gas. The injector and detector temperature were both set to 240 °C. Acetophenone and 1-phenylethanol were separated by a chiral capillary column Cyclodex-B (Agilent, Santa Clara, CA, USA, 30 m × 0.32 mm, 0.25 μm film thickness) at 120 °C, and the retention times were as follows: acetophenone (4.86 min), (*R*)-1-phenylethanol (7.04 min), and (*S*)-1-phenylethanol (7.41 min).

3.7. Molecular Simulation

The processor used was Intel(R)Core(TM)i5-7300HQ CPU @2.50 GHz. The display adapter was the NVIDIA GeForce GTX 1050. The wild-type structure of short-chain dehydrogenase/reductase EbSDR8 was obtained by homology modeling (MODELER) [32], and the modeling of the mutant was performed using Swiss-Pdb Viewer 4.03. The receptor model was built using MGLTools 1.5.6, and the ligand model was constructed using ChemOffice 2014. The docking tests of the wild-type and variant towards acetophenone were conducted with AutoDock Vina 1.1.2. During the process of molecular docking, the center is the binding site of small molecule ligands in the tripartite structure of the protein. The Grid Box size was set to 60, 60, and 60, the grid interval was the default value of 0.375 Å, and the other parameters were set to default. Using the parameter files generated by Autogrid, the default settings of the genetic algorithm were selected, and 300 configurations were run for selection. The optimal protein-substrate docking concomplex was selected based on the catalytic mechanism and conformational energy.

4. Conclusions

In this paper, we established a colorimetric HTS method for acetophenone bioreduction and used this method to facilitate the directed evolution of the short-chain dehydrogenase/reductase EbSDR8. A positive mutant, M190V, with a 74.8% improvement in specific activity towards 200 mM acetophenone was obtained. Aside from the enhanced acetophenone tolerance, M190V was also shown to have many other qualities regarding its practical application as a whole-cell biocatalyst, including a broad pH range, moderate reaction temperature, and low optimal cosubstrate concentration. A comparative molecular simulation study of the mutant and the wild-type indicated that the enhanced acetophenone tolerance could probably be attributed to the improved binding and release of coenzyme. Furthermore, directed evolution is proven to be helpful in creating mutants that are not predictable by the rational design of protein, and is therefore particularly suitable for modification of complex enzymatic properties such as substrate/product tolerance and environmental tolerance. However, mutants that may have more advantages in this random screening method were not detected, so they still need to be studied in the future. In future industrial production, attention should be paid to the catalytic capacity under the condition of a high concentration of chiral ketone, and the carbon source, nitrogen source, pH, inoculation capacity, fermentation temperature, fermentation speed, and fermentation cycle should still be optimized, so as to improve the expression of the enzyme protein to improve the catalytic activity of the unit cell.

Author Contributions: S.Y., H.Z. and L.Y. conceived of the study and its design, developed a chromogenic reaction-based high-throughput screening method, participated in employed directed evolution to enhance the acetophenone tolerance of EbSDR8 research, and drafted the manuscript. H.Z., B.W. and H.Y. carried out the directed evolution to enhance the acetophenone tolerance of EbSDR8, performed the statistical analysis, and helped to draft the manuscript. All authors have read and agreed to the published version of the manuscript.

Funding: This work was financially supported by the Natural Science Foundation of China (Grant Nos. 21406196 and 21576234), and the National High Technology Research and Development Program (863 Program) of China (Grant No. SS2015AA020601).

Data Availability Statement: Not applicable.

Acknowledgments: The authors would like to acknowledge the science and technology department of Zhejiang province. Great appreciation is given to all members of our laboratory for their enthusiastic participation in the research.

Conflicts of Interest: The authors declare no conflict of interest.

Ethics Approval and Consent to Participate: Not applicable.

Abbreviations

SDRs, short-chain dehydrogenases/reductases; PMS, phenazine methosulfate; MTT, thiazolyl blue tetrazolium bromide; HTS, High throughput screening; IPTG, isopropyl β-D-1-thiogalactopyranoside; GC, gas chromatography.

References

1. Li, A.P.; Ye, L.D.; Wu, H.P.; Yang, X.H.; Yu, H.W. Characterization of an excellent anti-Prelog short-chain dehydrogenase/reductase EbSDR8 from *Empedobacter brevis* ZJUY-1401. *J. Mol. Catal. B Enzym.* **2015**, *122*, 179–187. [CrossRef]
2. Morris, R.E.; Bu, X. Induction of chiral porous solids containing only achiral building blocks. *Nat. Chem.* **2010**, *2*, 353–361. [CrossRef]
3. Musa, M.M. Alcohol Dehydrogenases with anti-Prelog Stereopreference in Synthesis of Enantiopure Alcohols. *ChemistryOpen* **2022**, *11*, 4. [CrossRef]
4. Han, M.N.; Wang, X.M.; Pei, C.H.; Zhang, C.; Xu, Z.; Zhang, H.L.; Li, W. Green and scalable synthesis of chiral aromatic alcohols through an efficient biocatalytic system. *Microb. Biotechnol.* **2021**, *14*, 444–452. [CrossRef]
5. Goldberg, K.; Schroer, K.; Lutz, S.; Liese, A. Biocatalytic ketone reduction—A powerful tool for the production of chiral alcohols—Part I: Processes with isolated enzymes. *Appl. Microbiol. Biotechnol.* **2007**, *76*, 237–248. [CrossRef]
6. Huisman, G.W.; Liang, J.; Krebber, A. Practical chiral alcohol manufacture using ketoreductases. *Curr. Opin. Chem. Biol.* **2010**, *14*, 122–129. [CrossRef]
7. Garzón-Posse, F.; Becerra-Figueroa, L.; Hernández-Arias, J.; Gamba-Sánchez, D. Whole Cells as Biocatalysts in Organic Transformations. *Molecules* **2018**, *23*, 1265. [CrossRef]
8. Asako, H.; Shimizu, M.; Makino, Y.; Itoh, N. Biocatalytic reduction system for the production of chiral methyl (R)/(S)-4-bromo-3-hydroxybutyrate. *Tetrahedron Lett.* **2010**, *51*, 2664–2666. [CrossRef]
9. Hohne, M.; Schatzle, S.; Jochens, H.; Robins, K.; Bornscheuer, U.T. Rational assignment of key motifs for function guides in silico enzyme identification. *Nat. Chem. Biol.* **2010**, *6*, 807–813. [CrossRef]
10. Zheng, G.W.; Xu, J.H. New opportunities for biocatalysis: Driving the synthesis of chiral chemicals. *Curr. Opin. Biotechnol.* **2011**, *22*, 784–792. [CrossRef]
11. Alsafadi, D.; Paradisi, F. Effect of organic solvents on the activity and stability of halophilic alcohol dehydrogenase (ADH2) from *Haloferax volcanii*. *Extremophiles* **2013**, *17*, 115–122. [CrossRef]
12. Kallberg, Y.; Oppermann, U.; Persson, B. Classification of the short-chain dehydrogenase/reductase superfamily using hidden Markov models. *Febs J.* **2010**, *277*, 2375–2386. [CrossRef]
13. Li, Z.; Liu, W.; Chen, X.; Jia, S.; Wu, Q.; Zhu, D.; Ma, Y. Highly enantioselective double reduction of phenylglyoxal to (R)-1-phenyl-1,2-ethanediol by one NADPH-dependent yeast carbonyl reductase with a broad substrate profile. *Tetrahedron* **2013**, *69*, 3561–3564. [CrossRef]
14. Zhou, S.; Zhang, S.C.; Lai, D.Y.; Zhang, S.L.; Chen, Z.M. Biocatalytic characterization of a short-chain alcohol dehydrogenase with broad substrate specificity from thermophilic *Carboxydothermus hydrogenoformans*. *Biotechnol. Lett.* **2013**, *35*, 359–365. [CrossRef]
15. Shao, Z.H.; Su, B.M.; Yang, S.L.; Ye, L.D.; Yu, H.W. Rational design of the carbonyl reductase EbSDR8 for efficient biosynthesis of enantiopure (R)-3-chloro-1-phenyl-1-propanol. *Appl. Microbiol. Biotechnol.* **2020**, *104*, 9219–9228. [CrossRef]
16. Nezhad, N.G.; Rahman, R.N.Z.R.A.; Normi, Y.M.; Oslan, S.N.; Shariff, F.M.; Leow, T.C. Thermostability engineering of industrial enzymes through structure modification. *Appl. Microbiol. Biotechnol.* **2022**, *106*, 4845–4866. [CrossRef]
17. Basak, S.; Song, H.; Jiang, R.R. Error-prone PCR of global transcription factor cyclic AMP receptor protein for enhanced organic solvent (toluene) tolerance. *Process Biochem.* **2012**, *47*, 2152–2158. [CrossRef]
18. Chong, H.Q.; Huang, L.; Yeow, J.W.; Wang, I.; Zhang, H.F.; Song, H.; Jiang, R.R. Improving ethanol tolerance of *Escherichia coli* by rewiring its global regulator cAMP receptor protein (CRP). *PLoS ONE* **2013**, *8*, e57628. [CrossRef]
19. Chong, H.Q.; Yeow, J.W.; Wang, I.; Song, H.; Jiang, R.R. Improving acetate tolerance of *Escherichia coli* by rewiring its global regulator cAMP receptor protein (CRP). *PLoS ONE* **2013**, *8*, e77422.
20. Yang, Y.; Liu, J.; Li, Z. Engineering of P450pyr Hydroxylase for the Highly Regio- and Enantioselective Subterminal Hydroxylation of Alkanes. *Angew. Chem. Int. Ed. Engl.* **2014**, *53*, 3120–3124. [CrossRef]
21. Katz, M.; Frejd, T.; Hahn-Hagerdal, B.; Gorwa-Grauslund, M.F. Efficient anaerobic whole cell stereoselective bioreduction with recombinant *Saccharomyces cerevisiae*. *Biotechnol. Bioeng.* **2003**, *84*, 573–582. [CrossRef]
22. Borg, A.J.E.; Beerens, K.; Pfeiffer, M.; Desmet, T.; Nidetzky, B. Stereo-electronic control of reaction selectivity in short-chain dehydrogenases: Decarboxylation, epimerization, and dehydration. *Curr. Opin. Chem. Biol.* **2021**, *61*, 43–52. [CrossRef]
23. Bornscheuer, U.T.; Hauer, B.; Jaeger, K.E.; Schwaneberg, U. Directed Evolution Empowered Redesign of Natural Proteins for the Sustainable Production of Chemicals and Pharmaceuticals. *Angew. Chem. Int. Ed. Engl.* **2019**, *58*, 36–40. [CrossRef]
24. Zheng, R.C.; Ge, Z.; Qiu, Z.K.; Wang, Y.S.; Zheng, Y.G. Asymmetric synthesis of (R)-1,3-butanediol from 4-hydroxy-2-butanone by a newly isolated strain *Candida krusei* ZJB-09162. *Microbiol. Biotechnol.* **2012**, *94*, 969–976. [CrossRef]
25. Peters, J.; Minuth, T.; Kula, M.R. A novel NADH-dependent carbonyl reductase with an extremely broad substrate range from *Candida parapsilosis*: Purification and characterization. *Enzyme Microb. Technol.* **1993**, *15*, 950–958. [CrossRef]
26. Rao, P.V.; Krishna, C.M.; Zigler, J.S. Identification and characterization of the enzymatic activity of zeta-crystallin from guinea pig lens. A novel NADPH: Quinone oxidoreductase. *J. Biol. Chem.* **1992**, *267*, 96–102. [CrossRef]

27. Bradford, M.M. A rapid and sensitive method for the quantitation of microgram quantities of protein utilizing the principle of protein-dye binding. *Anal. Biochem.* **1976**, *72*, 248–254. [CrossRef]
28. Schneider, R.W. Effects of nonpathogenic strains of *Fusarium oxysporum* on celery root infection by *Fusarium oxysporum* f. sp. *apii* and a novel use of the Lineweaver-Burk double reciprocal plot technique. *Phytopathology* **1984**, *74*, 646–653. [CrossRef]
29. Li, A.P.; Ye, L.D.; Yang, X.H.; Yang, C.C.; Gu, J.L.; Yu, H.W. Structure-guided stereoselectivity inversion of a short-chain dehydrogenase/reductase towards halogenated acetophenones. *Chem. Commun.* **2016**, *52*, 6284–6287. [CrossRef]
30. Filling, C.; Berndt, K.D.; Benach, J.; Knapp, S.; Prozorovski, T.; Nordling, E.; Ladenstein, R.; Jornvall, H.; Oppermann, U. Critical residues for structure and catalysis in short-chain dehydrogenases/reductases. *J. Biol. Chem.* **2002**, *277*, 25677–25684. [CrossRef]
31. Luo, X.; Wang, Y.J.; Shen, W.; Zheng, Y.G. Activity improvement of a Kluyveromyces lactis aldo-keto reductase KlAKR via rational design. *J. Biotechnol.* **2016**, *224*, 20–26. [CrossRef]
32. Chen, W.W.; Ye, L.D.; Guo, F.; Lv, Y.; Yu, H.W. Enhanced activity of an alkaline phytase from *Bacillus subtilis* 168 in acidic and neutral environments by directed evolution. *Biochem. Eng. J.* **2015**, *98*, 137–143. [CrossRef]

catalysts

MDPI

Article

C058 and Other Functional Microorganisms Promote the Synthesis of Extracellular Polymer Substances in Mycelium Biofloc

Yiyong Li [1,2], Wanyi Luo [1,2], Wen Liu [1,2], Yongcong Yang [1,2], Zexiang Lei [1,2], Xueqin Tao [1,2] and Baoe Wang [1,2,*]

1 College of Resources and Environment, Zhongkai University of Agriculture and Engineering, Guangzhou 510225, China; yueguangzhuying@126.com (Y.L.); 13430372936@163.com (W.L.); lazliu@163.com (W.L.); yongcongyang@126.com (Y.Y.); lei-1965@163.com (Z.L.); xqtao@foxmail.com (X.T.)
2 Guangdong Provincial Engineering and Technology Research Center for Agricultural Land Pollution Prevention and Control, Guangzhou 510225, China
* Correspondence: baoewang@163.com

Abstract: The mycelium biofloc bioaugmented by *Cordyceps* strain C058 effectively purifies water, which may be related to the synthesis of extracellular polymer substances. To verify this conjecture, we analyzed the changes in extracellular polymer substances content in the mycelium biofloc under various hydraulic retention times (36 h, 18 h, and 11 h). The microstructure and microflora composition were analyzed using a scanning electron microscope and high-throughput sequencing. The ordinary biofloc without bioaugmentation was taken as a control. The results showed that under the above hydraulic retention time, the extracellular polymer substances contents of the mycelium biofloc were 51.20, 55.89, and 33.84 mg/g, respectively, higher than that of the ordinary biofloc (14.58, 15.72, and 18.19 mg/g). The protein content or the polysaccharide content also followed the same trend. Meanwhile, the sedimentation performance of the mycelium biofloc was better than that of the ordinary biofloc, attributed to the content of the extracellular polymer substances. It is worth noting that C058 is the main biofloc content, which promotes the synthesis of extracellular polymer substances in the mycelium biofloc. Other functional microorganisms in the mycelium biofloc were *Janthinobacterium*, *Phormidium*, *Leptolyngbya*, *Hymenobacter*, and *Spirotrichea*, which also promote the synthesis of extracellular polymer substances.

Keywords: C058; mycelium biofloc; extracellular polymeric substances; microcosmic structure; microflora composition

Citation: Li, Y.; Luo, W.; Liu, W.; Yang, Y.; Lei, Z.; Tao, X.; Wang, B. C058 and Other Functional Microorganisms Promote the Synthesis of Extracellular Polymer Substances in Mycelium Biofloc. *Catalysts* **2022**, *12*, 693. https://doi.org/10.3390/catal12070693

Academic Editors: Zhilong Wang and Tao Pan

Received: 21 April 2022
Accepted: 20 June 2022
Published: 24 June 2022

1. Introduction

Extracellular polymeric substances (EPS) are sticky polymers secreted by microorganisms through metabolism and cell autolysis [1]. They contain proteins (PN), polysaccharides (PS), humic acid, and nucleic acids [2,3]. PN and PS are the main EPS constituents, accounting for about 70–80% of the total EPS [4]. Because EPS promote adhesion and aggregation between microbial cells [5], and their content can affect the properties of flocs, such as flocculation capacity, settling performance, surface charge, and microbial community structure [6–8], EPS are key substances that determine the physical, chemical, and biological properties of flocs, and thus the water purification performance. Biofloc technology (BFT) is widely used in aquacultural tailwater purification. Zoogloea and filamentous bacteria are the core of ordinary biofloc (OBF). They adhere to bacteria, fungi, algae, protozoa, and organic polymers, forming structurally diverse flocs using secreted extracellular polymer substances [9]. OBF removes ammonium nitrogen, total nitrogen, and total phosphorus from water through microbial assimilation and nitrification [10] and removes organic matter from water through biological adsorption [11]. Recent studies have shown that the bioaugmented biofloc efficiently purifies water, but the mechanism is not clearly described, in particular, the EPS synthesis [12–14].

Cordyceps sp. belongs to phylum Ascomycota, class Sordariomycetes, order Hypocreales, and family Clavicipitaceae. Its ascospores germinate into and belong to large filamentous mycelium. Numerous studies have shown that fungi can efficiently remove contaminants in wastewater [15–18]. Our previous study showed that C058 and its bioaugmented biofloc named mycelium biofloc (MBF) effectively purified water [19,20]; however, the mechanism underlying this process is unclear. Zhang [21] used a mycelium ball as a biomass carrier to load Pseudomonas stutzeri T13, and found that the total nitrogen removal rate in the SBR reactor was about 10% higher than in the traditional activated sludge reactor, and the protein content in EPS secreted by mycelium ball is high, which enhances the solid–liquid separation performance of the system by improving the hydrophobicity of activated sludge. *Cordyceps* secretes large amounts of PS, PN, and other macromolecules [22,23]. Hence, we speculate that the water purification effect of MBF may be related to its EPS synthesis. To validate this hypothesis, we analyzed the changes in EPS content, the microstructure, and the microflora composition in MBF under various hydraulic retention times (HRT) using a scanning electron microscope (SEM) and high-throughput sequencing. The OBF without bioaugmentation was taken as a control. The findings of this study provide a theoretical basis for the water purification effect of MBF.

2. Results and Discussion

2.1. Changes in EPS Content in the Biofloc

Figure 1 shows that the EPS content was higher in the experimental reactor than that of the control reactor under various HRTs, which may be attributed to the C058 in the experimental reactor. C058 secretes a large amount of PS and PN [19], generating higher EPS content in MBF of 51.20, 55.89, and 33.84 mg/g under the above HRTs. Morgan et al. [24] proposed that the activity of functional microorganisms affects the EPS content. Accordingly, the EPS contents were different in the two reactors [25]. There was lower EPS content in OBF (14.58, 15.72, and 18.19 mg/g under the above HRTs). Meanwhile, It can be seen from Figure 1 that the PN content or PS content also followed the same trend. In addition, the flow rate influenced the HRT and thus the shear forces. Appropriated shear force is essential for optimal EPS secretion in aerobic granular sludge, but excessive shear force destabilizes the sludge [26]. This may explain the increase in EPS content in both reactors during the first and second operation stages (Figure 1). Filamentous bacteria are intolerant to shear forces, and HRT 11 h may cause autolysis of filamentous bacteria with a decrease in EPS content in MBF.

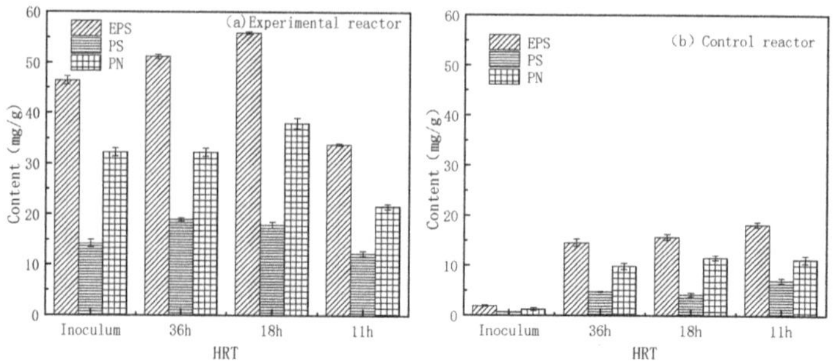

Figure 1. Contents of EPS and its components in biofloc from the two reactors under various HRTs.

2.2. Sludge Volume Index (SVI) Changes of Biofloc

The settling performance of flocs can indirectly reflect the content and composition of EPS. It can be seen from Figure 2 that the SVI was lower in the experimental reactor than that in the control reactor under various HRTs, and all the SVI in the experimental reactor were lower than 150 mL/g, implying that the settling performance of MBF occurred within a given favorable range. That of the control group was above 150 mL/g under various HRTs. The OBF was loose and was detrimental to sedimentation [27]; thus, the settling performance of MBF was better than that of OBF. Meanwhile, there was a positive correlation between EPS and SVI in the experimental reactor, consistent with a previous study [28]. Li et al. [29] found that microorganisms can secrete EPS to enhance flocculation and sedimentation. In the present study, the sedimentation performance of MBF was better than that of OBF, which indirectly proved that the EPS content of MBF was higher than that of OBF. In addition, the settling performance of flocs reflects the composition of EPS. Previous studies have shown that PN possesses hydrophobic groups, which improve the hydrophobicity of the cells and enhance the mutual attraction between bacteria, promoting their sedimentation [30,31]. In this study, the PN content of the experimental reactor was higher than that of the control reactor under various HRTs (Figure 1).

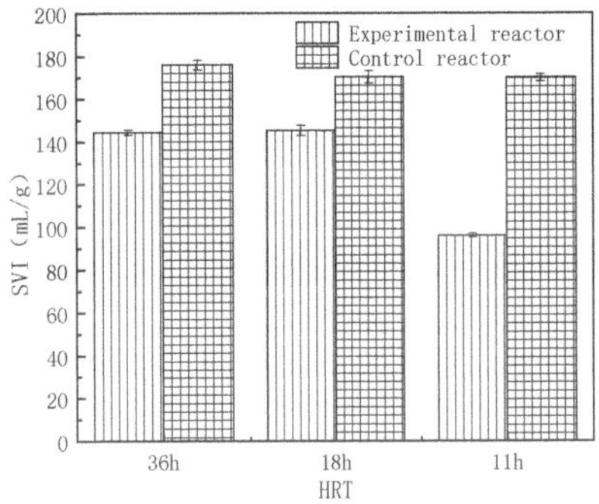

Figure 2. SVI changes of biofloc in both reactors under various HRTs.

2.3. *Changes in the Appearance of Biofloc*

The appearance of MBF and OBF were analyzed through a visual examination using an SEM. The morphological structure of C058 is shown in Figure 3. C058 is filamentous and supercoiled. The bioflocs in both reactors under various HRTs are shown in Figure 4. As shown in Figure 4a,c,e, bacteria aggregates on the MBF surface increased over time, forming a symbiotic structure with C058. Although MBF was constructed by fungi and bacteria together, fungi were still the dominant microorganism. As shown in Figure 4b,d,f, the OBF mainly comprised zoogloea, whose size decreased with time. It revealed that the shortened HRT strengthened the shear force and minimized the zoogloea size.

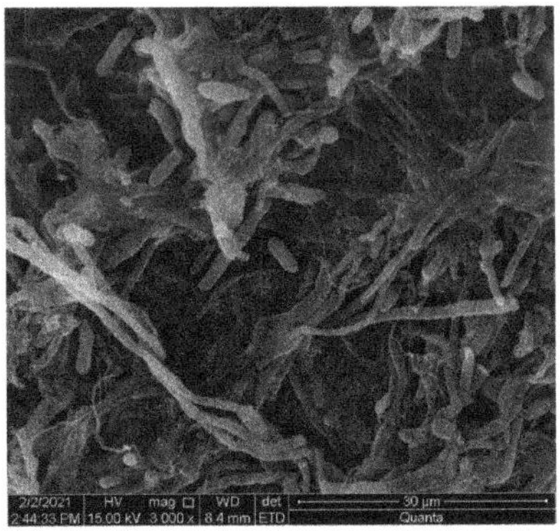

Figure 3. The morphological structure of *Cordyceps* strain C058.

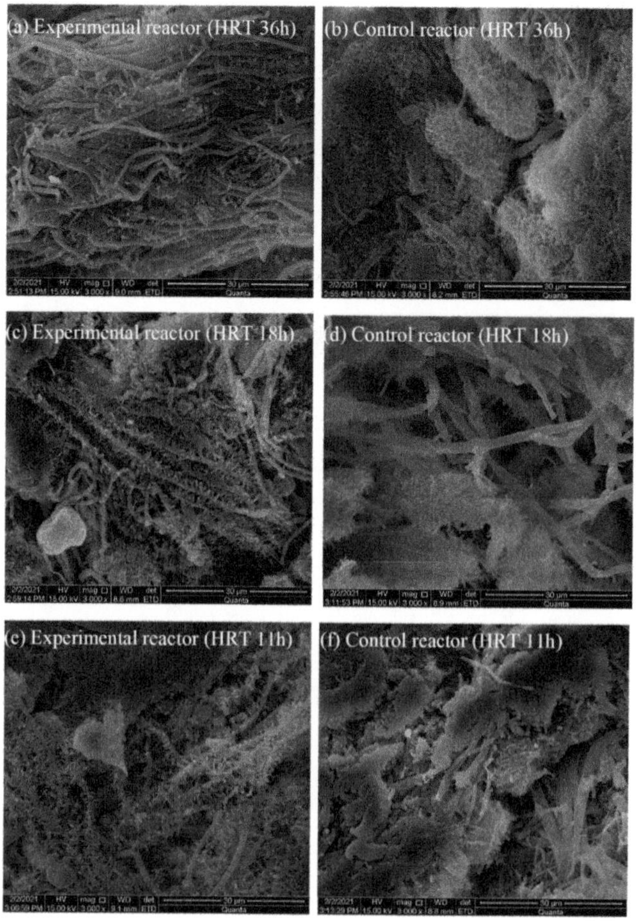

Figure 4. SEM of biofloc from both reactors under various HRTs.

2.4. *Changes in the Microbial Community Structure in Biofloc*

During the operation of both reactors, the biofloc was sampled when it had achieved stable operation under various HRTs. A total of eight samples were collected and included CKCB (inoculum of the experimental reactor), CB-1 (sample of the experimental reactor at HRT 36 h), CB-2 (sample of the experimental reactor at HRT 18 h), CB-3 (sample of the experimental reactor at HRT 11 h), CKWB (inoculum of the control reactor), BF-1 (sample of the control reactor at HRT 36 h), BF-2 (sample of the control reactor at HRT 18 h), BF-3 (sample of the control reactor at HRT 11 h).

The microbial community structure in biofloc was determined using 16S rRNA and 18S rRNA gene sequencing, and the sequencing process was completed in Guangdong Meilikang Biotechnology Co., Ltd., Guangzhou, China.

2.4.1. Changes in Prokaryote in Biofloc

(1) The diversity index of prokaryote

The diversity of the microbial community, including the species diversity and relative abundance of each group, was described based on the α-diversity index. We found that the Chao1 index in the experimental reactor negatively correlated with the HRT, and was at the lowest level (1906.67) at HRT 11 h. The change of the Chao1 index in the control reactor was the same as in the experimental reactor, and the lowest Chao1 index was 1489.73 at HRT 11 h (Table 1). The change of HRT had less of an effect on the bacteria abundance in the experimental reactor than in the control reactor. Meanwhile, the Shannon index and Simpson index also followed the same trend. Hence, the species diversity of microorganisms was higher in the experimental reactor than in the control reactor, implying that the microbial community in the experimental reactor was more stable.

Table 1. The alpha diversity index of biofloc from both reactors based on the 16S rRNA sequences.

	Chao1	Shannon	Simpson
CKCB	1701.59	6.16	0.969
CB-1	2868.92	8.42	0.988
CB-2	2720.18	8.27	0.987
CB-3	1906.67	6.16	0.942
CKWB	2477.06	7.80	0.988
BF-1	2242.12	7.82	0.981
BF-2	2059.89	6.67	0.915
BF-3	1489.73	3.97	0.745

(2) The relative abundance of prokaryotes at the phylum level

It can be seen from Figure 5 that the most dominant phylum in the experimental reactor at the first operation stage was Thermi, followed by Proteobacteria and Cyanobacteria. At HRT 36 h, Cyanobacteria became the most dominant phylum, with an abundance of 38.05%, followed by Proteobacteria (27.99%) and Bacteroidetes (13.34%). At HRT 18 h, the abundance of Proteobacteria increased to 53.34%, making it the most dominant phylum, followed by Cyanobacteria and Bacteroidetes. At HRT 11 h, the abundance of Proteobacteria was 37.35%, and it was still the most dominant phylum, followed by Cyanobacteria and Firmicutes. In the control reactor, the most dominant phylum at the first operation stage was Proteobacteria, followed by Actinobacteria and Bacteroidetes. The abundance of Proteobacteria decreased gradually from 27.21% (HRT 18 h) to 14.59% (HRT 11 h) with the shorting of HRT, and it was no longer the most dominant bacteria. The abundance of Actinobacteria also displayed the same trend. Instead, Thermi became the most dominant phylum at HRT 11 h; therefore, MBF was dominant by Proteobacteria. Proteobacteria plays an indispensable role in the material cycle, including nitrogen transformation in the water [32,33], enabling nitrogen removal in aquacultural wastewater [34]. Meanwhile, the abundance of Bacteroidetes and Cyanobacteria was also higher in the experimental reactor than in the control reactor under various HRTs. Bacteroidetes had the particularity of using nitrogen compounds for growth and Cyanobacteria had nitrogen fixation capacity [32,35]. These bacteria can promote the synthesis of EPS in MBF.

(3) The relative abundance of prokaryotes at the genus level

It can be seen from Figure 6 that the abundance of *Janthinobacterium* belonging to Proteobacteria in the experimental reactor reached 20.74% at HRT 11 h, significantly higher than that in the control reactor (0.68%). This may explain the effective water purification capability of MBF at short HRT, as reported in our previous study [18]. Other studies have shown that *Janthinobacterium* has nitrogen removal capacity in the water. Yang et al. [36] found that *Janthinobacterium* sp. M-11 still exhibited nitrite removal capacity under low temperature, and the removal rate of nitrite and nitrate even reached 93% and 98%, respectively, under anaerobic conditions, which was beneficial for the treatment of eutrophic water. Meanwhile, as dominant genera, the abundance of *Phormidium* and *Leptolyngbya* belonging to Cyanobacteria was higher in the experimental reactor than in the control reactor under various HRTs. Other studies have shown that *Phormidium* and *Leptolyngbya* had excellent nitrogen fixation activity and PN and PS production capacity. Through the $^{15}N_2$ gas tracer method combined with the stable isotope nucleic acid probe technique (DNA-stable isotope probing), it was confirmed that *Leptolyngbya* was one of the most important nitrogen fixation microorganisms in paddy soil [37]. Bar et al. [38] found that *Phormidium* secretes PS for flocculation. Kim et al. [39,40] found that *Leptolyngbya* sp. KIOST-1 produces a large amount of PN. In addition, the abundance of *Hymenobacter* belonging to Bacteroides was higher in the experimental reactor than in the control reactor under various HRTs. *Hymenobacter* possesses nitrogen fixation capacity [41]; therefore, in addition to C058, many bacteria that can synthesize EPS were present in the experimental reactor, which promoted the synthesis of EPS in MBF, enhancing the water purification efficiency of the system.

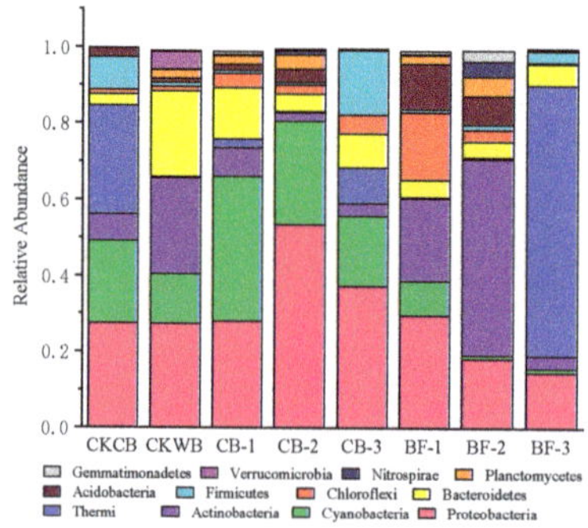

Figure 5. Composition of dominant prokaryote phyla.

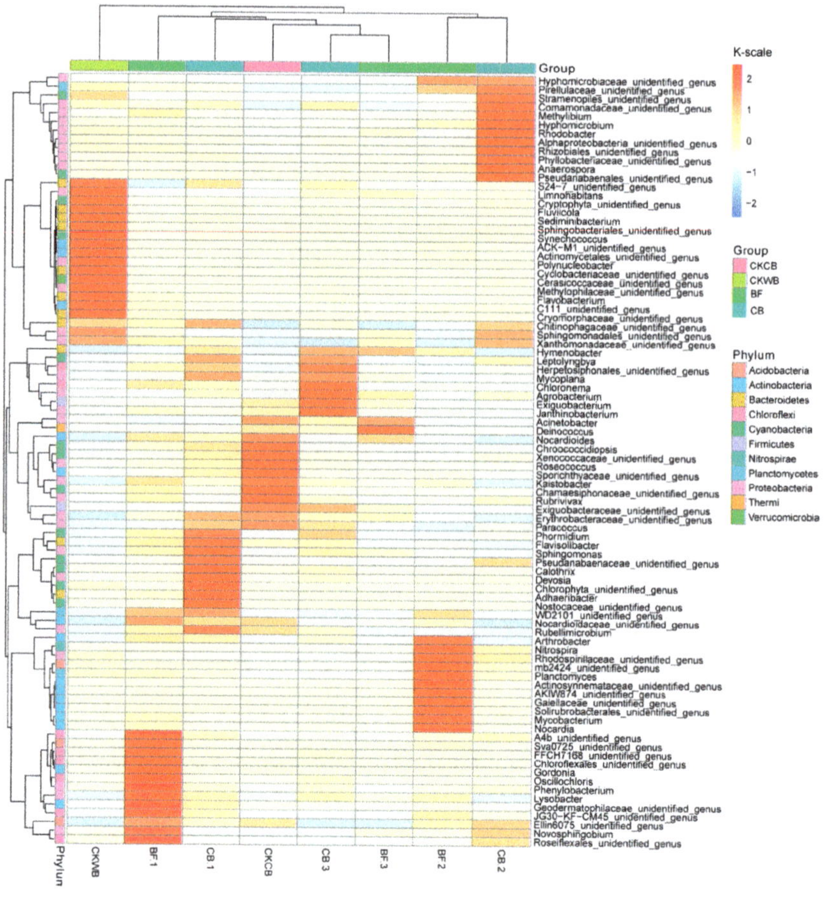

Figure 6. Heatmap diagram for the dominant prokaryote genera.

2.4.2. Changes in Eukaryotes in Biofloc

(1) The diversity index of eukaryotes

The microflora abundance in the experimental reactor was on the rise on the whole with the shortening of HRT, but the control reactor displayed an opposite trend (Table 2). Compared with the control reactor, the microflora diversity was lower in the experimental reactor, probably because C058 filamentous fungi dominated the experimental reactor, and competition limited the growth of the other eukaryotes.

Table 2. The alpha diversity index of biofloc from both reactors based on the 18S rRNA sequences.

	Chao1	Shannon	Simpson
CKCB	290.25	4.43	0.92
CB-1	387.69	2.94	0.63
CB-2	259.12	2.21	0.52
CB-3	393.45	3.17	0.75
CKWB	492.11	4.06	0.87
BF-1	481.27	4.93	0.93
BF-2	372.07	4.16	0.88
BF-3	353.57	3.75	0.86

(2) The relative abundance of eukaryotes at the phylum level

It can be seen from Figure 7 that the most abundant eukaryote was Opisthokonta at the first operation stage of the experimental reactor, achieving an abundance of 46.72%. In the latter stages, SAR became the most abundant eukaryote, at 88.36% (HRT 36 h), 86.19% (HRT 18 h), and 90.49% (HRT 11 h), respectively; however, the proportion of SAR in the control reactor declined. Studies have shown that protozoa caused flocs formation by excreting gelatinous mucus before ingesting the bacteria [42]. After the formation of the floc, ciliated protozoa become the main protozoan group, and the secretions during its predation further promote the accumulation of microorganisms in the sludge [43]. Hence, the abundance of SAR remained higher in the experimental reactor, which promoted the flocs' formation and EPS synthesis.

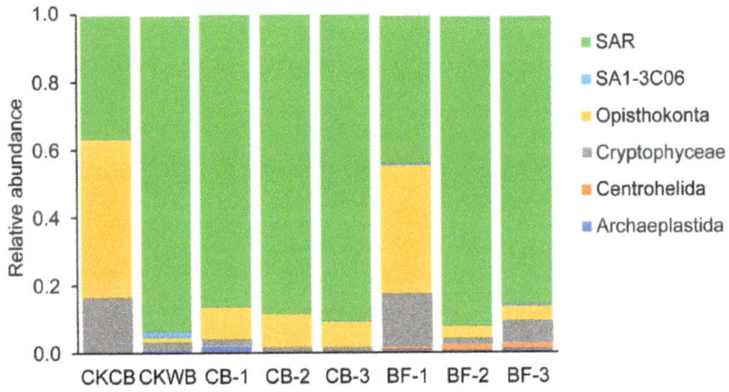

Figure 7. Composition of dominant eukaryote phyla.

(3) The relative abundance of eukaryotes at the genus level

It can be seen from Figure 8 that *Spirotrichea* was the most dominant group in both reactors, but their abundance changed with time. The abundance of *Spirotrichea* in the experimental reactor increased from 18.67% (inoculum) to 79.02% (HRT 36 h), 84.67% (HRT 18 h), and 62.26% (HRT 11 h), but decreased in the control reactor from 70.64% (inoculum) to 21.24% (HRT 36 h), 40.82% (HRT 18 h), and 56.42% (HRT 11 h). *Spirotrichea* belonging to Ciliophora is a complex and diverse group of ciliates. It is an indispensable part of the microfood ring and is the main contributor to the energy cycle. Its abundance in marine water was negatively correlated with the N and P contents [44–46]. Ciliates stimulate the formation and adhesion of bacteria colonies, which are efficient for nutrient uptake [47]; Ciliates appeared on the surface of aerobic granular sludge and secreted the viscous substances that can absorb suspended particles and bacteria, and the dead remains were used as the skeleton to form granular sludge [48]; therefore, the increase in the abundance of *Spirotrichea* in the experimental reactor promoted flocs' formation, and thus improved the efficiency of water purification.

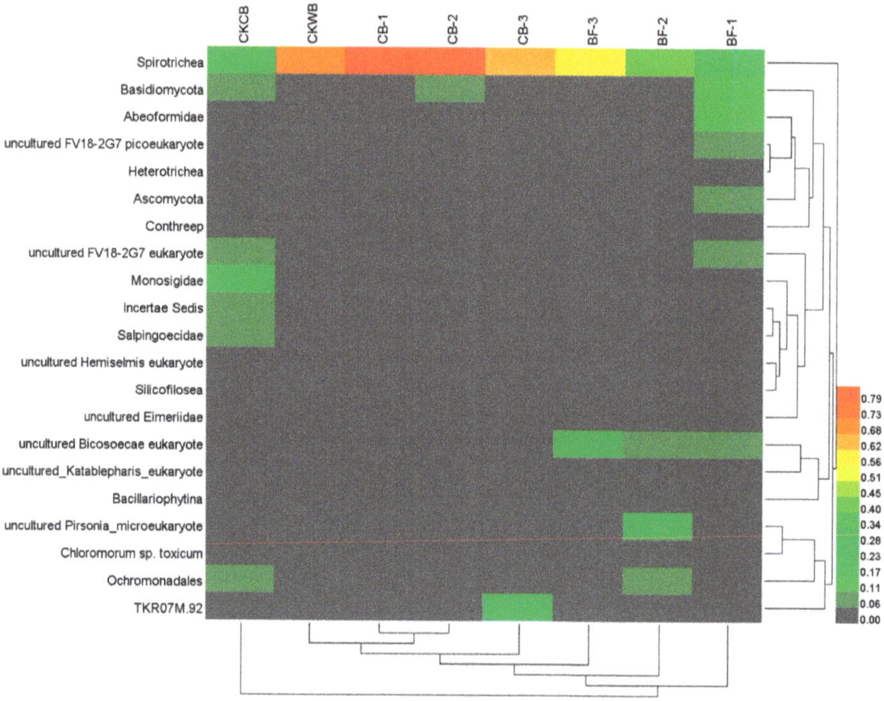

Figure 8. Heatmap diagram for the dominant eukaryote genera.

3. Materials and Methods

3.1. Experimental Materials

The Cordyceps strain C058 was donated by the Medicinal Fungi Research Group of the Institute of Microbiology, Guangdong Academy of Sciences. The formula of the C8 medium was referred to by Li et al. [22]. The formula of the simulated wastewater was referred to by Schryver and Verstraete [49]. The size and structure of the baffled reactor were described by Yang [20].

3.2. Experimental Methods

3.2.1. Culture and Collection of C058

The mycelium was taken from the C058 slope and inoculated into a C8 culture medium. After a large number of new mycelium balls were grown out at 28 °C, 150 rpm in a rotating incubator, they were collected by centrifugation under the condition of 2000 rpm, 10 min, and were washed three times with sterilized water. The washed mycelium balls were used for inoculation.

3.2.2. Operation of the Reactors

The experiment was carried out in two sets of identical baffled reactors. After filling with the simulated wastewater, one reactor, as an experimental reactor, was inoculated with 1% of C058 and 1% of Pearl River water (23°06′37.9″ N, 113°16′51.6″ E) to form mycelium biofloc (MBF), and the other reactor as a control reactor was only inoculated with 1% of Pearl River water to form OBF. Here, Pearl River is chosen as a typical natural water body that is rich in natural microflora. The reactors were operated at room temperature and contained above 5 mg/L of dissolved oxygen by continuous aeration.

After a consideration of the effect of HRT on the EPS content [26], the reactor was operated under three stages: HRT 36 h (v = 4.7 L/h), 18 h (v = 9.2 L/h) and 11 h (v = 15.3 L/h). The first operation stage was HRT 36 h. When the reactor became stable, sampling for latter index determination was carried out. Then, the influent velocity was changed to control the HRT of 18 h for a second operation stage; so did the third operation stage at HRT 11 h.

3.3. Analysis Method

3.3.1. Extraction and Determination of EPS

The EPS was extracted using the centrifugal heating method [50]. In this study, the main components of EPS, PN, and PS, were used to characterize EPS. The PN content was determined using Coomassie brilliant blue spectrophotometry, and the PS content was determined using phenol-sulfuric acid spectrophotometry [51].

3.3.2. Determination of SVI

Sludge settling velocity in 30 min (SV_{30}) and mixed liquid suspended solids (MLSS) content were determined according to "Monitoring of Water and waste Water" [52], and $SVI = SV_{30}/MLSS$.

3.3.3. SEM Analysis

Samples were dried (Tousimis Autosamdri-815, American) and sprayed with gold (EMS 150T, American) before the electron microscope scanning (Tungsten Filament SEM Q25, American). The specific operation was referred to by Shen et al. [53].

3.3.4. Analysis of Microbial Community Structure

(1) The 16S rRNA sequence analysis

Total DNA was extracted using the soil strong DNA extraction kit (DNeasy PowerSoil Kit, QIAGEN, Hilden, Germany). The DNA was diluted to 10 ng/μL for PCR amplification. According to Tamaki [54], the V4-V5 hypervariable region of the 16S rRNA gene of the experimental sample was amplified by primers 515F(5′-GTGYCAGCMGCCGCGGTA-3′) and 909R (5′-CCCCGYCAATTCMTTTRAGT-3′). Reaction system: every 25 μL PCR reaction solution contains 1× PCR buffer, 1.5 mM $MgCl_2$, 0.2 mM dNTP (Transgen, Beijing, China), 1.0 μM primers, 0.25U Ex Taq (TaKaRa, Beijing, China), and 10 ng DNA template. The PCR reaction procedure was pre-denatured at 94 °C for 3 min, then extended for 10 min at 72 °C after 30 cycles of conventional amplification (denatured at 94 °C for 40 s, annealing at 56 °C for 60 s, extension at 72 °C for 60 s). The same sample was amplified twice, and the two PCR products obtained from the same sample were mixed. After 1.2% agarose gel electrophoresis, the gel was cut and purified by sanPrep DNA gel recovery kit (Raw engineering, China). After all the purified DNA was mixed in the same amount, PE250 sequencing was carried out using the Illumina Miseq platform [55].

The original sequencing data were spliced using the FLASH1.2.8 software and screened using the QIIME1.9.0 software after splicing [29,56,57], using the Uchime program [58] to detect and remove chimera sequences, then using the QIIME1.9.0 software to divide OTUs (Operational Taxonomic Units) according to 97% similarity of sequences, and using the RDP classifier [46] to annotate each species.

(2) The 18S rRNA sequence analysis

The 18S rRNA was extracted as 16s rRNA. According to Lejzerowicz et al. [59], the V4 hypervariable region of 18s rRNA gene of the experimental sample was amplified by primer TAReuk454WD1 (5′-CCAGCASCYGCGGTAATTCC-3′) and primer TAReukREV3 (5′-ACTTTCGTTCTTGATYRA-3′). PCR reaction system: every 25 μL PCR reaction solution contains 1× PCR buffer, 1.5 mM $MgCl_2$, 0.2 mM dNTP (Transgen, China), 1.0 μM primers, 0.25 U TransFast Taq DNA polymerase (Transgen, China), and 10 ng DNA template. The PCR reaction procedure was the same as that of 16S rRNA. After the completion of the PCR reaction, two PCR products amplified from the same sample were mixed, and the gel was cut using 1.2% agarose gel electrophoresis and purified by the AxyPrep DNA gel recovery kit (Axygen, Hangzhou, China). After all the purified DNA was mixed in the same amount, PE250 sequencing was carried out using the Illumina Miseq platform [29].

The original sequencing data were analyzed as 16S rRNA.

3.4. Data Processing

Microsoft Excel 2019 was used for data processing and analysis, and Origin 2018 was used for drawing.

4. Conclusions

(1) The EPS contents in MBF were 51.20 mg/g (HRT 36 h), 55.89 mg/g (HRT 18 h), and 33.84 mg/g (HRT 11 h), respectively, higher than the EPS content of OBF under the corresponding HRTs. PN content or PS content also followed the same trend.
(2) The sedimentation performance of MBF was better than that of OBF, attributed to higher EPS PN contents.
(3) MBF was constructed by fungi and bacteria together, and C058 was the main component, promoting the synthesis of EPS.
(4) Compared with OBF, MBF bioaugmented by *Cordyceps* strain C058 had higher diversity and abundance of microorganisms, realizing a more stable operation of the experimental reactor. More importantly, C058 promoted the growth of some functional bacteria, including *Janthinobacterium* belonging to Proteobacteria, *Phormidium* and *Leptolyngbya* belonging to Cyanobacteria, and *Hymenobacter* belonging to Bacteroides, which participate in nitrogen fixation and PN and PS production, promoting the EPS synthesis. In addition, C058 also promoted the growth of *Spirotrichea* belonging to Ciliophora, which benefited floc formation and enhanced the water purification.

5. Patents

Chinese patent: A MPBR reactor suitable for sewage purification, CN212833060U. Applicant: Zhongkai University of Agriculture and Engineering. Inventor: Yiyong Li, Yongcong Yang, Kangchun Peng, Baoe Wang, Xueqin Tao, Chong Lin, Zexiang Lei, Jianjun Du.

Author Contributions: Conceptualization, Y.L. and B.W.; methodology, Y.L. and Y.Y.; software, W.L. (Wanyi Luo); validation, W.L. (Wen Liu), Z.L., X.T. and B.W.; formal analysis, W.L. (Wanyi Luo); data curation, B.W.; writing—original draft preparation, W.L. (Wanyi Luo); writing—review and editing, Y.L.; supervision, B.W.; funding acquisition, Y.L. All authors have read and agreed to the published version of the manuscript.

Funding: This research was funded by the Science and Technology Planning Project of Guangzhou, China, grant numbers 201803030039, 201704020187, and B318221315.

Institutional Review Board Statement: Not applicable.

Informed Consent Statement: Not applicable.

Data Availability Statement: The data presented in this study are available on request from the corresponding author. The data are not publicly available due to confidentiality issues.

Acknowledgments: The authors would like to thank the Medicinal Fungi Research Group, Institute of Microbiology, Guangdong Academy of Sciences for the donation of *Cordyceps* sp. C058 strain. The authors would also like to thank Guangdong Meilikang Biotechnology Co., Ltd. for the sequencing process and all the staff who contributed to this experiment.

Conflicts of Interest: The authors declare no conflict of interest.

References

1. Wang, Z.C.; Gao, M.C.; Wei, J.F.; Yang, Y.S.; Zhang, J.; Yu, S.P.; Wang, Y.J. Effect of salinity change on extracellular polymer of anaerobic sludge. *Acta Sci. Circumst.* **2016**, *36*, 3273–3281.
2. Sheng, G.P.; Yu, H.Q.; Li, X.Y. Extracellular polymeric substances (EPS) of microbial aggregates in biological wastewater treatment systems: A review. *Biotechnol. Adv.* **2010**, *28*, 882–894. [CrossRef]
3. More, T.; Yadav, J.; Yan, S.; Tyagi, R.; Surampalli, R. Extracellular polymeric substances of bacteria and their potential environmental applications. *J. Environ. Manag.* **2014**, *144*, 1–25. [CrossRef]
4. Dignac, M.; Urbain, V.; Rybacki, D.; Bruchet, A.; Snidaro, D.; Scribe, P. Chemical description of extracellular polymers: Implication on activated sludge floc structure. *Water Sci. Technol.* **1998**, *38*, 45–53. [CrossRef]
5. Jagaba, A.H.; Kutty, S.; Noor, A.; Haruna Birniwa, A.; Affam, A.; Lawal, I.M.; Usman Kankia, M.; Usman Kilaco, A. A systematic literature review of biocarriers: Central elements for biofilm formation, organic and nutrients removal in sequencing batch biofilm reactor. *J. Water Process Eng.* **2021**, *42*, 102178. [CrossRef]
6. Wang, Z.C.; Gao, M.; Wang, S.; Xin, Y.J.; Ma, D.; She, Z.L.; Wang, Z.; Chang, Q.B.; Ren, Y. Effect of hexavalent chromium on extracellular polymeric substances of granular sludge from an aerobic granular sequencing batch reactor. *Chem. Eng. J.* **2014**, *251*, 165–174. [CrossRef]
7. Basuvaraj, M.; Fein, J.; Liss, S.N. Protein and polysaccharide content of tightly and loosely bound extracellular polymeric substances and the development of a granular activated sludge floc. *Water Res.* **2015**, *82*, 104–117. [CrossRef]
8. Zhu, N.; Liu, L.; Xu, Q.; Chen, G.W.; Wang, G. Resources availability mediated EPS production regulate microbial cluster formation in activated sludge system. *Chem. Eng. J.* **2015**, *279*, 129–135. [CrossRef]
9. Ray, A.J.; Seaborn, G.; Leffler, J.W.; Wilde, S.B.; Lawson, A.; Browdy, C.L. Characterization of microbial communities in minimal-exchange, intensive aquaculture systems and the effects of suspended solids management. *Aquaculture* **2010**, *310*, 130–138. [CrossRef]
10. Zhao, D.H.; Pan, L.Q.; Wang, C. Cleaning effect of biofloc on culture environment and effects of physiological indexes of Penaeus vannamei. *Trans. Oceanol. Limnol.* **2014**, 67–73.
11. Jagaba, A.H.; Kutty, S.R.M.; Noor, A.; Affam, A.C.; Ghfar, A.A.; Usman, A.K.; Lawal, I.M.; Birniwa, A.H.; Kankia, M.U.; Afolabi, H.K.; et al. Parametric optimization and kinetic modelling for organic matter removal from agro-waste derived paper packaging biorefinery wastewater. *Biomass Convers. Biorefin.* **2022**, 1–18. [CrossRef]
12. Zhang, H.; Wang, X.; Li, C. Isolation and identification of a bacillus sp. strain and its role in bioflocs for the shrimp culture system. *Prog. Fish. Sci.* **2016**, *37*, 111–118.
13. Gao, G.; Zhu, K.L.; Zhang, Q.Q.; Wang, Z.J.; Huang, J. Simplified fermentation of a functional probiotics and the application in prawn (*Litopenaeus vannamei*) bio-floc breeding. *Prog. Fish. Sci.* **2017**, *38*, 140–147.
14. Wang, S.H.; Tan, H.X.; Luo, G.Z.; Sun, D.C. Effects of temperature, pH and dissolved oxygen on the construction of biofloc system by *Lactobacillus plantarum*. *Genom. Appl. Biol.* **2020**, *39*, 2628–2637.
15. Baldrian, P. Interactions of heavy metals with white-rot fungi. *Enzyme Microb. Technol.* **2003**, *32*, 78–91. [CrossRef]
16. Crini, G. Non-conventional low-cost adsorbents for dye removal: A review. *Bioresour. Technol.* **2006**, *97*, 1061–1085. [CrossRef]
17. Zhao, L.H.; Jin, R.F.; Sun, H.J.; Gao, Y.J. Study on treatment of cotton pulp black liquor by white rot fungi and activated sludge. *Ecol. Environ.* **2009**, *18*, 28–31.
18. Wang, L.; Chen, G.Q.; Zeng, G.M.; Zhang, W.Q.; Chen, Y. Research progress on extracellular polymers of fungi and their interaction with heavy Metals. *Environ. Pollut. Control* **2010**, *32*, 74–80.
19. Li, Y.; Yang, Y.C.; Wang, B.E.; Lin, C.; Tao, X.Q.; Du, J.J. Optimization and effect of purification of aquaculture wastewater by Cordyceps sinensis. *Adv. Environ. Prot.* **2020**, *10*, 10.
20. Yang, Y.C. *Purification Effect and Mechanism of Cordyceps Sinensis on Aquaculture Wastewater*; Zhongkai University of Agriculture and Engineering: Guangzhou, China, 2020.
21. Zhang, R.Y. *Occurrence Regularity of Mycelium Pellets in Activated Sludge System and Its Bioaugmentation Application*; Hebei University of Engineering: Handan, China, 2020.
22. Li, Y.; Chen, Y.C.; Li, D.L.; Zhang, Y.X.; Zhang, W.M. Study on antibacterial and antitumor activity of several fermented extracts of Cordyceps sinensis. *Sci. Technol. Food Ind.* **2010**, *31*, 88–90.
23. Zhou, X.; Huang, M.Q.; Qiu, R.; Li, Z.Q.; Guo, S.H.; Cui, Y.H. Research progress on composition, production and application of Cordyceps sinensis. *Priv. Sci. Technol.* **2018**, 98–99.
24. Morgan, J.W.; Forster, C.F.; Evison, L. A comparative study of the nature of biopolymers extracted from anaerobic and activated sludges. *Water Res.* **1990**, *24*, 743–750. [CrossRef]

25. Ge, L.Y.; Wang, H.W.; Ma, L.M.; Deng, H.H.; Liu, Y. Study on influencing factors of extracellular polymers in aerobic activated sludge. *Environ. Sci. Technol.* **2007**, *02*, 8–9.
26. Gu, Y.N. *Study on the Efficiency and Stability of Aerobic Granular Sludge in the Treatment of Domestic Wastewater*; Harbin Institute of Technology: Harbin, China, 2020.
27. Zhang, X.; Zhang, Q.S.; Xu, Y.; Liu, X. Study on sedimentation and dewatering performance of activated sludge from a sewage treatment plant. *Environ. Prot. Sci.* **2011**, *37*, 20–22.
28. Zhang, L.H.; Li, J.; Guo, J.B.; Jia, Y.P.; Zhang, H.F. Effect of EPS on flocculation settling performance and surface properties of activated sludge. *J. Chem. Ind. Eng.* **2012**, *63*, 1865–1871.
29. Li, H.; Qu, J.; Li, T.; Li, j.; Lin, Q.; Li, X. Pika population density is associated with the composition and diversity of gut microbiota. *Front. Microbiol.* **2016**, *7*, 758. [CrossRef]
30. Tsuneda, S.; Jung, J.; Hayashi, H.; Aikawa, H.; Hirata, A.; Sasaki, H. Influence of extracellular polymers on electrokinetic properties of heterotrophic bacterial cells examined by soft particle electrophoresis theory. *Colloids Surf. B Biointerfaces* **2003**, *29*, 181–188. [CrossRef]
31. Satoshi, T.; Hirotoshi, A.; Hiroshi, H.; Atsushi, Y.; Akira, H. Extracellular polymeric substances responsible for bacterial adhesion onto solid surface. *FEMS Microbiol. Lett.* **2010**, *223*, 287–292.
32. Cardona, E.; Gueguen, Y.; Magré, K.; Lorgeoux, B.; Piquemal, D.; Pierrat, F.; Noguier, F.; Saulnier, D. Bacterial community characterization of water and intestine of the shrimp *Litopenaeus stylirostris* in a biofloc system. *BMC Microbiol.* **2016**, *16*, 157. [CrossRef]
33. Addo, F.G.; Zhang, S.; Manirakiza, B.; Ohore, E.O.; Yuan, S.D. The impacts of straw substrate on biofloc formation, bacterial community and nutrient removal in shrimp ponds. *Bioresour. Technol.* **2021**, *326*, 124727. [CrossRef]
34. Yang, C.; Wei, Z.; Liu, R.; LI, Q.; Li, B.; Wang, S.; Song, C.; Qiao, C.; Aulchandani, A. Phylogenetic diversity and metabolic potential of activated sludge microbial communities in full-scale wastewater treatment plants. *Environ. Sci. Technol.* **2011**, *45*, 7408–7415. [CrossRef] [PubMed]
35. Lee, I.S.; Parameswaran, P.; Rittmann, B.E. Effects of solids retention time on methanogenesis in anaerobic digestion of thickened mixed sludge. *Bioresour. Technol.* **2011**, *102*, 10266–10272. [CrossRef]
36. Yang, M.; Lu, D.W.; Qin, B.D.; Liu, Q.L.; Zhao, Y.M.; Liu, H.L.; Ma, J. Highly efficient nitrogen removal of a coldness-resistant and low nutrient needed bacterium, *Janthinobacterium* sp M-11. *Bioresour. Technol.* **2018**, *256*, 366–373. [CrossRef] [PubMed]
37. Ma, J.; Bei, Q.C.; Wang, X.J.; Lan, P.; Liu, G.; Lin, X.W.; Liu, Q.; Lin, Z.B.; Liu, B.J.; Zhang, Y.H.; et al. Impacts of Mo application on biological nitrogen fixation and diazotrophic communities in a flooded rice-soil system. *Sci. Total Environ.* **2019**, *649*, 686–694. [CrossRef] [PubMed]
38. Bar-Or, Y.; Shilo, M. Characterization of macromolecular flocculants produced by *Phormidium* sp. strain J-1 and by Anabaenopsis circularis PCC 6720. *Appl. Environ. Microbiol.* **1987**, *53*, 2226–2230. [CrossRef]
39. Kim, J.H.; Choi, W.; Jeon, S.M.; Kim, T.; Park, A.; Kim, J.; Heo, S.J.; Oh, C.; Shim, W.B.; Kang, D.H. Isolation and characterization of *Leptolyngbya* sp. KIOST-1, a basophilic and euryhaline filamentous cyanobacterium from an open paddle-wheel raceway Arthrospira culture pond in Korea. *J. Appl. Microbiol.* **2015**, *119*, 1597–1612. [CrossRef]
40. Kim, J.H.; Kang, D.H. Draft genome sequence of *Leptolyngbya* sp. KIOST-1, a filamentous cyanobacterium with biotechnological potential for alimentary purposes. *Genome Announc.* **2016**, *4*, e00984-16. [CrossRef]
41. Lu, X.P.; Tang, K.; Li, H.; Cheng, Y.L.; Yang, S.S.; Guo, H.L.; Guo, H.Q.; Yun, X.Y.; Feng, F.Y. Isolation and Identification of thin layer strain L28 and its promoting effect on the growth of Potato seedlings. *Microbiol. China* **2020**, *47*, 4050–4058.
42. Tenney, M.W.; Stumm, W. Chemical flocculation of microorganisms in biological waste treatment. *J. Water Pollut. Control Fed.* **1965**, *37*, 1370–1388.
43. Chen, Y.H. *Composition and Function Analysis of Extracellular Polymers during Activated Sludge Aggregation*; Chongqing University: Chongqing, China, 2019.
44. Andersen, P. The quantitative importance of the microbial loop in the marine pelagic: A case study from the North Bering/Chukchi seas. *Arch. Hydrobiol. Beih.* **1988**, *31*, 243–251.
45. Nielsen, T.G.; Hansen, B. Plankton community structure and carbon cycling on the western coast of Greenland during and after the sedimentation of a diatom bloom. *Mar. Ecol. Prog. Ser.* **1995**, *125*, 239–257. [CrossRef]
46. Wang, Q.; Garrity, G.M.; Tiedje, J.M.; Cole, J.R. Naïve Bayesian classifier for rapid assignment of rRNA sequences into the new bacterial taxonomy. *Appl. Environ. Microbiol.* **2007**, *73*, 5261–5267. [CrossRef] [PubMed]
47. Weerman, E.J.; Van Der Geest, H.G.; Van Der Meulen, M.D.; Manders, E.M.M.; Koppel, J.V.D.; Herman, P.M.J.; Admiraal, A.W. Ciliates as engineers of phototrophic biofilms. *Freshw. Biol.* **2011**, *56*, 1358–1369. [CrossRef]
48. Weber, S.D.; Ludwig, W.; Schleifer, K.H.; Fried, J. Microbial composition and structure of aerobic granular sewage biofilms. *Appl. Environ. Microbiol.* **2007**, *73*, 6233–6240. [CrossRef]
49. Schryver, P.D.; Verstraete, W. Nitrogen removal from aquaculture pond water by heterotrophic nitrogen assimilation in lab-scale sequencing batch reactors. *Bioresour. Technol.* **2009**, *99*, 1162–1167. [CrossRef]
50. Zhang, X.; Bishop, P.L.; Kinkle, B.K. Comparison of extraction methods for quantifying extracellular polymers in biofilms. *Water Sci. Technol.* **1999**, *39*, 211–218. [CrossRef]
51. Bo, F.; Palmgren, R.; Keiding, K.; Nielsen, P.H. Extraction of extracellular polymers from activated sludge using a cation exchange resin. *Water Res.* **1996**, *30*, 1749–1758.

52. Eaton, A.D.; Clesceri, L.S.; Greenberg, A.E.; Franson, M.A.H. Standard methods for the examination of water and wastewater. *Am. J. Public Health Nations Health* **1966**, *56*, 387–388.
53. Shen, P.; Chen, X.D. *Microbiology Experiment*; Higher Education Press: Beijing, China, 2007.
54. Tamaki, H.; Wright, C.L.; Li, X.; Lin, Q.; Hwang, C.; Wang, S.; Thimmapuram, J.; Kamagata, Y.; Liu, W.T. Analysis of 16S rRNA amplicon sequencing options on the Roche/454 Next-Generation Titanium Sequencing Platform. *PLoS ONE* **2011**, *6*, e25263. [CrossRef]
55. Caporaso, J.G.; Kuczynski, J.; Stombaugh, J.; Bittinger, K.; Bushman, F.D.; Costello, E.K.; Fierer, N.; Peña, A.G.; Goodrich, J.K.; Gordon, J.I.; et al. QIIME allows analysis of high-throughput community sequencing data. *Nat. Methods* **2010**, *7*, 335–336. [CrossRef]
56. Li, S.; Peng, Y.Z.; Wang, R.D. The role of extracellular polymers substance in biological treatment of wastewater. *J. Nat. Sci. Heilongjiang Univ.* **2016**, *33*, 515–520.
57. Ni, J.J.; Li, X.J.; He, Z.L.; Xu, M.Y. A novel method to determine the minimum number of sequences required for reliable microbial community analysis. *J. Microbiol. Methods* **2017**, *139*, 196–201. [CrossRef] [PubMed]
58. Edgar, R.C.; Haas, B.J.; Clemente, J.C.; Quince, C.; Knight, R. UCHIME improves sensitivity and speed of chimera detection. *Bioinformatics* **2011**, *27*, 2194–2200. [CrossRef]
59. Lejzerowicz, F.; Esling, P.; Pillet, L.; Wilding, T.A.; Black, K.D.; Pawlowski, J. High-throughput sequencing and morphology perform equally well for benthic monitoring of marine ecosystems. *Sci. Rep.* **2015**, *5*, 13932. [CrossRef] [PubMed]

MDPI
St. Alban-Anlage 66
4052 Basel
Switzerland
Tel. +41 61 683 77 34
Fax +41 61 302 89 18
www.mdpi.com

Catalysts Editorial Office
E-mail: catalysts@mdpi.com
www.mdpi.com/journal/catalysts

www.ingramcontent.com/pod-product-compliance
Lightning Source LLC
Chambersburg PA
CBHW061448210926
51850CB00008B/1651